AF556015

Fruit Breeding

About the Authors

Dr. Anil Kumar Shukla obtained his B.Sc. (Ag) and M.Sc Ag (Hort.) from N.D. University of Agriculture and Technology Kumarganj, Faizabad during 1993 and 1996 respectively. He completed his PhD from IARI, Pusa, New Delhi in the year 2000. He has been awarded with Gold Medal in B.Sc (Agri.) and he has second rank in ICAR in JRF (1993). He was topper of ARS and SRF (1998). He worked as Scientist and Sr. Scientist at CIAH, Bikaner. He also served as Associate Professor at Maharana Pratap University of Agriculture and Technology, Udaipur. Dr. Shukla also worked as Principal Scientist (Horticulture) at Directorate of Research on Women in Agriculture, Bhubaneswar, Odisha. Presently, he is head, ICAR-CAZRI, RRS, Pali (Rajasthan). He is the recipient of N.E. Borlaug Fellowship for year 2008, He also received Best worker Award by Vice-Chancellor, MPUAT, Udaipur. Fellowship Award 2012, from Indian Society of Horticultural Research and Development, Uttarakhand (Society). Fellowship Award 2012, from Hi-Tech Horticultural Society, Meerut (Society), Fellowship Award 2012, from Confederation of Horticulture Association of India, New Delhi (Society), Fellowship Award 2018, from Indian Society of Arid Horticulture, Bikaner. Fellowhip Award 2019, International Society for Noni Science Chennai, Fellowship Award 2019, Indian Academy of Horticultural Science, Eminent Scientist Award 2016, from SVWS, Lucknow, Himadri Young Scientist Award 2010 from GBPUAT, Pant Nagar, Uttarakhand and Indian Society of Horticulture Research and Development, Uttarakhand.

Dr. Anil Kumar Shukla immensely contributed in developing two varieties of Ber (Thar Sevika and Thar Bhubraj) and two varieties of guava MPUAT-S-1 and MPUAT-S-2. He has developed several Agro techniques in Ber and Guava crops. He has guided 02 Ph.D. and 01 M.Sc. (Ag) students as Major Advisor and 05 Ph.D. and 04 M Sc. (Ag.) students as Co-Advisor. He has published 70, research papers, 38 book chapters, 76 research abstracts, 68 popular articles, 11 bulletins and 03 books.

Dr. Arun Kumar Shukla graduated from T.D. College, Jaunpur (Purvanchal University), M.Sc. (Ag.) Horticulture and PhD from Department of Horticulture, Institute of Agricultural Sciences, BHU, Varanasi. He has been recipient of National Scholarship, ICAR-JRF, UGC-JRF/SRF, Young Scientist-2015, Outstanding Scientist-2016, Scientist of the Year Awards-2017 by Academic Research Foundation, APJ Abdul Kalam Scientist, Dr BP Pal Scientist Awards-2019 by Society of Tropical Agriculture. Awarded with fellow-ISAH 2018 and Fellow-STA 2018. ICAR-NAIP Team award for outstanding achievement of NAIP Project (Component-3) entitled "Integrated farming system ..... Dhar districts of Madhya Pradesh"(As CoPI-IGFRI, Jhansi) during Regional committee meeting, Goa, on 09.11.2012. Besides production technologies, he is actively involved in genetic improvement of temperate fruits He has published more than 180 publications including research paper, popular articles, book, book chapter, training manual, leaflet/extension bulletins etc. At present he is working as Principal Scientist (Fruit Science) at ICAR-IARI, Regional Station, Amartara Cottage, Shimla.

Dr. M.B. Noor mohamed, is presently working as Scientist (Agroforestry) at ICAR- Central Arid Zone Research Institute, RRS, Pali Marwar (Rajasthan) with more than 5 years of experience in tree improvement, agroforestry, carbon sequestration and breeding works in shrubs and trees. He completed his UG, PG and Ph.D. in Forestry at Forest College and Research Institute, TNAU, Mettupalayam (Tamil Nadu). He has been awarded with Gold medal in PhD (Forestry). He is actively involved in germplasm collection, evaluation and improvement in trees and fruit crops. He has published 20 research papers in international as well as national journals, 2 popular articles, 8 book chapters, 2 books, 1 training compendium and more than 20 abstracts. He has been honoured with Young Scientist Award, Innovative Scientist Award and Best oral presentation award by various societies.

Dr. Akath Singh, Principal Scientist (Horticulture) at ICAR- Central Arid Zone Research Institute with more than 16 years of experience in production and improvement of various fruit crops. He has remarkable research achievements in strawberry, guava, peach and pomegranate production and improvement. He contributed in development of two guava cultivars, and one each in *lasora* and *karonda*. He was actively participated in finalizing national DUS guidelines for pomegranate. Selected as Member of Indian scientist's delegation visited to China. He has been honoured with Young Scientist Associate Award, SARP Associate Award, SDSH fellow by various societies and also granted ITS by DST for Spain visit. He has around 85 publications in his credit including 35 research papers in high impact national and international journals and author for three edited books, 18 book chapters and 7 technical bulletins. He has also guided two M.Sc. students.

Dr. Divya Tiwari completed her higher studies from Narendra Deva University of Agriculture & Technology, Faizabad, U.P. (India) and acquired B.Sc. (Ag.), M.Sc. (Ag.) and Ph.D. in Horticulture (Fruit Science) Degrees in continuation during 2000-2010. She has also Awardee of The Best Teacher Award (2016) and Young Scientist award (2017) and got recognition as CHAI Fellow in 2019. She has published several research papers, authored/edited books, book chapters, popular articles, folders and manuals. She is also developed, introduced and popularized several technologies. Among which fertigation and irrigation scheduling for fruit crops, development of modified BBCH scale for mango and methodology for recording the phenological observations was the most significant. Moreover, she has also the experience of handling several projects which are either University funded or externally funded.

2nd Fully Revised and Enlarged Edition

Fruit Breeding

Anil Kumar Shukla
Arun Kumar Shukla
M.B. Noor mohamed
Akath Singh
Divya Tiwari

NEW INDIA PUBLISHING AGENCY
New Delhi – 110 034

NEW INDIA PUBLISHING AGENCY
101, Vikas Surya Plaza, CU Block, LSC Market
Pitam Pura, New Delhi 110 034, India
Phone: + 91 (11)27 34 17 17 Fax: + 91(11) 27 34 16 16
Email: info@nipabooks.com
Web: www.nipabooks.com

Feedback at feedbacks@nipabooks.com

ISBN : 978-93-89130-46-1

Composed and Designed by NIPA

Dedicated to
Late Shree Yamuna Prashad Shukla
Agriculture Extension Teacher

डॉ गौतम कल्लू
उप महानिदेशक
(बागवानी एवं फसल विज्ञान)
DR. G. KALLOO
Deputy Director General
(Horticulture & Crop Science)

भाकृअनुप
ICAR
भारतीय कृषि अनुसंधान परिषद
कृषि अनुसंधान भवन–II पूसा
नई दिल्ली – 110012
Indian Council of Agricultural Research
Krishi Anusandhan Bhawan-II
Pusa, New Delhi-110012
Ph.:25842068 (O) 25840191 (R)
Fax: 25841976, Email: kallog@icar1.nic.in

Foreword to the First Edition

There are many challenges in fruit production to meet out the daily requirement of fruits for growing population in the new millennium because of faster rate of population growth, limited area under fruit crop, reduction in cultivable land, poor yield potential, and susceptibility to insect pests and diseases. However, for nutritional security of growing population, development of precocious, prolific, biotic and abiotic stress resistant fruit varieties are essentially required.

There have been conscious efforts in fruit improvement through planned hybridization and selection. Consequently, most of the outstanding present day cultivars have resulted from chance seedling selection, some of these selections are quite old. In response to new challenges need for better tree canopy resistant to biotic and abiotic stresses, better shelf life and high productivity per unit area still requires more attention. I am happy that concerted efforts made by the authors to compile the work done in this direction are under one umbrella. I congratulate the authors for bringing out such an useful publication. I hope the book will be a valuable source of reference and text for students, teachers and scientists associated with the discipline.

(G. Kalloo)

Preface of the Second Edition

The pressure of an ever-increasing population and periodic famine due to unexpected flood and drought has forced and awakened the horticultural scientist, to evolve new plant types for diversified use. Besides, some limitations in the improvement of fruit crop such as long juvenile phase, high heterozygosity, limited information on inheritance pattern, excessive fruit drop, parthenocarpy and lesser number of seeds per fruit, hybridization, selection, mutation and other tools of fruit breeding have resulted in the development of a number of varieties in mango, grape, papaya, banana and guava for various purposes.

A vast literature is available on every aspects of fruit cultivation and many excellent books and manuals have been written from time to time howcvcr, compiled information on fruit breeding are not available under a single head. The present book entitled " Fruit Breeding Approaches and Achievements" Revised and enlarged edition is ventured with the objective to provide latest possible information on basic approaches in fruit breeding, breeding for biotic stresses resistance, use of plant growth regulators in fruit improvement, improvement of important fruit crops such as mango, banana, papaya, grape, guava, citrus, ber, aonla, pomegranate, date palm, litchi, coconut, cashewnut, pineapple, few under utilized fruits and temperate fruits in a broad spectrum.

It is with more than mere formality we thank Dr. A.K. Singh, Deputy Director General (Horticultural Sciences), ICAR, New Delhi for inspiration given by them in writing this book.

Authors express their deep sense of gratitude to Dr. O.P. Yadav, Director, Central Arid Zone Research Institute, Jodhpur for giving more thrust to increase quality publications. We are also thankful to Dr. Dipak Kumar Gupta and Dr. R.S. Mehta for their critical comments and suggestions. Authors are thankful to All Scientist and workers of Central Arid Zone Research Institute, Regional Research Station, Pali (Rajasthan) for their help during writing this book.

Authors books, book chapters, research papers, popular articles and technical papers used in writing of this book have duly been acknowledged in the text.The authors will appreciate receiving suggestions for further improvement of this book. We hope that information presented here in the book would receive due response from the teaches, students, researchers and other readers

.

Date: August, 2019
Place: Pali Marwar (Rajasthan)

Anil Kumar Shukla
Arun Kumar Shukla
M.B. Noor mohamed
Akath Singh
Divya Tiwari

Preface of the First Edition

The pressure of an ever-increasing population and periodic famine due to unexpected flood and drought has forced and awakened the horticultural scientist, to evolve new plant types for diversified use. Besides, some limitations in the improvement of fruit crop such as long juvenile phase, high heterozygosity, limited information on inheritance pattern, excessive fruit drop, parthenocarpy and lesser number of seeds per fruit, hybridization, selection, mutation and other tools of fruit breeding have resulted in the development of a number of varieties in mango, grape, papaya, banana and guava for various purposes.

A vast literature is available on every aspects of fruit cultivation and many excellent books and manuals have been written from time to time however, compiled information on fruit breeding are not available under a single head. The present book entitled " Fruit Breeding Approaches and Achievements" Volume 1 is ventured with the objective to provide latest possible information on basic approaches in fruit breeding, breeding for biotic stresses resistance, use of plant growth regulators in fruit improvement, improvement of important fruit crops such as mango, banana, papaya, grape, guava, citrus, ber, aonla, pomegranate, date palm, litchi, coconut, cashewnut, pineapple and few underutilized fruits in a broad spectrum.

It is with more than mere formality we thank Dr. G. Kalloo, Deputy Director General (Horticultural and Crop Sciences), ICAR, New Delhi and Prof. R.K. Pathak, Director, Central Institute for Subtropical Horticulture, Lucknow for the inspiration given by them in writing this book. Authors are highly grateful to Dr. S.N. Pandey, ADG (Horticulture and Plantation crops), Indian Council of Agricultural Research, new Delhi for his encouragement and keen interest.

Authors express their deep sense of gratitude to Dr. D.G. Dhandar, Director, Central Institute of Arid Horticulture, Bikaner for giving more thrust to increase quality publications. We are also thankful to Dr. O.P. Awasthi, Dr. R.S. Singh, Dr. D. Singh and Dr. D.K. Samdia for their critical suggestions. Authors are thankful to Mr. D.C. Joshi, Mr. Sanjay Patil, Mr. M.K. Jain, B.R. Khatri and Mr. Akhil Thukral of CIAH, Bikaner for their help during writing this book. Thanks are due to Dr. J. Singh, Assistant Professor (Horticulture), Rajasthan Agriculture University, Bikaner for thorough checking of the manuscript.

Authors books, book chapters, research papers, popular articles and technical papers used in writing of this book have duly been acknowledged in the text.

The authors will appreciate receiving suggestions for further improvement of this book. We hope that information presented here in the book would receive due response from the teaches, students, researchers and other readers.

Date: April, 2004
Place: Bikaner

Anil Kumar Shukla
Arun Kumar Shukla
B.B. Vashistha

Contents

Abbreviations

2,4 – D	2,4 Dichlorophenoxy Acetic Acid
ABM	*Agrobacterium* mediated
AES	Agriculture Experiment Station
AFLP	Amplified Fragment Length Polymorphism
AICRP	All India Coordinated Research Project
APAU	Andhra Pradesh Agricultural University
AP-PCR	Arbitrarily Primered Polymerase Chain Reaction
BAC	Bihar Agricultural College
BCKV	Bidhan Chandra Krishi Vishwavidyalaya
CAP	Cleaved Amplified Polymorphism
CHES	Central Horticulture Experiment Station
CIAH	Central Institute for Arid Horticulture
CISH	Central Institute for Subtropical Horticulture
CPCRI	Central Plantation Crop Research Institute
DBT	Department of Biotechnology
DNA	Deoxyribonucleic Acid
E M	East Malling
EC	Exotic Collection
EES	Ethyl Methane Sulphonate
EP	Electroporation
FRS	Fruit Research Station
FW-450	1,2 Dichloro- iso – butyrate
GA_3	Gibberellic Acid
GAU	Gujarat Agricultural University
GBPUAT	Govind Ballabh Pant University of Agriculture and Technology
HARP	Horticulture and Agroforestry Research Programme
HAU	Haryana Agricultural University
HETC	Horticultural Experiment and Training Centre
HS	Highly Susceptible
IARI	Indian Agricultural Research Institute
IBA	Indole Butyric Acid
IC	Indigenous Collection
IIHR	Indian Institute of Horticulture Research
IITA	International Institute for Tropical Agriculture
IPM	Integrated Pest management
IW	Indigenous Wild Collection
KAU	Kerala Agricultural University
LD	Lethal Dose
MAS	Marker Assisted Selection
MH	Maleic Hydrazide
MMC	Megaspore Mother Cell
MS	Moderately Susceptible

NAA	Naphthalene Acetic Acid
NBPGR	National Bureau of Plant Genetic Resources
NDUAT	Narendra Dev University of Agriculture and Technology
NMU	N-Nitrose-N-Methyl-Urea
NMUT	N-Nitroso-N-Methyl Urethane
NRC	National Research Centre
PB	Particle Bombardment
PB+ AMB	Particle Bombardment followed by *Agrobacterium* Co-cultivation
PEG	Polyethylene Glycol
PGM	Polyethylene Glycol Mediated
PGR	Plant Growth Regulator
PHB	Polyhydroxy Butyrate
PMC	Pollen Mother Cell
PRSV	Papaya Ring Spot Virus
R	Resistant
RAPD	Random Amplified Polymorphic DNA
RAU	Rajendra Agricultural University
RFLP	Restriction Fragment Length Polymorphism
RFRS	Regional Fruit Research Station
S	Susceptible
SADH	Dimethylamino Succinamic Acid
SCAR	Sequence Characterized Amplified Regions
SPAR	Single Primer Amplification Reaction
SSCP	Single-Strand Conformation Polymorphism
SSRP	Single Sequence Repeat Polymorphism
STS	Sequenced Tagged Sites
TIBA	2,3,5, Tri iodobenzoic Acid
TNAU	Tamil Nadu Agricultural University
TSS	Total Soluble Solids

1

Basic Approaches in Fruit Breeding

India is bestowed with a wide range of agro-climatic and soil conditions. Therefore, almost all types of fruits can be grown in one or the other parts of the country. India is the second largest producer of fruits next to China. In India, fruit crops occupied 6.5 million hectares and it produces 96.75 million metric tonnes of fruits. India has exported fruits worth Rs. 4817.35 crores (692.01 USD Millions) (National Horticulture Board).

Improvement of fruit crops is difficult because of long gestation period, high heterozygosity, scanty information on inheritance pattern, often cross pollination, excessive fruit drop, parthenocarpy and less number of seeds restricting the availability of hybrid seedlings for evaluation. Even though, planned hybridization and clonal selections have been attempted in a number of fruit crops and these efforts have resulted in the development of promising varieties (mango, grape, guava, papaya, sapota, banana, etc). However, systematic and dedicated efforts are still required for the development of ideal varieties through modern tools in fruit crops.

In future, search for desired gene, critical study of inheritance pattern and use of biotechnological tools require due attention in combining ideal characteristics in varietal improvement programme of fruit crops.

1.1. Definition

Fruit breeding is the manipulation of a biological system that requires many generations to achieve results. It is also a dynamic, exciting and challenging profession, operating under continually changing conditions.

1.2. Major problems in fruit breeding

- It has long generation cycle of 2-10 years depending upon species and cultivars.
- It is not possible to have more recombination.
- Fruit crop has long juvenile period due to that early assessment of strains are not possible e.g. mango, *Madhuka latifolia*, jack fruit etc.

- Majority of the fruit species are highly heterozygous, therefore, large population is required for selection.
- Most of the fruit species are polyploid in nature e.g. ber, banana etc.
- Polyembryony e.g. citrus, mango.
- Parthenocarpy and seedlessness e.g. banana, pineapple etc.
- Presence of sexual incompatibility e.g. mango, apple, pear, loquat etc.
- More number of chromosomes, creating a big problem in genetical analysis e.g. ber, mulberry.
- Excessive fruit drop e.g. mango, citrus, grape etc.
- Presence of single seed in single fruit, requiring more number of crossing e.g. mango, litchi, mahua etc.

1.3. Objectives of fruit breeding

The objectives of fruit breeding vary with the fruit plants in relation to different location and requirements. Even though, the main objective of fruit breeding is to get maximum quality production per unit area with low cost and free from biotic and abiotic stresses. The objectives may also vary with respect to breeding for rootstock and scion.

1.3.1 For rootstock

- Wide geographical adaptability.
- Easily propagated, preferably through asexual means.
- It should be compatible with most of the scion cultivars having strong scion-stock union and more longevity.
- It should be resistant to biotic and abiotic stresses.
- For high density planting, it should have dwarfing effect without affecting the productivity of scion cultivars.
- Plants should not have brittle root system e.g. EM 9 rootstock of apple.
- It should be free from suckering habit.

1.3.2 For scion cultivars

- Dwarf growth stature.
- Regular, precocious and prolific bearer per unit canopy area.
- High productivity with good quality fruits.

- Resistant to biotic and abiotic stresses.
- Attractive fruit colour with pleasant aroma.
- Suitable for processing and export.
- Good keeping and transport quality.

1.4. Steps in fruit breeding

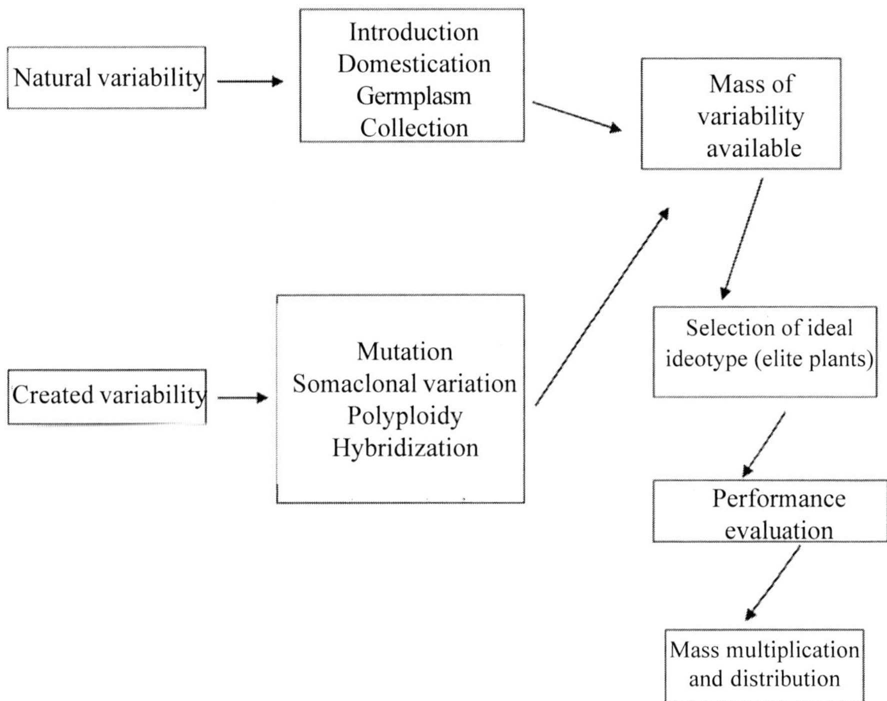

1.5. Modes of reproduction in relation to fruit breeding

The reproduction system is principally responsible for perpetuation and preservation of a particular genotype. Further, the mode of reproduction also determines the genetic constitution of fruit crop whether it is homozygous or heterozygous (Singh, 1990).

1.6. Types of reproduction in fruits

1.6.1 Asexual reproduction

- By using apomictic seeds e.g. citrus, mango, apple.
- By using vegetative parts of the plants e.g. mango, banana, guava, aonla, bael, date palm, apple, peach, plum, etc.

1.6.1.1 Reproduction by apomictic seed (apomixis)

Apomixis refers to the development of embryo without the usual process of meiosis and fertilization. The plants developed from apomictic seeds are true to type (Raghaven,1986, Naumova, 1993, Nyrgen, 1954).

Obligate apomictic seeds develop with or without pollination but without fertilization. e.g. *Malus* spp. However, in case of facultative apomictic seeds both types of embryo may develop i.e. zygotic as well as nucellar embryo (Maheshwari and Sachar, 1963, Bhatnagar and Johri, 1972) e.g. citrus and polyembryonic cultivars of mango.

1.6.1.1.1 Classification of apomixis

1.6.1.1.1.1 Recurrent apomixis

The embryo sac (female gametophyte) develops from the megaspore mother cell where meiosis is disturbed (sporogenesis failed) or from adjoining cell (megaspore mother cell disintegrates). The egg cell is diploid and embryo develops directly from the diploid egg cell without fertilization. Generally, somatic apospory, diploid parthenogenesis and diploid apogamy fall under recurrent apomixis.

Example: *Rubus* sp. (Raspberry), *Malus hupehensis, Malus sikkimensis, Malus sargenti* and *Malus toringoides* (Mitra, 1991, Vashishtha *et al*., 2004).

1.6.1.1.1.2 Non recurrent apomixis

The development of embryo takes place from haploid egg cell without fertilization. Such type of apomixis rarely occurs. Generative apospory, haploid parthenogenesis, haploid apogamy and androgamy fall under this category.

1.6.1.1.1.3 Adventitive embryony

This is also known as nucellar embryony, nucellar budding or polyembryony. In this case more than one embryo develops in a single seed. In the seed both types of embryo develops i.e. nucellar embryo from nucellar cell and zygotic embryo from egg cell with the result of syngamy.

Example: Olour, Goa, Kurukkan, Bappakai, Vellaikolamban, Nileswar dwarf, Salem, Bellary, Goakasargod, Mazagaon, Chandrakaran etc cultivars of mango (Majumder and Sharma, 1991) and most of the species of citrus except *Citrus medica* (citron), *Citrus grandis* (pummelo or shaddock) and *Citrus latifolia* (Ghosh, 1991, Vashishtha *et al*., 2004).

1.6.1.1.1.4 Vegetative apomixis

This is not common in fruit crops. However, in some cases like *Poa bulbosa* and some *Allium*, *Agave* and grass species produce vegetative buds or bulbils instead of flower in the inflorescence.

1.6.1.1.1.5 Apospory

Some times when embryo sac develops from archisporial or from the nucellus or from other cell. If it develops from haploid megaspore cell, it is known as generative or haploid apospory. On the other hand, if it develops from diploid cell i.e. nucellus of other cells it is termed as somatic or diploid apospory.

1.6.1.1.1.6 Parthenogenesis

It can be defined as development of embryo from egg cell with or without pollination but without fertilization. Depending upon the ploidy levels of egg cell, parthenogenesis can be halploid (non recurrent) and diploid (recurrent type) e.g. mangosteen (*Garcinea mangostana*).

1.6.1.1.1.7 Apogamy

Development of embryo from synergids or antipodal cells within the embryo sac with or without pollination but without fertilization is termed as apogamy. This type of apomixis is also grouped into haploid and diploid apogamy depending upon the ploidy level of cell. Diploid apogamy is recurrent type whereas, haploid aopgamy is non recurrent type.

1.6.1.1.1.8 Androgamy

Development of the embryo from male gametes inside or out side of embryo sac is known as androgamy. Since the cells are haploid in nature therefore, it comes under non recurrent type.

1.6.1.1.1.9 Genetics of apomixis

Stebbins (1958) stated that, as a rule, the apomictic condition is recessive to sexuality, although polyploid apomicts show tendency towards dominance. However, this recessiveness is not usually due to a monogenic difference. Since there is frequent reversion of apomicts to normal sexuality or sterility or some abnormal genetic behavior in crosses involving an apomict and an amphimictic or involving two apomicts of diverse origins, it appears that a successful apomict cycle is the result of an interaction of many gene which tend to break on hybridization. It is only in the relatively simple type of apomixis like adventive embryony and vegetative reproduction that simple genetic behaviour can be expected. Recently, Vardy *et al.* (1989) recorded three recessive genes with additive effects which are responsible for parthenocarpy.

1.6.1.1.1.10 Advantages of apomixis

1. Faster multiplication of genetically uniform individual can be done without any problem of segregation.
2. Heterosis or hybrid vigour can permanently be fixed in plants.
3. Efficient exploitation of maternal effect (if present) is possible from generation to generation.

1.6.1.1.1.11 Disadvantages of apomixis

Apomixis is nuisance when the breeder desires to obtain sexual progeny i.e. self or hybrid.

1.6.1.1.1.12 Reproduction by the vegetative parts of the plants

In this method of reproduction, a new plant develops from a portion of the plant body. Based on the different plant parts used it can be classified in to following groups

1.6.1.2. Reproduction through specialized parts

Offshoot : date palm

Sucker : banana, pineapple

Runner : strawberry

Others : e.g. tuber, stolon, bulb, rhizome and corms etc. These are not common in fruit crops.

1.6.1.2.1 Cutting

Stem cutting	Hard wood cutting : grape, fig, mulberry, guava, mango, quince, currant, pomegranate etc.
	Semi hard wood cutting : citrus, olive etc.
	Soft wood cutting : crab apple, apple, peach, pear, plum, apricot, cherry etc.
	Herbacious cutting : Most common in ornamental plants.
Root cutting	apple, crab apple, blackberry, raspberry
Leaf cutting	Not common in fruit plants.
Layering	Air layering: litchi
	Stool layering: guava, apple rootstock, pear etc.
	Trench layering: apple and pear rootstocks.

	Compound layering: muscadine grapes.
	Tip layering: blackberry, raspberry, gooseberry.
Grafting and budding	mango, guava, cashewnut, aonla, bael, ber, citrus, grape, apple, pear, plum, almond etc.

1.6.2 Sexual reproduction

It is also known as amphimixis. Multiplication is done through seed that develop after normal meiosis followed by syngamy e.g. phalsa, karonda, papaya etc.

There are four important steps involved in reproductive process.

(i) Development of gametophytes and gametes

(ii) Pollination

(iii) Fertilization or syngamy

(iv) Post fertilization development, i.e. formation of embryo, seed and fruit.

1.6.2.1 Floral biology and development of gametophytes

A normal flower consists of calyx, corolla, androecium and gynoecium. The last two parts of flowers have great significance in entire process of reproduction. When androecium and gynoecium both are present in same flower it is known as hermaphrodite or bisexual flower e.g. ber, citrus, guava, sapota, etc.

When male (androecium) and female (gynoecium) parts of flower are present in separate flower but on same plant it is known as monoecious condition e.g. aonla, walnut, hazelnut, jackfruit etc.) but when they are present on different plants separately it is known as dioecious condition e.g. papaya, date palm, kiwi fruit etc. Andromonoecious condition is found in mango where male and hermaphrodite flower are present in same panicle. Futher, the gynodioecious condition is found in some cultivars of papaya (Pusa Delicious, Pusa Majestic etc.)

On the basis of occurrence of flower, there are different types of inflorescence found in fruit crops.

Racemose :

i) Catkin: mulberry, chestnut, walnut, pecannut.

ii) Corymb: pear.

iii) Raceme: wild cherry, currents, gooseberry, blackberry, blueberry, raspberry.

Solitary: orange, quince, guava, peach, apricot, almond.

Cymose inflorescence: citrus, phalsa, sapota, strawberry, persimmon.

i) Fasicle: cherry, plum, ber.

ii) Panicles: grape, mango, litchi, loquat, pistachionut.

Anomolus inflorescence: apple.

Special types inflorescence:

i) Spadix: date palm, banana.

ii) Hypanthodium: fig.

1.6.2.2 Microsporogenesis and microgametogenesis

Production of microspore from pollen mother cell (PMC) is termed as microsporogenesis. Generally, each anther has four pollen sacs which contain numerous pollen mother cells (PMC's). After meiosis, each PMC results into four haploid cells or microspores. After thickening of the cell wall of microspore, it is termed as the pollen grain.

On other hand, microgametogenesis refers to the production of male gametes. During maturation of the pollen, the microspores nucleus divides mitotically to produce a generative and a vegetative or tube nucleus. The pollen is generally released in this binucleate stage. When the pollen reaches the stigma of flower it is termed as pollination. Shortly after pollination, pollen grain starts germination. The pollen tube enters in to stigma and grows through the style. The generative nucleus now undergoes a mitotic division to produce two male sperms. The pollen alongwith pollen tube is known as microgametophyte. The process between microspore to microgametophyte is termed as microgametogenesis. Pollen tube finally enters in to ovule through a small pore, micropyle, and discharges the two sperms in to the embryo sac.

1.6.2.3 Megasporogenesis and megagametogenesis

Megasporogenesis can be defined as production of megaspore from megaspore mother cells (MMC's). Megasporogenesis occurs in ovule which are present inside the ovary. A single cell in each ovule differentiates into a megaspore mother cell. The megaspore mother cell undergoes meiosis to produce four haploid megaspores. Three of them degenerate leaving one functional megaspore per ovule.

Under the process of megagametogenesis, the nucleus of a functional megaspore divides mitotically to produce four or more nuclei. The exact number of nuclei and their arrangement varies considerably from one species to another species.

In most of the cases megaspore nucleus undergoes three mitotic divisions to produce eight nuclei. Three of these nuclei move to one pole and produce central egg cell and two synergid cell. One synergid is situated on either side of the egg cell. Another three nuclei migrate to the opposite pole to give rise an antipodal cell. Remaining two nuclei in the centre are known as polar nuclei and after fusion forms the secondary nucleous. Megaspore develops into a mature megagametophyte or embryo sac and the development of embryo sac from a megaspore is known as megagametogenesis.

1.6.2.4 Fertilization (syngamy)

Fusion of the one of the two sperms with egg cell, producing a diploid zygote, is known as fertilization or syngamy. The fusion of remaining sperm with the secondary nucleus leading to the formation of a triploid primary endosperm nucleus is termed as triplet fusion. When triplet fusion and syngamy takes place simultaneously it is termed as the double fertilization.

1.6.2.5 Post fertilization development

A series of changes in the ovule follows as the fertilization is over. Synergids and antipodal cells become disorganized, and the egg cell becomes covered with a cell wall, known as oospore. This follows the development of seed. The various parts of gynoecium develop into ovary-fruit, ovule-seed, integument outer-testa (outer seed coat), integument inner tegmen (inner seed coat), secondary nucleus after fertilization-endosperm (triploid), egg cell (after fertilization) embryo, nucellus-perisperm (nutritive tissue like endosperm).

1.6.2.6 Significance of sexual reproduction

Sexual reproduction makes it possible to combine genes from two parents into a single hybrid. Recombination of these genes produces a large number of genotypes. This is an essential step in creating variation through hybridization. Almost the entire fruit breeding is based on sexual reproduction except clonal selection which is perpetuated by the vegetative means.

1.7 Modes of pollination

Transfer of pollen grains from anther to stigma is termed as pollination. When the pollen grain from an anther falls on the stigma of same flower on same plant (variety) it is termed as self-pollination or autogamy. But if pollen grain from flower of a plant is transferred to the stigma of flower of another plant (variety), it is known as cross pollination or allogamy. Third situation is known as geitonogamy when pollen from flower of a plant falls on the stigma of other flowers of the same plant (variety). The genetic consequences of geitonogamy are the same as those of autogamy.

1.7.1 Conditions and mechanism of autogamy (self-pollination)

In self-pollinated species generally 5% cross pollination takes place. The degree of cross pollination in self-pollinated species is affected by cultivars, location and environmental conditions like temperature, humidity etc. Hermaphrodite nature of flower, cleistogamy, and homogamy mechanism of flower favours to self-pollination.

1.7.1.1 Hermaphrodite or perfect flower

In such type of flowers both male and female parts are present in a single flower. It leads to self-pollination. Although in some cases perfect flower does not allow the self-pollination due to self-incompatibility, sterility and dichogamy. Self-pollinated fruits are guava, citrus etc.

1.7.1.2 Cleistogamy

Dehiscence of the pollen grains takes place before opening of the flower e.g. grape, sapota.

1.7.1.3 Homogamy

The male and female parts of the flowers mature at same time which favours to self pollination. e.g. citrus, peach and apricot.

1.7.2 Conditions and mechanism of allogamy (cross pollination)

In cross pollinated species 5-10% self pollination takes place.

The mechanisms which facilitate the cross pollination are as under.

1.7.2.1 Unisexuality

It is also known as dicliny.

i) Monoecious plants: jackfruit, aonla, walnut, hazelnut, pecannut

ii) Dioecious: papaya, kiwifruit, date palm etc.

iii) Andromonoecious: mango

1.7.2.2 Dichogamy

Stamens and pistils of hermaphrodite flowers may mature at different time facilitating cross pollination. It is of two types (a) protandry in which androecium matures first than gynoecium, e.g. walnut and coconut (b) protogyny where gynoecium matures first than androecium e.g. annona, banana, plum, pomegranate, avocado and fig etc.

1.7.2.3 Self incompatibility

It refers to the failure of pollen from a flower to fertilize the same flower or other flowers on the same plant e.g. almond, apple, mango, pear, pineapple, cherry, and apricot etc. The sporophytic incompatibility is found in mango whereas gametophytic incompatibility is found in loquat.

1.7.2.4 Male sterility

Presence of nonfunctional pollen grain in the flower favours to cross pollination e.g. triploid banana, J.H. Hale cutivar of peach, triploid apple varieties such as Bramley's Seedling, Blenheim Orange, King of Topkins etc and pear varieties Beurre-De-Amanlis, Marguerite Marillat are mostly sterile (Crane and Lawrence,1956).

1.7.2.5 Heterostyly

Presence of variable length of style refers to heterostyly condition. Which favours to cross pollination. Flowers of pomegranate possess this condition.

1.7.2.6 Heteromorphism

Dimorphism and trimorphism also facilitate the cross pollination. e.g. pin type of flowers are found in sapota and pomegranate whereas thrumb type of flowers are found in almond.

1.7.2.7 Environmental factors

Strawberries grown in rich soils have been reported to produce sterile pollens (Darrow,1957). Pear was found to be self-fertile in rich soil of California. (Gourley and Howlett, 1955). Similarly, Bose pear is self fruitful in New York and regularly parthenocarpic in South Africa (Reinicke, 1930).

1.7.3 Agencies of pollination

Based on the pollinating agent fruits can be grouped into (i) Anemophillus (by wind), (ii) Hydrophillus (by water), (iii) Entomophillus (by insect). The wind and insect are the major pollinating agents.

Fruits and their pollinating agents are as under

S. No.	Pollinating agents	Fruit crops
1.	House fly	mango
2.	Hony bee, wasp and other insects	apple, pear, plum, almond
3.	Wind	papaya, date palm, walnut, aonla
4.	Bird	banana and pineapple

1.8 Self incompatibility

It is the inability of a functional male and female gametes of the hermaphrodite flowers to set seeds on self pollination.

1.8.1 Genetic control of self incompatibility

Incompatibility is generally controlled by a special gene at S-locus represented by multiple allelic series in the population, each of these alleles control the formation of a specific substance that determines the incompatibility reactions, both in the pistil and pollen. Identical substances specified by identical genes in pollen and pistil favours to prevent fertilization. Based on the timing and mode of S-gene activity, the incompatibility reaction among homomorphic angiosperm is categorized into two group.

(A) Gametophytic control of pollen reaction

(B) Sporophytic control of pollen reaction

1.8.1.1 Gametophytic self incompatibility

In this type of incompatibility, pollen is binucliate and pollen behaviour is determined by the S allele present in each pollen and in this type of incompatibility stigma is wet type. It means the incompatibility reaction of pollen is determined by its own genotypes, and not by the genotype of the plant on which it is produced. Generally, incompatibility reaction is determined by a single gene having multiple alleles. Sometimes, polyploidy may lead to the loss of incompatibility due to a competition between the two S alleles present in diploid pollen. Important examples are pineapple, loquat, apple, pear, plum, cherry, almond, apricot, some citrus and members of Solanaceae family.

1.8.1.2 Sporophytic incompatibility

The incompatibility reaction of pollen is governed by the genotype of plant on which the pollen is produced, not by the genotype of the pollen. It means the incompatibility is imposed by the maternal genotype, due to that all the pollen grains from a given plant behave similarly. Incompatibility occurs at the stigmatic surface resulting in the inhibition of pollen germination. Pollens are trinucleate and the stigmatic surface is dry e.g.*Mangifera indica*.

1.8.2 Mechanism of self incompatibility

Based on the various phenomenon observed in self incompatible mating it can be grouped in to three :

i) Pollen stigma interaction.

ii) Pollen tube style interaction.

iii) Pollen tube ovule interaction.

1.8.2.1 Pollen stigma interaction

This interaction occurs just after the pollen grains reach the stigma and generally prevent pollen germination. In the gametophytic system, stigma surface is plumose having elongated receptive cells and is commonly known as wet stigma. Incompatibility reaction occurs at a later stage. There are clear cut serological differences among the pollen grains with different S genotypes, such differences have not been observed in sporophytic system.

In sporophytic system, stigma is papillate and dry covered with a hydrated layer of proteins known as pellicle. There is evidence that the pellicle is involved in incompatibility reaction. There are striking differences in the stigma antigens related to the S allele composition. Within few minutes of reaching the stigmatic surface, the pollen releases an exine exudates which is either protein or glycoprotein in nature. This exudate induces immediate callose formation in papilae (which are in direct contact with the pollen) of incompatible stigma. Often callose is also deposited on the young protruding pollen tubes preventing any further germination of the pollen. Thus, in the sporophytic system, stigma is the site of incompatibility reaction.The incompatibility reaction of pollen is probably due to the deposition of some compounds from anther tapetum on to the pollen exine.

1.8.2.2 Pollen tube style interaction

In most of the gametophytic system, pollen grains germinate and pollen tubes penetrate the stigmatic surface. But in the incompatible combinations, the growth of pollen tube is retarded within the stigma.

1.8.2.3 Pollen tube ovule interaction

In some cases, pollen tube reaches the ovule and affect fertilization. However, in incompatible combinations, embryo degenerate at early stage of development.

1.8.3 Methods of overcoming self-incompatibility

One of the following methods can be used for bringing partial fertility by temporarily suppressing the incompatibility reaction.

- **Bud pollination** - Application of mature pollens to immature non-receptive stigma i.e. 1-2 days prior to anthesis.

- **Surgical technique** - Removal of stigmatic surface.
- **High temperatue** - Exposure of pistils to temperature up to 60^0C.
- **Irradiation** - With x rays or g rays for single locus gametophytic incompatibility.
- **Double pollination** - Incompatible pollen is applied as mixture with a compatible pollen.
- Pollination at the end of season.
- Arora and Singh (1988) observed that in low chilling plum and peach cultivars, methanol killed the mentor pallen and not helpful in overcoming incompatibility barriers, however, frozen and thawed mentor pollen (one which, if alive would be fully compatible with style receiving it) improved fruit set in both intra and interspecific incompatibility.

In case of sporophytic incompatibility system the break down is comparatively easy because the incompatibility reaction takes place between stigmatic surface and pollen wall in comparison to gametophytic incompatibility in which reaction starts when the pollen tubes has already travelled 1/3 to ½ length of styler tissue (Arora, 1993).

1.8.4 Advantages of self-incompatibility

1. Where male sterility is nonexistent, self-incompatibility can alternatively facilitate the production of F_1 hybrids.
2. Self fertility can be induced temporarily or permanently by mutation of S allels to S_f through artificial irradiation in clonally propagated orchard species like cherry and apple (Lewis, 1956).
3. Seedless varieties, such as in pineapple, grape etc. can be evolved if self-incompatibility is present.

1.8.5 Disadvantages

- Variations in seed set due to poor fertility.
- Poor preservation of genetic purity of improved varieties since cross pollination is non-restricted.
- Difficulties in development and maintenance of homozygous lines (inbreds) which can be utilized for hybridization.
- Uneven quality of fruits because of mixed planting of different varieties based on their cross compatibility.

1.9 Male sterility

Male sterility is characterized by non-functional pollen grains, while female gametes are functional. Male sterility can be classified into three groups genetic male sterility, cytoplasmic male sterility, and cytoplasmic genetic male sterility.

1.9.1 Genetic male sterility

Like any other morphological traits, particularly mono and oligogenic, this type of male sterility occurs in plant due to mutation of the fertility locus, situated on a chromosomes within the nucleus. In this case cytoplasm is not involved in bringing the sterility. There could be three possible genotypes for this locus and only one of them is male sterile.

Fertile (R-line) = RR

Fertilie (B-line) = Rr

Sterile (A-line) = rr

1.9.1.1 Sterility maintenance

By crossing AxB lines sterile and fertile progenies are produced in equal proportions.

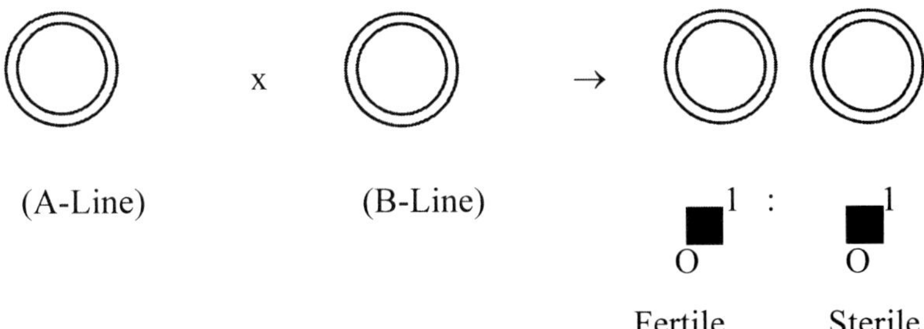

For the maintenance of sterile line the fertile plants need to be quickly removed before the shedding of the pollen grains. The fertile plants can be removed in early stage of plant growth by using marker gene.

1.9.1.2 Fertility restoration

Fertile line can be obtained by crossing A-line with R-line.

e.g.

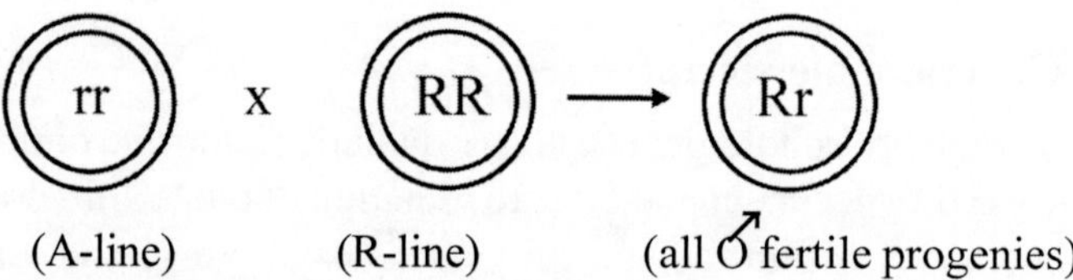

Uses: It can be used in hybrid seed production and genetical studies or for the preservation of variability.

1.9.2 Cytoplasmic male sterility

It occurs due to the mutation of mitochondria or to some other cytoplasmic factors outside the nucleus, resulting in the transformation of the fertile cytoplasm in to a sterile one. Nuclear genes are not involved. Further, with two types of cytoplasm i.e. sterile and fertile. At the most only two kinds of genotypes are possible, one of them is sterile and another fertile. The fertile cytoplasm is denoted by (F – B Line) and sterile cytoplasm is denoted by (f - A-Line)

1.9.2.1 Sterility maintenance

Due to two different types of genotype cytoplasmic sterility can be maintained as under:

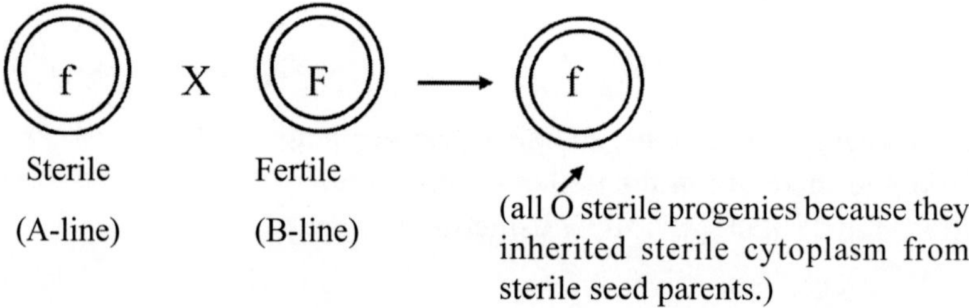

1.9.2.2 Fertility restoration

Since there is no third type of genotype which can act as R-line, as such restoration of fertility is not feasible. However, this does not exhaust all the possibilities of the use of cytoplasmic sterile lines.

1.9.2.3 Uses

As restoration is not possible, this type of sterility is useful only in crops where the seed is not desired end product. This is important for horticultural crops where vegetative parts are of economic value.

1.9.3 Cytoplasmic-genetic male sterility

Such sterility arises from the interaction of nuclear gene (s) and conditioning sterility with sterile cytoplasm. The cytoplasmic genetic sterility is essentially a cytoplasmic sterility with a provision for restoration of fertility. The fertility is restored by (R) gene present in the nucleus. The combination of both nuclear gene (s) and cytoplasmic factors determine the fertility or sterility in such plants. Based on these combinations, there can be maximum of six types of genotype and only one of them is sterile.

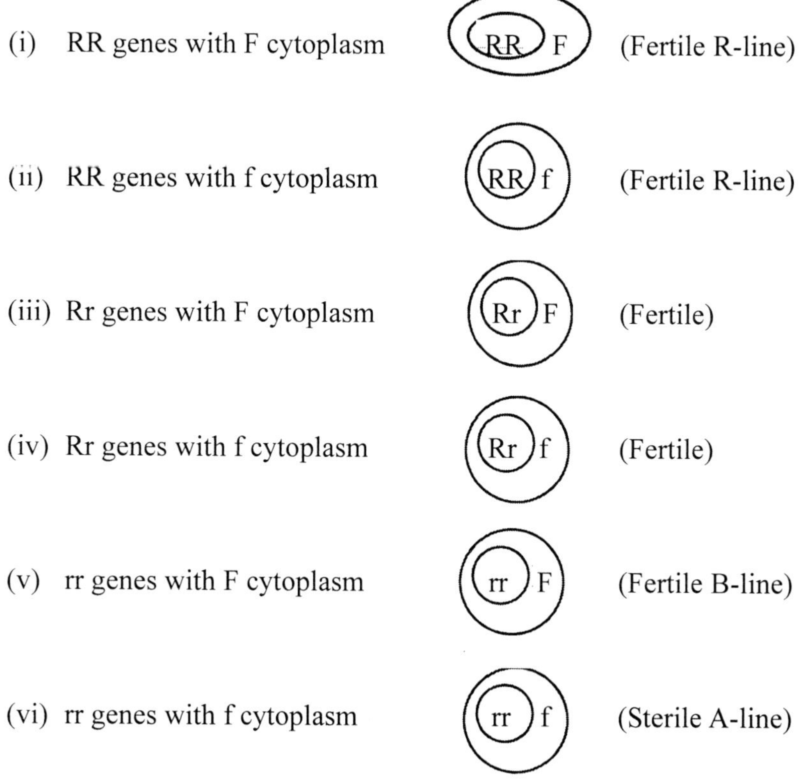

1.9.3.1 Sterility maintenance

As visualized by their genetic composition and cytoplasm, only [(rr) f] genotype can maintain the sterility of A-line.

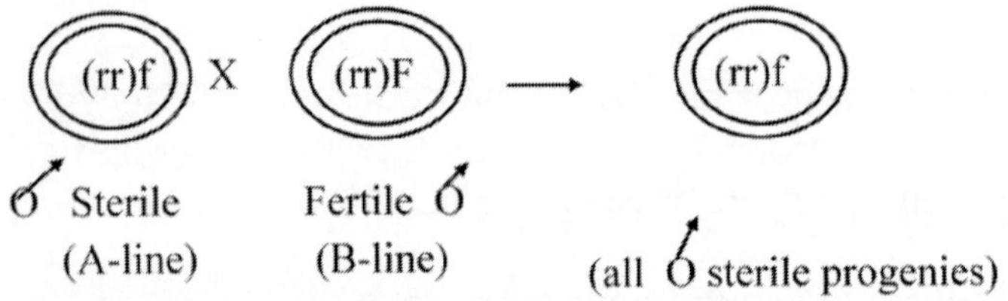

(rr)f X (rr)F → (rr)f

♂ Sterile (A-line) Fertile ♂ (B-line) (all ♂ sterile progenies)

1.9.3.2 Fertility restoration

This is achieved by suitable restorer lines which can give rise to all fertile progenies on crossing with A-line. Among the possible six genotypes, only [(RR) F] and [(RR)f] are such restorer or R-line. They produced all fertile progenies.

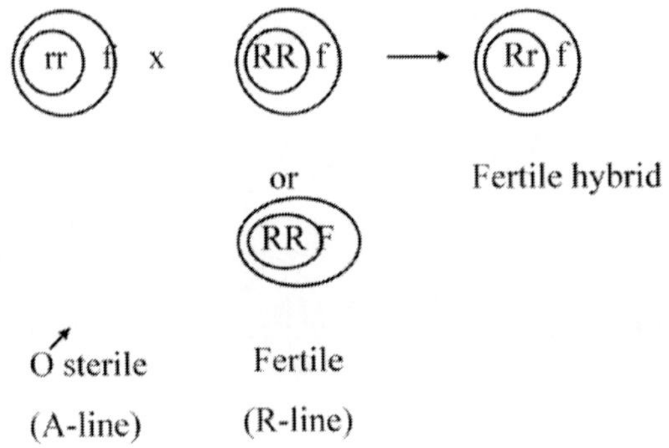

(rr) f x (RR) f → (Rr) f

or (RR) F

Fertile hybrid

♂ sterile (A-line) Fertile (R-line)

1.9.3.3 Uses

Cytoplasmic-genetic male sterile lines are of immense importance in exploitation of hybrid vigour in crops where seed is desired end product.

1.10 Centres of origin (primary and secondary)

The concept of centres of origin was given by N. I. Vavilov based on his studies of a vast collection of plants at the institute of plant industry, Laningrad. He was the director of this institute from 1916 to 1936. According to Vavilov, crop plants evolved from wild species in the areas showing great diversity and termed them as primary centres of origin. But in some areas, certain crop species show considerable diversity of forms although they did not originate there, such areas are known as secondary centre of origin of these species. Eight main centres of origin are recognized as proposed by Vavilov.

1.10.1 China centre of origin

This is one of the largest and oldest centres of origin. It includes mountaineous parts of Central and Western China besides, neighbouring lowlands.

Examples

Pears (*Pyrus communis*)

Peach (*Prunus persica*)

Apricot (*Prunus armeniaca*)

Plum (*Prunus divaricata*)

Orange (*Citrus nobilis*)

1.10.2 Hindustan centre of origin

This centre includes Burma, Assam, Malaya Archipelago, Java, Borneo, Sumatra and Philippines. But this centre does not include North West India, Punjab and North Western Frontier Provinces. Later on this centre of origin is devided into Indo-Burma and Siam-Malaya-Java centre of origin.

Examples

Mango (*Mangifera indica)*

Sour lime (*Citrus aurantifolia*)

Mandarin*(Citrus reticulata*) Some other citrus species

Coconut (*Cocos nucifera*)

Banana (*Musa sapientum*)

1.10.3 Central Asia centre of origin

It is also known as the Afghanistan centre of origin. It includes North West India (Punjab, North-West Frontier Provinces and Kashmir), all Afghanistan, Soviet Republics of Tadjikistan and Uzbekistan and Tian-Shan.

Examples

Pistachionut (*Pistacia vera*)

Almond (*Prunus amygdalus*)

Grape (*Vitis vinifera*)

Apple (*Malus* sp.)

Some species of apricot and pear

1.10.4 Asia minor centre of origin

It includes the interior of Asia minor, the whole of Transcaucasia, Iran, and high lands of Turkmenistan. This centre is also known as the Near East or Persian centre of origin.

Primary centre of origin

Fig (*Ficus carica*)

Pomegrante (*Punica granatum*)

Some species of apple, P*yrus*, *Prunus* and grape

Secondary centre of origin

Chestnut and pistachionut

1.10.5 Mediterranean cetnre of origin Example Pippermint (*Mentha* sp.)

Abyssinian centre of origin It includes Ethiopia and hilly country of Eritrea.

Example : Coffee

1.10.6 Central American centre of origin

This includes region of South Mexico and Central America. It is also known as Mexican centre of origin.

Examples

Papaya (*Carica papaya*)

Guava (*Psidium guajava*)

Avocado (*Persia americana*)

1.10.7 South American centre of origin

This centre includes the high mountainous regions of Peru, Bolivia, Ecuador, Colombia, Parts of Chilie and Brazil and whole of Paraguay. Further, this centre was sub divided into three centres i.e. Peru, Chile and Brazil-Paraguay centre of origin.

Examples

Pineapple

Some species of guava

1.11 Methods of fruit breeding

1.11.1 Introduction

Plant introduction pertains to taking a genotype or a group of genotypes in to a new environments, where they were not being grown before. The systematic introduction and exchange of plant genetic resources of horticultural plants started in 1946 by Division of Botany, IARI, New Delhi, which was replaced by the plant introduction and exploration organization in 1956 and then by the Division of Plant Introduction in 1961. Later in 1976 National Bureau of Plant Genetic Resources (NBPGR) was established in New Delhi. The important cultivars and species of different fruit crops introduced from time to time for various purposes are as under: (Singh and Rana, 1993)

Banana - Cultivar Lady Finger (EC 160160) from Australia as resistant/tolerant to bunchy top disease, Grand Nain MS (EC27237) from France and West Indies.

Citrus - Torocco Blood from USA, Sunramon from Peru, Vainiglia Sanguigno (pigmented) from USA, Tha and Thai-1 from Thailand.

Date palm - Tayar, Hatemi, Mejnaj, Khalas, Ruziz and Khesab from Saudi Arabia. Samani, Zaghlool, Hayani, Samy, Ari, Agolani, Saklote and Amhat from Egypt.

Fig - Genoa white, 0030Genoa, 0009 Flanders,Candria (Adriatic type Fig) from USA. Mission, Ficus Rocdin-3 and Ficus standard from USA.

Grape - Thompson Seedless, Perlette and Beauty Seedless from USA. Kishmish Beli and Kishmish Charni from USSR. Totlocha from Brazil and Dogridge from USA.

Guava - Beaumont G-35 and Indonesia Seedless from Australia. Verdie and M 25988 from USA.

Mango - Tomy, Zilets, Haden, Sensation Julie from USA.

Papaya - PI41436, Soniyimma, Malinchly from Nigeria, Sunrise, Wilder, Sunset T88/211, *C. cauliflora* from USA.

Pomegranate - Wonderful from USA, Rannij G1-8-23, and Rannyj G-1-3-34 from formerly USSR.

1.11.2 Procedure of introduction

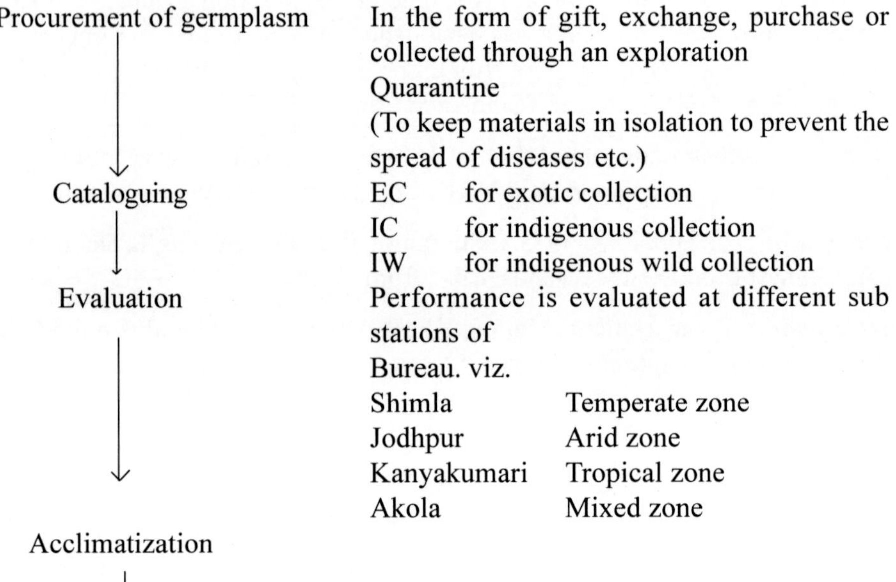

Multiplication and distribution

1.11.3 Merits of introduction

- It provides entirely new crop plants i.e. kiwi fruit.
- It may provide superior varieties either directly, or after selection or hybridization.
- Introduction and exploration are the only feasible means of collecting germplasm and to protect variability from genetic erosion.
- It is quick and economical method of crop improvement.
- Plants may be introduced in new disease free area to protect them from damage.

1.11.4 Demerit

The disadvantanges of plant introductions are associated with the introduction of foreign weeds, diseases and pests.

1.12 Selection

Most of the present day cultivars of fruits are the result of open pollinated seedling selection from the mixed population. For the selection there must be:

Large population

Variation must be present in the population

Variation must be heritable

1.12.1 Achievements

S.No.	Name of the Fruit	Name of the cultivars
1.	Mango	Langra , S.B. Chausa, Sukul, Dashehari, Bombay Green etc. most of the commercial cultivar of mango
2.	Guava	L-49, Arka Mridula, Allahabad Safeda, Pant Prabhat, Lalit.
3.	Litchi	Swarn Roopa
4.	Aonla	Kanchan (NA4), Krishna (NA5), NA6, NA7
5.	Papaya	Co-1, Co-2, Co-5, Co-6, Pusa Delicious, Pusa Majesty, Pusa Giant, Pusa Dwarf, Coorg honey dew.
6.	Kagzi-lime	Vikram, Pramalini, Saisarbati, Jaidevi
7.	Lemon	Pant lemon
8.	Bael	NB-5, NB-9, CISHB-1, CISHB-2, Pant Aparna, Pant Shivani, Pant Urvashi, Pant Sujata
9.	Sapota	Co-2 and PKM-1

1.13 Clone and clonal selection

Clone

A clone is a group of plants produced exclusively from a single individual plant through asexual reproduction. Most of the fruit plants are propagated asexually which consists of large number of clones that is why these plants are known as a group of plant derived from a single plant by vegetative means. In other word all the vegetative progenies of a single plant make a clone.

1.13.1 Characteristics

- **Clones are stable**- Clones are retain their original traits just like pure line veriey.
- **Theoretically clones are immortal**- A clone can be maintained indefinitely by asexual reproduction but these are very much susceptible to diseases or insect pest depending upon the species and cultivar.
- **Homogeneous**- Individual plant of a clone is a mitotic derivative of the same plant and therefore, homogeneity in phenotypy is major feature of clones. Same tissue of the original plant which has multiplied into a group of plants and each plant has the same genotype. Phenotypic variation in clones is due to environmental impact.

- Individual clone is heterozygous in nature.
- The phenotype of a clone is due to effect of gene (G), environment (E) and G x E interaction over the population mean (h). Therefore P=h+G+E+GE.
- Clones are vegetatively propagated plants.

1.13.2 Genetic variation within clones

Genetic variation within clones may be due to mutation, mechanical mixture and sexual reproduction.

Clonal degeneration

The loss in vigour and productivity of clones with the passing of time is known as clonal degeneration and it may be due to mutation, infection of virus and bacteria.

Clonal selection

Clonal selection is the selection of devisable clone from the mixed population of asexually propagated plant.

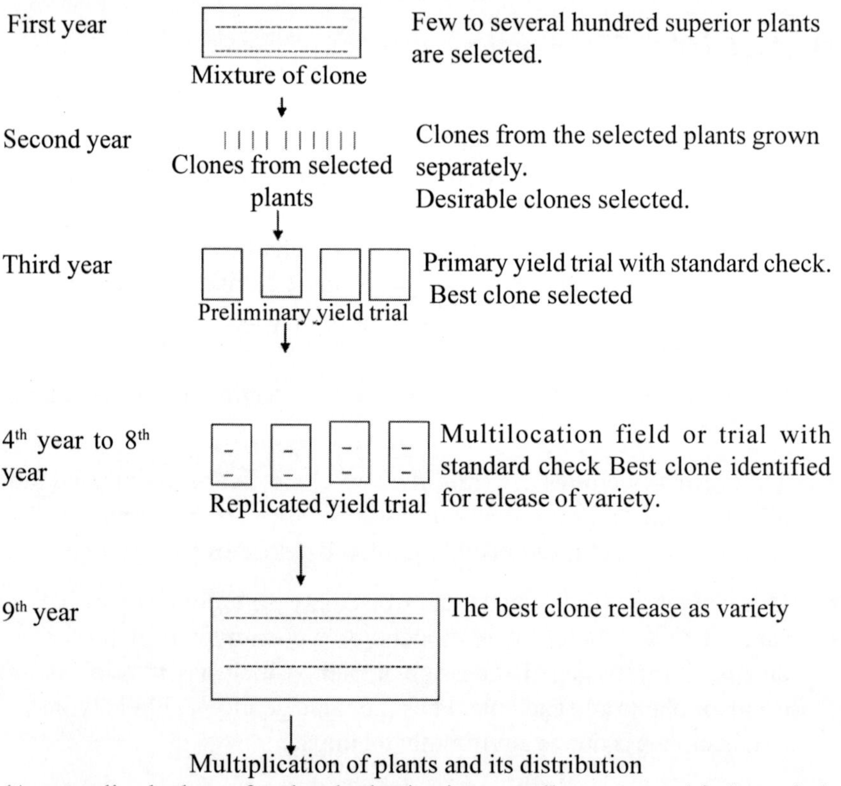

(A generalized scheme for clonal selection in asexually propagated fruit species)

Generally clonal selection involves

i) Collection of clonal variability.

ii) Evaluation of clone with respect to yield, quality and biotic stress resistance. The high yielding and disease free clones are selected and further evaluated and multiplied for distribution of planting material to the farmers.

1.13.3 Advantages

i) Clonal selection is easy and less time-consuming method.

ii) Easy maintenance because there is no problem of outcrossing and loss of seed viability. Variation occurs due to somatic mutation only, which can be managed by removal of undesired plants.

iii) Permanent hybrid - Heterotic clones on selection may be used as permanent hybrid. Heterosis can be exploited for longer time without production of hybrid seed every year (for vegetatively propagated vegetable crops).

iv) Clonal selection is only the method of breeding in vegetatively propagated fruit plants.

1.13.4 Limitations

- There is limited chance of getting new and useful type of variability.
- Low multiplication rate.
- It is only useful for vegetatively propagated plants.

1.13.5 Achievements

Clone No. 51 from Dashehari, MA-1 from Alphanso, Tomy Atkin from Haden, Pusa Surya from Elden in mango, Pusa Seedless from Thompson Seedless of grape etc.

1.14 Mutation

Sudden heritable change in characteristic of the plants is termed as mutation. It may be spontaneous (without any treatment by man) or induced (artificially induced by a treatment with certain physical or chemical agents) in plant population.

1.14.1 Procedure of mutation breeding

Treating a biological material with a mutagen in order to induce mutation is known as mutagensis. When mutations are induced for crop improvement the entire operation of induction and isolation of mutants is termed as mutation breeding. The various steps involved in mutation breeding are as under

- Objectives of programme- Objective should be clear cut and well defined.
- Selection of variety for mutagen treatment –Locally accepted best veriety in which improvement is needed either in polygenic or monogenic trait.
- Part of plant to be treated- Seeds, pollen grains or vegetative propagule (buds and cuttings) may be used for mutagenesis. Selection of plant part for mutagenic treatments based on mode of multiplication / reproduction. In sexually propagated fruit plants, seed treatment is common, pollen grain may be used but it has some limitations such as it is difficult to collect large amount of pollen grains and pollen survival life is short. In case of a sexually propagated fruit plant, buds or cuttings are used for mutagenic treatment.
- Dose of mutagen-An optimum dose is that one which produces the maximum frequency of mutation and causes the minimum killing i.e. LD 50. It is that dose of mutagen which would kill 50% of the treated individual. Dose of mutagen depends upon intensity and time of treatment.
- Mutagen treatment- Selected plant part is exposed to the desired mutagen dose. In case of irradiation, the plant are immediately planted to raise M_1 plants from them. In case of seed treatment, they are pre-soaked for a few hours to initiate metabolic activities and then exposed to mutagen. Treated seeds are sown immediately in field to raise M_1 generation. When pollen grains are treated, germination resulting from the seeds produced by treated pollen grain would be the M_1 generation. M_2, M_3, M_4 etc. are subsequent generations derived from M_1, M_2, M_3 etc plants through selfing or clonal propagation.
- Handling of mutagen treated populations.

1.14.2 General characteristics of mutation

Mutations are generally recessive but dominant mutations also occur.

Mutations are generally harmful to the organism.

Mutations are random.

Mutations are recurrent.

Induced mutations commonly show pleiotrophy, often due to mutations in closely linked gene.

1.14.3 Mutagens

Agent used for induction of mutations, are known as mutagens. The different mutagens may be grouped as follows

1.14.3.1 Physical mutagens

1. Ionizing radiations

a) Particulate radiations - a-rays, b-rays, fast neutron, thermal neutrons.

b) Non particulate radiations - X-rays, g-rays.

2. Non ionizing radiation - Ultra violet radiation

1.14.3.2 Chemical mutagens

1. Alkylating agents - Sulphur mustard, nitrogen mustards, EMS (Ethylmethane sulphonate) etc.
2. Acridine dyes - acriflavin, proflavin, acridine orange, acridine yellow, ethedium bromide.
3. Base analogues - 5-bromouracil, 5-Chlorouracil.
4. Others - Nitric acid, hydroxyl amine.

1.14.4 Achievements

Mango - Rosica from Peruvian variety Rosadodelca.

Papaya - Pusa Nanha from local type

Grape - Marvel Seedless from Delight

Banana - Highgate from Gros michel, Motta poovan from Poovan

Orange - Washington Navel

Grapefruits - Marsh and Thompson

1.15 Polyploidy breeding

The individuals carrying chromosome number other than diploid (2x and not 2n) number are known as polyploid and the situation is known as heteroploidy. Polyploidy breeding seems to have prospects of obtaining large sized fruit with dwarf plant type (Illaichi cultivar of ber is octaploid but size of fruit is very small is exception). Production of triploid by hybridizing tetraploids with diploids may be useful in obtaining seedless varieties. Change in chromosome number may contribute in crop evolution as polyploidy significantly contributes to the evolution of the genus *Psidium* (Subramanyam and Iyer, 1993). Majumder and Mukherjee (1972a and 1972b) made the crosses between triploid (seedless) and diploid (Allahabad Safeda). Out of 73 hybrid seedlings, 26 were diploids, 9 trisomic, 5 double trisomic and 14 tetrasomic. They showed distinct variations in the tree habit, leaf and fruit characters. The three trisomic plants had dwarf growth habit and normal shape and size of fruits with few seeds.

1.16 Hybridization

The mating or crossing of two plants or lines of dissimilar genotypes is known as hybridization.

1.16.1 Types of hybridization

1.16.1.1 Intervarietal hybridization

The parents involved in hybridization belong to the same species. They may be two strains, varieties or races of the same species. It is also known as intraspecific hybridization. The intervarietal crosses may be simple or complex depending upon the number of parents involved.

a. **Simple cross** In a simple cross, two parents are crossed to produce the F_1

A x B
↓
F_1

b. **Complex cross** More than two parents are crossed to produce the hybrid.

example

Three parent cross (A,B,C)

A x B

↓

(A B) x C

↓

(A B C) (Complete hybrid)

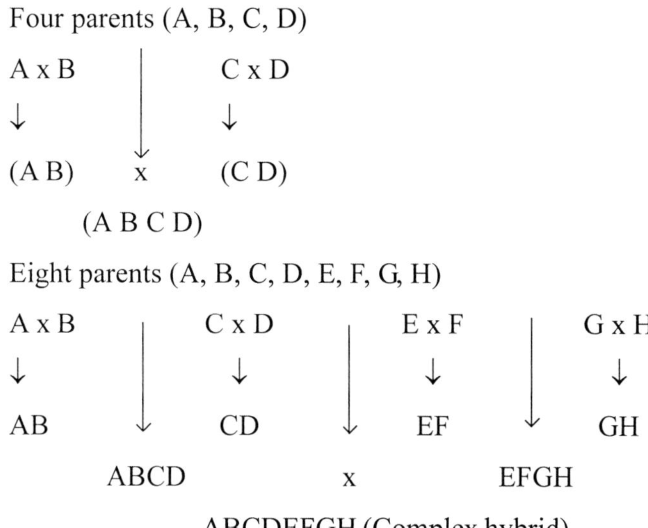

1.16.1.2 Distant hybridization

Distant hybridization includes crosses between different species of the same genus or of different genera. When two species of the same genus are crossed, it is known as inter specific hybridization, but when they belong to two different genera, it is intergeneric hybridization.

1.16.2 Procedure of hybridization programme

There are seven steps involved in hybridization.

1.16.2.1 Choice of parents

It mainly depends upon the objective of breeding programme. In addition to other objectives, increased yield is always an objective of the breeder.

1.16.2.2 Evaluation of parents

If the performance of parents used in breeding is known, evaluation is not necessary. But if their performance is not known, it should be evaluated, particularly for the characters to which they are expected to contribute.

1.16.2.3 Emasculation

The removal of the stamens or anthers or the killing of pollen grains of a flower without affecting in any way to female reproductive organs is known as emasculation. The purpose of emasculation is to prevent self-fertilization in the flowers of female parent.

1.16.2.3.1 Types of emasculation

1. Hand emasculation
2. Suction emasculation
3. Hot water emasculation
4. Alcohol treatment
5. Cold treatment
6. Genetic emasculation e.g. male sterility

1.16.2.4 Bagging

Immediately after emasculation, the flowers or the inflorescence are closed in suitable bags of appropriate size to prevent random cross pollination.

1.16.2.5 Tagging

Emasculated flowers are tagged just after bagging. The following information is recorded on the tags with a carbon pencil.

1. Date of emasculation
2. Date of pollination
3. Name of the female and male parents. The name of female parent written first, and that of male parent is written later.

1.16.2.6 Pollination

In case of pollination, mature, fertile and viable pollen should be placed on a receptive stigma to bring fertilization.

1.16.2.7 Harvesting and storage of F_1 seeds

The crossed fruits along with seed should be harvested. For recalcitrant seeds sowing is done immediately after extraction of the seeds. However, orthodox seeds of some fruit plants can be stored for longer time.

1.16.3 Achievements

Fruit	Varieties
Mango	Mallika, Amrapali, Pusa Arunima, Arka Anmol, Arka Puneet, Arka Aruna, Arka Neelkiran, Ratna, Sindhu, PKM-1, PKM-2
Guava	Arka Amulya, Safed Jam, Kohir Safed
Litchi	Sabohar Madhu, Sabohar Priya
Papaya	CO-3, CO-2
Sapota	CO-1, DHS-1, DHS-2, Hybrid 214, Hybrid - 711

1.17 Biotechnological approach

Realizing the importance of biotechnology in National development, the Govt. of India set up a full-fledged Department of Biotechnology (DBT) in 1986 to coordinate and oversee priority areas. DBT has initiated a number of programmes to promote fruit industries. As a result of this, biotechnological revolution, has taken place in horticulture. Biotechnological tools are appropriate for accelerating the productivity. Application of biotechnological tools in plant improvement has been found effective in three ways (i) rapid multiplication of existing allied clones and varieties, (ii) speeding up the process of conventional breeding and (iii) conservation of germplasm and evolving novel genotypes through genetic engineering technology. The different biotechnological processes are as under:

1.17.1 Micropropagation

Superior selections and hybrids developed at various research centres failed to reach the orchardists due to lack of sufficient planting materials. It leads to non realization of the potential of improved cultivars, thus making the efforts of fruit improvement programme unfruitful. In this case micropropagation can be a powerful tool for large scale propagation of fruit crops. This is also an ideal system for production of disease free plants. Among the fruits, micropropagation has been most successful in banana, papaya and datepalm multiplication. Rapid clonal multiplication through culture of apical shoots and nodal explants has been reported in mulberry (Pattnaik and Chand, 1997). Long term micropropagation of passion fruit by formation of multiple shoot primordial initiated from leaf explants (Kawata *et al*., 1995). *In vitro* propagation of grape vine is also possible (Heloir *et al.,* 1997, Gray and Fisher, 1985).

1.17.2 Conservation of germplasm

The potential importance of natural gene pool to meet the future need of crop improvement by introducing specific characters such as biotic and abiotic stress resistance cannot be under estimated. However, the number of wild species and their natural habitats are disappearing rapidly, as a result of introduction of highly bred modern varieties, growing urbanization and an increased pressure on land for cultivation. This leads to the erosion of the natural germplasm to such extent that, there is a general fear that potentially valuable germplasm is being lost irretrievably. In plant improvement it is necessary to facilitate the assimilation of germplasm collection in working collection for the use of the breeders. The process of genetic erosion necessitates measure that germplasm must be conserved in such a manner that there should be minimal losses of gentic variability of a population. The conventional methods of germplsam conservation are proven to possible catastrophic losses due to (i) Attack by pest and pathogens (ii) Climatic disorders (iii) Natural disasters and (iv) Political and economic causes. Besides this, the seeds of many important fruit plants such as mango, litchi etc. may lose their viability in a short time under conventional storage system. NBPGR, New Delhi is maintaining large *in-vitro* germplasm collection of bananas, phalsa, bael, jackfruit, pomegranate etc. There are two basic approaches followed to maintain the germplasm *in-vitro*.

(i) Minimal growth

It is most applicable for short and medium term storage. It can be achieved by reduced temperature and light, incorporation of sub-lethal levels of growth retardants, induction of osmotic stress and maintenance of culture of a reduced nutritional status particularly reduced carbon, reduction of gas pressure over the culture. Advantage of this approach is that culture can be readily brought back to normal culture conditions to produce plants on demand. But the disadvantage is that culture may subjected to contamination and somaclonal variation.

(ii) Cryopreservation

Cryopreservation at ultra low temperature of liquid nitrogen (-196° C), offers the possibility for long term storage with maximum phenotypic and genotypic stability.

1.17.3 Anther culture

In vitro androgenesis holds a myriad of possibilities for improvement of horticultural crops. This technology has been extended to a number of horticultural crops. The pupose of anther and pollen culture is to produce haploid plants by the induction of embryogenesis from repeated divisions of monoploid spores, either microspore or immature pollen grains.

The major interest in haploids is based upon there being on intermediate in the production of homozygous plants produced by breeding as an alternative to repeated cycles of inbreeding in self pollinated crops. In cross pollinated species, double haploids are more likely to be used as parents in the production of single or double cross hybrids. The advantage of haploids in fruit plants are as follows:

- As a result of haploid induction, followed by chromosome doubling, homozygosity is attained in a very short time. This is particularly useful in heterozygous and self-incompatible crops like mango, etc.
- With the use of homozygous parents in crossing programme the production of pure F_1 hybrids become possible.

- Haploid cell lines have great advantages in studies on mutant selection in *in vitro*.
- Plant transformation through *Agrobacterium* mediated gene transfer and direct DNA transfer is the fore front areas of plant biotechnology.

1.17.4 Overcoming crossing barriers (embryo culture)

This technique pertains to cultivation of excised plant embryo in artificial medium. Embryo culture technique has found its application both in the applied and basic research. In the conventional plant breeding programme, breeder often faces problem in transfering resistance from wild species to the cultivated in getting the desirable interspecific hybrids (Yeung *et al.*,1981). Application of embryo rescue can overcome some of the pre and post fertilization barriers in fruit crops. Further, most useful and popular application of zygotic embryo culture has been in raising hybrids. Embryo culture technique has important role in haploid production, shortening of breeding cycle (Lammerts, 1942), rapid seed viability test and propagation of rare plants.

1.17.5 Somaclonal variation

Somaclonal variation explores the naturally occuring or *in vitro* induced variability of somatic cells following plant regeneration. Somaclonal variation is an excellent method for shortening breeding programmes. Somaclonal variation may be due to variation in chromosomes number, structural variation of chromosomes due to deletions, duplication, translocation, genetic and cytoplasmic mutation etc.

Hwang and Ko (1987) identified somaclonal variations in the cultivars Giant Cavendish with Putative field resistance to Fusarium wilt (race 4) but inferior in agronomic characters. A somaclonal variant of Cavendish expressing resistance to yellow sigatoka disease with satisfactory yield has been reported (Chadha and Sahijram, 2000).

1.17.6 Somatic hybridization

It is an approach of immense value in the area of fruit breeding. Somatic hybridization provides a method where sexual incompatibility in the plants can be by-passed. Protoplast culture includes a series of operations such as isolation of the protoplasts from cells, culturing them in a suitable medium, inducing them to divide and then regenerating plantlets from them. Fusion of protoplasts may occur spontaneously or they may be induced to fuse in the presence of fusigenic agents. The polyethylene glycol (PEG) is the most widely used fusigenic agent (Chadha *et al.*, 2000)

Fruit plants in which androgenesis has been induced

S.No.	Name of the Species	Mode	Response	Chromosome constitution	References
1	*Ananas comosus*	Indirect androgenesis	Callus	-	Banega *et al.*, 1996
2	*Annona squamosa*	Indirect androgenesis	Callus, Plants	n	Nair *et al.*,1983
3	*Carica papaya*	Direct androgenesis	Plants	n	Litz and Conover,1978
		Indirect androgenesis	Callus, embryo,plants	n	Su and Tsay,1985, Litz and Conover, 1986
		Direct/indirect androgenesis	Callus, embryo, Plants	n	Tsay and Su, 1985
4	*Citrus aurantium*	Direct androgenesis	Embryo, Plants	-	Hidaka *et al.*, 1982
		Indirect androgenesis	Callus, embryo	-	Hidaka, 1984a
5	*Litchi chinensis*	Indirect androgenesis	Callus, embryo, plants		n Fu and Tang, 1983
6	*Malus* x *domestica*	Indirect androgenesis	Callus, Plants	-	Kubicki *et al.*,1975
		Indirect androgenesis	Callus, embryo, plants	n, mixoploids	Fei and Xue, 1981
7	*Passiflora edulis*	Indirect androgenesis	Callus	-	Tsay *et al.*, 1984
8	*Poncirus trifoliata*	Direct androgenesis	Plants	n	Hidaka *et al.*, 1979
		Indirect androgenesis	Callus, embryo	-	Hidaka, 1984b
		Direct androgenesis	Embryo, plants	-	Chen, 1992
9	*Prunus amygdalus*	Indirect androgenesis	Callus	n	Michellon *et al.*, 1974
10	*Prunus armeniaca*	Indirect androgenesis	Callus	-	Harn and Kim,1972
11	*Prunus persica*	Indirect androgenesis	Callus	-	Todorovic *et al.*,1992
12	*Psidium guajava*	Indirect androgenesis	Callus	-	Babbar and Gupta, 1986
13	*Punica granatum*	Indirect androgenesis	Callus, embryo, Plants	2n	Moriguchi *et al.*, 1987
14	*Vitis vinifera*	Indirect androgenesis	Callus,Plants	n	Gresshoff and Doy, 1974

Important fruit plants in which protoplast fusion is practised are as under :

Name	Method of fusion
Citrus (Tangelo) + *Murraya paniculata*	Electrofusion
(*Citrus reticulata* x *Citrus paradisi*) + *Citrus jambhiri*	Electrofusion
Citrus sinensis + *Citrus reticulata*	Commercial method-PEG mediated
Citrus sinensis + *Clausena lansium*	Electrofusion

1.17.7 Molecular biology techniques

Morphological characters, chemical composition and cytological information have been used over the years for the classification of plants. However, these techniques have certain limitations as they could be influenced by environmental and developmental effects. The molecular markers are now being increasingly used in the areas of plant classification and breeding. Polygenic characters which are very difficult to analyse using traditional plant breeding methods can be easily analysed using molecular markers.

1.17.7.1 Uses

Molecular markers can be used for -

- Characterization of germplasm
- Varietal identification and clonal fidelity testing.
- Assessment of genetic diversity.
- Validation of genetic relationship.
- Marker assisted selection.
- Developing linkage map in fruit crops. (Helentjaris,1987, Bernatzsky and Tanksley, 1986)

1.17.7.2 Types of molecular marker technique

S.No.	Name of the marker	Specific characters
1	Biochemical marker (enzyme based) Isozyme patterns	Isozymes are different forms of an enzyme exhibiting the same catalytic activity but differing in charge and electrophoretic mobility. In isozyme analysis, crude plant extracts are subjected to electrophoresis using starch or polyacrylamidegels. Following electrophoresis, the enzymes of interest are detected by treating the gels with specific activity strains. Variations in banding patterns obtained between individual sample can be used to sort out the varieties of interest.Advantage: Technique is most simple and less expensive.Disadvantages: i) Availability of limited number of enzyme loci. ii) Developmental and seasonal dependent expression of activity. Example: In mango, enzyme polymorphism has been used to distinguish nucellar from zygotic seedlings (Chemda *et al.*, 1990). In citrus for classification (Dequin *et al.*, 1994, Zheng *et al.*, 1996) and in grape for root stock identification (Walker and Liu, 1995) isozyme pattern was used.
2	DNA based markers	DNA based molecular markers allow direct comparison of the genetic material of two individual plants avoiding any environmental influences on gene expression. DNA based markers are more stable. DNA is the ideal molecule for such analysis (Caetano-Anolles *et al.*, 1991). Polymorphism in the nucleotide sequence is sufficient for it to function as a molecular marker in maping. The important DNA based molecular markers are as under. Retriction Fragment Length Polymorphism (RFLP) Arbitrarily Primered Polymerase Chain Reaction (AP-PCR) Random Amplified Polymorphic DNA (RAPD) Sequence Characterized Amplified Regions (SCAR) Single Primer Amplification Reaction (SPAR) Amplified Fragment Length polymorphism (AFLP) Single Sequence Repeat Polymorphism (SSRP) Single-Strand Conformation Polymorphism (SSCP) Cleaved Amplified Polymorphism (CAP) Sequenced Tagged Sites (STS). **Examples:**

(i) DNA marker in the gentic diversity assesment

Name of the fruits	References
Avocado	Lavi *et al.*, 1991, Mhameed *et al.*, 1997
Banana	Bhat and Jarret 1995, Gawel *et al.,* 1992
Blueberry	Rowland and Levi, 1994
Citrus	Liou *et al.*, 1996, Durham *et al.*, 1992
Grapes	Bourquin *et al.*, 1993, Xianping *et al.,* 1996
Mango	Adato *et al.*, 1995, Schnell *et al.*, 1995
Strawberry	Graham *et al.*, 1996
Walnut	Fjellstrom and Parfitt, 1994, Fjellstrom *et al.*, 1994

(ii) DNA marker in varietal identification

Crop	Marker	Reference
Annona sp.	RARD	Ronning *et al.*, 1995
Apple Minisatellite	RAPD	Koller *et al.*, 1993
Grape cvs.	Repetitive Sequence	Thomas *et al.*, 1993
Grape (rootstock)	RAPD	Hong *et al.*, 1995
Lemon	RAPD	Deng *et al.*, 1995
Mango	RAPD	Schnell e*t al.*, 1995

(iii) Marker assisted selection (MAS) is being increasingly adopted in breeding programmes. Since this approach permits the breeder to make early decisions about his selection while examining fewer plants.The essential requirements for marker-assisted selection in plant breeding programmes are –

a)Marker (s) should co-segregate or be closely linked with the desired trait. (ICM or less)

b) An efficient means of screening large populations for the molecular markers should be available. PCR based technology meets this requirements.

c) Screening technique should have high reproducibility across laboratories, be economical to use and very user friendly.

Crop	Trait of interest	References.
Apple	-Columnar growth habit	Minon *et al.*, 1997
	Scab resistance	Hong *et al.*, 1997
		Tartarin, 1996
	Morphological and development traits	Darlene *et al.*, 1995
	Avocado - Skin colour	Lavi *et al.*, 1995
Citrus	- Differentiation of zygotic and nucellar progeny	Orford *et al.*, 1995
	Dwarfing	Cheng and Roose, 1995
	CTV resistance	Gmitter *et al.*, 1996
Papaya	- Sex determination	Sondur *et al.*, 1996

1.18 Genetic engineering

The advent of recombinant DNA technology has opened tremendous possibilities for transforming almost any plant by transferring any gene from any organism across, taxonomic barriers. The recombinant DNA technology involves following major steps.

- Isolation of gene of desired characters.
- Insertion of the isolated gene in a suitable vector (making it a recombinant vector).
- Transformation - Insertion of the recombinant vector into a suitable host (organism/cell).
- Selection of the transformed host.
- Multiplication followed by expression of the introduced gene into the host.

1.18.1 Gene transfer technology

Important gene transfer methods used for production of transgenic plants are as under

- *Agrobacterium*-mediated transformation (Hohn *et al.*, 1989)
- Microprojectile bombardment- mediated transformation (Sanford, 1990)
- Protoplast-mediated transformation (Paszkowski *et al.*, 1989)
- In-planta electroporation (Chowrira *et al.*, 1996)
- Intact tissue electroporation (D' Halluin *et al.*,1992).
- Silicon carbide fibres (Songstad *et al.*, 1995)
- Electrophoresis (Songstad *et al.*, 1995)
- Microinjection (Neuhaus and Spangenburg, 1990)
- Sonication (Joersbo and Brunstedt, 1992)
- Laser mediated gene transfer (Guo *et al.*, 1995)

1.18.2 Important fruit crops in which transgenic plants are reported

S.No.	Name of the fruits	Technology used and references
1	Kiwi fruit (*Actinidia diliciosa)* Variety Hayward	PGM, Oliveira *et al.*, 1991ABM, Uematsu *et al.*, 1991, Janssen and Gardner, 1993
2	Papaya (*Carica papaya* L.) variety Sunrise SoloKapoho, Solo, Sunset, Maradol, Sunrise	ABM, Yang *et al.*, 1996,Pang and Sanford, 1998PM, Fitch *et al.*, 1990, Fitch *et al.*, 1992
3	Pecannut (*Carya illinoensis*) Elliot 6, Schley S/3, Wichita 9	ABM, Mcgranahan *et al.*, 1993
4	Lime (*Citrus aurantifolia*) Mexican lime seedling	ABM, Pena *et al.*, 1997
5	Rough lemon (*Citrus jambhiri*, Luch)	PGM, Vardi *et al.*, 1990
6	Mandarin (*Citrus reticulata*) Ohta, Ponkan	Electroporation, Hidaka and Omura, 1993
7	Tangelo	PB, Yao *et al.*, 1996
8	Strawberry (*Fragaria x ananassa* Duch)Redicoat, Rapella Line 77101	ABM, Nehra *et al.*, 1990 a and Nehra *et al.*, 1990 bJames *et al.,* 1990EP, Nyman and Wallin ,1992
9	Walnut (*Juglans regia* L. *J. hindsi* Jeps and hybrid paradox, Scharch, Franquette, Sunland	AMB, Dandekar *et al.*, 1988 Dandekar *et al.*, 1994, Mcgranahan *et al.*, 1988, Mcgranahan *et al.*,1990
10	Apple (*Malus* sp) Greensleeves, M-26, Granny Smith, Delicious, Royal Gala.	AMB, James *et al.*, 1989 Dandekar, 1991Lambert and Tepfer, 1992Yao *et al.*, 1995

Contd.

11	Banana (*Musa* species) Bluggoe (ABB group) Cavendish Clone Grand Nain	PB, Sagi *et al.*, 1995EP Sagi *et al.*, 1994PB + *Aqrobacterium cocultiration*) May *et al.*, 1995
12	Tamarillo(*Cyphomandra betacea*) var. Sendter	ABM, Atkinson and Gardner,1993
13	Passion fruit (*Passiflora edulis*) Flavicarpa, Degener	ABM, Manders *et al.*, 1994
14	Avocado (*Persea americana* Mill)	ABM, Cruz-Hernandez *et al.*, 1998
15	Sweet orange (*Citrus sinensis* L. Osb.)	PGM, Kobayashi and Uchimiya, 1989
16	Grape (*Vitis vinifera* L.)	PB+ABM, Scroza *et al.*, 1995ABM, Baribault *et al.*, 1989, Baribault *et al.*, 1990, Nakano *et al.*, 1994
17	Peach (*Prunus persica*)	PB, Xiaojian *et al.*, 1994ABM, Smigocki and Hammerschlag, 1991
18	Plum (*Prunus domestica* L.)	ABM, Scorza *et al.*, 1994
19	Apricot (*Prunus armeniaca* L.)	ABM, Machado *et al.*, 1992, Machado *et al.*, 1994
20	Pear (*Pyrus communis*)	ABM, Merkulov *et al.*, 1998
21	Almond (*Prunus amygdalus*)	ABM, Archilletti *et al.*, 1995
22	Trifoliate orange (*Poncirus trifoliata* L.)	ABM, Kaneyoshi *et al.*,1994
23	Citrange (*Citrus sinensis* x *Poncirus trifoliata*)	ABM, Moore *et al.*, 1992,
24	Cranberry (*Vaccinium macrocarpon* Ait)	PB, Serres and Stang, 1992
25	Red raspberry (*Rubus* sp.) Blackberry (*Ribes* sp.)	ABM, Graham and Mcnicol 1990, Graham and Mcnicol 1991

ABM – *Agrobacterium* mediated, (PGM) –Polyethylene glycol mediated
PB- Particle bombardment, (EP) Electroporation
PB+ AMB- Particle bombardment followed by *Agrobacterium* co-cultivation.

1.18.3 Role of Genetic engineering

Biotechnology in improving shelf life of fruits.

Name of the crops	Gene
Apple	ACC synthase
	Attacin esterase
	Polygalacturonase
Banana	ACC synthase
Strawberry	Acetolactate synthase
	Chitinase
	Glucanase
	PG inhibitor protein
	SAM transferase

Edible vaccines

Transgenic banana is one of the most recent examples,where genes of antigenic proteins of many deadly disease causing pathogens have been expresed in banana fruits. Thus, children can be immunized only by feeding with these transgenic fruits, instead of painful injections for immunization.

Biodegradable plastic

Polyhydroxy butyrate (PHB) is used as a substrate for the production of biodegradable plastic. By transforming the PHB genes into the plants, this could be produced on large scale and at a very low cost, which may reduce the threat imposed by polyethylene to the ecosystem.

Pest management

Fruit crops suffer from a variety of insect pests. It is possible to implement biotechnological approaches to manage insect pests in a rational, durable and ecofriendly manner. Therefore, novel insecticidal proteins and their respective genes need to be identified and used in conjunction with Bt to prevent development of resistant insect. In addition, IPM will have to play a central role in sustainable horticulture.

Disease resistance, herbicides resistance, abiotic resistance etc. are the another areas where genetic engineering can play an important role in imparting resistance in fruit crops. Example: In apple gene attacin (from *Hyalophora cecropia*) lysozyme (fram chicken) and cecropin B (from *H. cecropia*) can be used for disease resistance against *Eriwinia amylovora.*

1.19 References

Adato, A.D., Sharon, V., Lavi, V.S., Hillel, J. and Gazit, S. (1995) Application of DNA finger-prints for identification and genetic analysis of mango (*M. indica*) genotypes. *J. Amer. Sco. Hort. Sci.*, **120**:259-264.

Archilletti, T., Lauri, P. and Damiano, C. (1995) *Agrobacterium* mediated transformation of almond leaf pieces. *Pl. Cell. Rep*., **14**: 267-272.

Arora, R. L. (1993) Incompatibility system in fruits. In: Advances in Horticulture Vol. I (Eds. K. L. Chadha and O. P. Pareek), Malhotra Publishing House, New Delhi, pp 497-514.

Arora, R.L. and Singh R. (1988) Break down of incompatibility in *Prunus salicina* and *Prunus persica. Indian J. Hort*., **45** : 29-44.

Atkinson, R.G. and Gardner, R.C. (1993) Regeneration of transgenic tamarillo plants. *Pl. Cell Rep*., **12**:347-351.

Babbar, S.B. and Gupta, S.C. (1986) Induction of androgensis and callus formation in *in vitro* cultured anthers of a myrtaceous fruit tree (*Psidium guajava* L.). *Bot. Mag. Tokyo*, **99**:75-83.

Banega, R., Isidron, M., Grneros, A., Arias, E., Daquinta, M., Companioni, L. and Martinez, J. (1996) Induction of callus in pineapple anthers. *Cult. Trop*., **17**:72-74.

Baribault, T.J., Skene, K.G.M., and Scott, N.S. (1989) Genetic transformation of grape vine cell. *Pl. Cell Rep.*, **8**:137-140.

Baribault, T.J., Skene, K.G.M., Cain, P.A. and Scott, N.S. (1990) Transgenic grape vines : regeneration of shoots expressing â - glucuronidase. *J. Exptl. Bot.,* **41**:1045-1049.

Bernatzsky, R. and Tanksley, S.D. (1986) Towards a saturated linkage map in tomato based on isozymes and random cDNA sequences. *Genet.*, **112**:887-898.

Bhat , K.V. and Jarret, R.L. (1995) Random amplified polymorphic DNA and genetic diversity in Indian *Musa* germplasm. *GRACO* , **42**:107-118.

Bhatnagar, S.P. and B.M. Johri (1972) Development of angiosperm seeds. In: Seed Biology Vol 1 (Ed T.T. Kozlowski), New york Academic Press, pp 77-149.

Bourquin, J.C., Sonko, A., Otten, L. and Walter, B. (1993) RFLP molecular taxonomy in *Vitis vinifera* L. *Theor. Appl. Genet.*, **87**:431-438.

Caetano-Anolles, G., Bassam, B.J. and Gresshoft, P.M. (1991) DNA amplification finger printing: a strategy for genome analysis. *Plant Mol. Biol. Rep.*, **9**:294-307.

Chadha, K.L. and Sahijram, L. (2000) Application of biotechnology to *Musa*. In: Biotechnology in Horticultural and Plantation Crops (Eds. K.L. Chadha, P.N. Ravindran and Leela Sahijram), Malhtora Publishing Hosue. New Delhi, pp 232-247.

Chadha, K.L., Ravindran, R.N. and Sahijram, L. (2000) Biotechnology in Horticulture. In: Biotechnology in Horticulture and Plantation Crops (Eds. K.L. Chadha, P.N. Ravindran and Leela Sahijram), Malhtora Publishing House, New Delhi, pp 1-25.

Chemda, D., Ruth, E.B. and Gazit, S. (1990) Enzyme polymorphism in mango. *J. Amer. Soc. Hort. Sci.*, **115**:844-847.

Chen, L.G. (1992) Studies on increasing the induction frequency of embryoids in anther culture of citrus. *Sci. Agric.Sin.*, **25**:90-101.

Cheng, F.S.A. and Roose, M.L. (1995) Origin and inheritance of dwarfing by the citrus root stock *Poncirus trifoliata* flying dragon. *J. Amer. Soc. Hort. Sci.*, **120**:286-291.

Chowrira, G.M., Akella, V., Fuerst, P.E. and Lurqin, P.F. (1996) Transgenic grain legumes obtained by in planta electroporation mediated gene transfer. *Mol. Biotech.*, **5**:85-95.

Crane, M. B. and Lawrence, W. J. (1956). The genetics of garden plants. Mac Millam and Co., London.

Cruz-Hernandez, A.W., Litz, R.E. and Lim, M.G. (1998) *Agrobacterium tumefaciens*- mediated transformation of embryogenic avocado cultures and regeneration of somatic embryos. *Pl. Cell Rep.*, **17**:497-503.

D'Halluin, E., Bonne, K.E., Bossut, M., Beuckeleer, M.D. and Leemans, J. (1992) Transgenic maize plants by tissue electroporation. *Pl. Cell*, **4**: 1495-1505.

Dandekar, A.M. (1991) To conduct a planned release of genetically engineered apple plants. *Biotechnology Permit Number* 91-218-03.

Dandekar, A.M., Martin, L.A. and Mcgranahan, G.H. (1988) Genetic transformation and foriegn gene expression in walnut tissue. *J. Amer. Soc. Hort. Sci.*, **113**:945-949.

Dandekar, A.M., Mcgranahan, G.H., Vail, P.V., Uratsu, S.L., Leslie, C.A. and Tebbet, J.S. (1994) Low levels of expression of wild type *Bacillus thuringiensis* var. Kurstakicry, (c) sequences in transgenic walnut somatic embryos. *Plant Sci.*, **96**:151-162.

Darlene, M.L., Minon, H. and Nounan, F.W. (1995) The use of molecular markers to analyse the inheritance of morphological and development traits in apple. *J . Amer. Soc. Hort. Sci.*, **120**:532-537.

Darrow, G. M. (1957) Sterility and fertility in strawberry. *J. Agric. Res.*, **34**: 393-411.

Deng, Z.N., Gentile, A., Nicolosi, E., Domina, F., Vardi, A. and Tribulato, E. (1995) Identification of *in vivo* and *in vitro* lemon mutants by RAPD markers. *J. Hort. Sci.*, **70**:117-125.

Dequin, F., Wencai, Z. and Shienyuan, X. (1994) Isozymes and classification of citrus in China. *Acta Bot. Sin.*, **36**:124-138.

Durham, R.E., Liou, P.C., Gmitter, F.G. and Moore, G.A. (1992) Linkage of restriction fragment length polymorphisms and isozymes in citrus. *Theor. Appl. Genet.*, **84**:39-48.

Fei, K.W. and Xue G.R. (1981) Induction of haploid plantlets by anther culture *in vitro* in apple cv. Delicious. *Sci. Agric. Sin.*, **4**:41-44.

Fitch, M.M.M., Manshardt, R.M., Gonsalves, D., Slightom, J.L. and Sanford, J.C. (1990) Stable transformation of papaya via micro projectile bombardment. *Pl. Cell Rep.*, **9**:189-94.

Fitch, M.M.M., Manshardt, R.M., Gonsalves, D., Slightom, J.L. and Sanford, J.C. (1992) Virus resistant papaya plants derived from tissues bombarded with the coat protein gene of papaya ringspot virus. *Bio / Tech.*, **10**:1466-1472.

Fjellstrom, R.G. and Parfitt, D.E. (1994) RFLP inheritance and linkage in walnut. *Theor. Appl. Genet.*, **89**:665-670.

Fjellstrom, R.G., Parfitt, D.E. and Mcgranahan, G.H. (1994) Genetic relationship and characterization of Persian walnut (*Juglans regia* L.) cultivars using RFLPs. *J. Amer. Soc. Hort. Sci.,* **119**:833-839.

Fu, L.F. and Tang, D.Y. (1983) Induction of pollen plants of litchi tree (*Litchi chinensis* Sonn.). *Acta Genet. Sin.*, **10**:369-374.

Gawel, N., Jarret, R.L. and Whittemore, A.T. (1992) Restriction fragment length polymorphism (RFLP) based phylogenetic analysis of *Musa. Theor. Appl. Genet.*, **84**:286-290.

Ghosh, S.P. (1991) Citrus. In: Fruits Tropical and Subtropical (Eds. T.K. Bose and S.K. Mitra), Naya Prokash, Culcutta, pp 69-73.

Gmitter, F.G.(Jr.), Yiao, S.Y., Huang, S., Hu, X.L., Garnsey, S.M. and Deng, Z. (1996) A localized linkage map of the citrus tristeza virus resistance gene region. *Theor. Appl. Genet.*, **92**:688-695.

Gourley, J. H. and Howlett, F. S. (1955). Modern fruit production. The Mac Millan Co., New York.

Graham, J. and Mcnicol, R.J. (1990) Plantlet regeneration and genetic transformation in soft fruit species. *Acta Hort.*, **280**:517-522.

Graham, J. and Mcnicol, R.J. (1991) Regeneration and transformation of *Ribes. Plant Cell, Tiss. Org. Cult.*, **24**:91-95.

Graham, J., Mcnicol, R.J. and Mcnicol, J.W. (1996) A comparison of methods for the estimation of genetic diversity in strawberry cultivars. *Theor. Appl. Genet.*, **93**:402-406.

Gray, D.J. and Fisher, L.C. (1985) *In vitro* shoot propagation of grape species, hybrids and cultivars. *Proc.Florid.StateHortl.Soc.*, **98**: 172-174.

Gresshoff, P.M. and Doy, C.H. (1974) Derivation of a haploid cell line from *Vitis vinifera* and the importance of the stage of meiotic development of anthers of haploid culture of this and other genera. *J. Pflanzen physiol.*, **73**:132-141.

Guo, Y., Liang, H. and Beins, W. (1995) Laser-mediated gene transfer in rice. *Physiol. Pl.* **93**:19-24.

Harn, C. and Kim , M.Z. (1972) Induction of callus from anthers of *Prunus armeniaca. Kor. J. Breed.*, **4**:49-53.

Helentjaris, T. (1987) A genetic linkage map for maize based on RFLPs trends. *Genet.*, **3**: 217-221.

Heloir, M.C., Fournious, J.C., Oziol, L. and Bessis, R. (1997) An improved procedure for the propagation *in vitro* of grapevine (*Vitis vinifera* cv. Pinotnoir) using axillary-bud microcuttings. *Pl.Cell Tiss.Org. Cult.*, **49**:223-225.

Hidaka, T. (1984 b) Induction of the plantlets from anthers of Trovita orange. *J. Jap. Soc. Hort. Sci.*, **53**:1-5.

Hidaka, T. (1984a) Effect of sucrose concentration, pH of media and culture temperature on anther culture of citrus. *Jap. J. Breed.*, **34**:416-422.

Hidaka, T., Yamada, Y. and Schichijo, T. (1979) *In vitro* differentiation of haploid plants by anther culture in *Poncirus trifoliata* L. *Raf. Jap. J. Breed.*, **29**:248-254.

Hidaka, T., Yamada, Y. and Schichijo, T. (1982) Plantlet formation from anthers of *Citrus aurantium* L. Proc. Intl. Soc. Citricult. Japan, pp. 153-155.

Hidaka, T.and Omura, M. (1993) Transformation of citrus protoplast by electroporation. *J.Japanese Soc. Hort. Sci*., **62**:371-376.

Hohn, B., Koukolikova-Nicola, Z. and Bakkeren, G. (1989) *Agrobacterium* mediated gene transfer to monocots and dicots. *Genome*, **31**:987-993.

Hong, Xu., Diane, J.W., Aulsekar, S. and Alan, T.B. (1995) Sequence specific PCR markers derived from RAPD markers for finger printing grape(*Vitis*) rootstocks. *J. Amer.Soc. Hort. Sci*., **120**:714-720.

Hong, Y.Y., Schuyler, S.K., Jutta, K. and Hanna, S. (1997) A RAPD marker tightly linked in to the scab-resistance gene V_f in apple. *J. Amer. Soc. Hort. Sci*., **122**: 47-52.

Hwang, S.C. and Ko, W.H. (1987) Somaclonal variation of bananas and screening for resistance to *Fusarium* wilt. Proc. Banana and Plantain Breeding Strategies (Eds. G.J. Persley and E.A. Decanghe), ACIAR., pp. 151-156.

James, D.J., Passey, A.J. and Barbara, D.J. (1990) *Agrobacterium*- mediated transformation of the cultivated strawberry (*Fragaria* x *ananassa* Duch.) Using disarmed binary vectors. *Pl. Sci*.,**69**:79-94.

James, D.J., Passey, A.J., Barbara, D.J. and Bevan, M. (1989) Genetic transformation of apple (*Malus pumila* Mill.) using a disarmed Ti-binary vector. *Pl. Cell. Rep*., **7**:658-661.

Janssen, B.J. and Gardner, R. (1993) The use of transfer GUS expression to develop an *Agrobacterium*-mediated gene transfer system for kiwi fruits. *Pl.Cell Rep*, **13**:28-31.

Joersbo, M. and Brunstedt, J. (1992) Sonication: A new methods for gene transfer to plants. *Physiol. Pl*., **85**:230-234.

Kaneyoshi, H.J., Kobayashi, S., Nalcamura, Y., Shigemoto, N. and Doi, Y. (1994) A simple and efficient gene transfer system of trifoliate orange (*Poncirus trifoliata* Raf.). *Pl Cell rep.*, 13:541-545.

Kawata, K., Ushida, C., Kawai, F., Kanamori, M. and Kuriyama, A. (1995) Micropropagation of passion fruit from sub cultured multiple shoot premordia. *J. Pl. Phy*., **147**:281-284.

Kobayashi, S. and Uchimiya, H. (1989) Expression and integration of a foreign gene in orange (*Citrus sinensis* Osb.) protoplast by direct DNA transfer. *Japanese J. Genet*., **64**:91-97.

Koller, B., Lehman, A., Mcdermott, J.M. and Gessier, C. (1993) Identification of apple cultivars using RAPD markers.*Theor. Appl. Genet*., **85**:901-904.

Kubicki, B., Telezynska, J. and Milewoska,- Pawliczuk, E. (1975) Induction of embryoid development from apple pollen grains. *Acta Soc. Bot. Pol*., **44**:631-635:

Lambert, C. and Tepfer, D. (1992) Use of *Agrobacterium* rhizogenes to create transgenic apple trees having an altered organogenic response to hormones. *Theor. Appl. Genet*., **85**:105-109.

Lammerts, W.E. (1942) Embryo culture an effective technique for shortening the breeding cycle of deciduous trees and increasing germination of hybrid seed. *Amer*. J. *Bot*., **29**:166-171.

Lavi, U., Adato, A., Kaufman, D., Mhameed, S., Sharon, D., Lahav. E., Hillel, J. and Gregan, P.B. (1995) Generation and application of VNTR DNA markers to fruit trees: Induced mutations and molecular technique for crop improvement. *Proc. Vienna, Austria*, pp 265-273.

Lavi, U., Millel, J., Vainstein, A., Lahav, E. and Sharon, D. (1991) Application of DNA fingerprints for identification and genetic analysis of avocado. *J. Amer. Soc. Hort. Sci*., **116**:1078-1081.

Lewis, D (1956) Incompatibility and plant breeding. *Brookhaven Symp. Biol*., **9**: 89-100.

Liou, P.C., Gmutter,F.G. and Moore, G.A. (1996) Characterization of the citrus genome through analysis of RFLP. *Theor. Appl. Genet*., **92**:425-35.

Litz, R.E. and Conover, R.A. (1978) Recent advances in papaya tissue culture. *Proc. Flo. St. Hort. Soc.*, **91**:180-182.

Litz, R.E. and Conover, R.A. (1986) *In vitro* improvement of *Carica papaya* L. *Proc. Trop. Reg. Amer, Soc. Hort. Sci.*, **23**:157-159.

Machado, A.C., Katinger, H. and Machado, M.L. C. (1994) Coat protein-mediated protection against plum pox virus in herbaceous model plants and transformation of apricot and plum. *Euphytic*, **77**:129-134.

Machado,M.L.C., Machado, A.C., Hanzer, V., Weics, H., Regner, F., Steinkellner, H. Mattanovich, D., Plail, R., Knapp, E., Kalthaff, B. and Katinger, H. (1992) Regeneration of transgenic plants of *Prunus armeniaca* containing the coat protein gene of plum-pox virus. *Pl. Cell Rep.*, **11**:25-29.

Maheshwari, P. and R.C. Sachar (1963) Polyembryony. In: Recent Advances in the Embryology of Angiosperm (Ed. P. Maheshwari), University of Delhi Inter Soc. of Plant Morph, pp 265-296.

Majumder, P. K. and Mukherjee, S. K. (1972a). Aneuploidy in guava. I Mechanism of variation in number of chromosomes. *Cytologia*, **37**: 541-548.

Majumder, P. K. and Mukherjee, S. K. (1972b). Aneuploidy in guava. II The occurrence of trisomics, tetrasomics and higher aneuploids in the progeny of triploid. *Nucleus*, **13**: 42-47.

Majumder, P.K. and Sharma, D.K. (1991) Mango. In: Fruits Tropical and Subtropical (Eds. T.K. Bose and S.K. Mitra), Naya Prokash, Calcutta, pp 6-7.

Manders, G., Otoni, W.C., Vaz, F.B.U., Blackhall, N.W., Power, J.B. and Davey, M.R.(1994) Transformation of passion fruit (*Passiflora edulis* cv. Flavicarpa, Degener.) using *Agrobacterium tumefacines*. *Pl. Cell Rep.*, **13**:697-702.

May, G.D., Afza, R., Mason, H.S., Wiecko, A., Novak, F.J. and Arntzen, C.J.(1995) Generation of transgenic banana (*Musa acuminata*) plants via *Agrobacterium*–mediated transformation. *Bio/Tech.*, **13**:486-492.

Mcgranahan, G.H., Leslie, C.A., Dandekar, A.M., Uratsu, S.L. and Yates, I.E.(1993) Transformation of pecan and regenerating transgenic plants. *Pl. Cell Rep.*, **12**:634-638.

Mcgranahan, G.H., Leslie, C.A., Uratsu, S.L. and Dandekar, A.M.(1990) Improved efficiency of the walnut somatic embryo gene transfer system. *Pl. Cell Rep.*, **8**:512-516.

Mcgranahan, G.H., Leslie, C.A., Uratsu, S.L., Martin, L.A. and Dandekar, A.M. (1988) *Agrobacterium*-mediated transformation of walnut somatic embryos and regeneration of transgenic plants. *Bio/Tech.*,**6**:800-804.

Merkulov, S.M., Bartish, I.V., Dolgov, S.V., Pasternak, T.P. and Mchugen, A.(1998) Genetic transformation of pear (*Pyrus communis* L.) mediated by *Agrobacterium tumefaciens*. *Genet.*, **34**:373-378.

Mhameed, S., Sharon, D., Kaufman, D., Lahav, E., Hillek, J., Degani, C., and Lavi, U. (1997) Genetic relationship with in avocado (*P. americana*) cultivars and between *Persea* species. *Theor. Appl. Genet.*, **94**:279-286.

Michellon, R., Hugard, J. and Jonard, R. (1974) Sur I isolement de colonies tissularies de pecher (*Prunus persica* Batsch. Cultivars Dixred et Nectared iv) et d' Amandier (*Prunus amygdalus* stokes, Cultivar Ai) a partir d" anthers cultivees in vitro. *C.R. Acad. Sci. Paris*, **278**:1719-1722.

Minon, M., Norman, F.W., Patric, J.C. and Brown, S.K. (1997) A DNA marker for columnar growth habit in apple contains a simple sequence repeat. *J.Amer. Soc. Hort. Sci.*, **122**:347-349.

Mitra, S.K. (1991) Apple. In: Temperate Fruits (Eds. S.K. Mitra, T.K. Bose and D.S. Rathore), Horticulture and Allied Publishers Calcutta, pp 15.

Moore, G.A., Jacono, C.C., Neidigh, J.L., Lawrence, S.D. and Cline, K. (1992) *Agrobacterium*-mediated transformation of citrus stem segments and regeneration of transgenic plants. *Pl. Cell Rep.*, **11**-238-242.

Moriguchi, T., Omura, M., Matsuta, N. and Kozaki, I. (1987) *In vitro* adventitious shoot formation from anthers of pomegranate. *Hort. Sci.*, **22**:94

Nair, S., Gupta, P.K. and Mascarenhas, A.F. (1983) Haploid plants from *in vitro* anther culture of *Annona squamosa* Linn. *Pl. Cell Rep.*, **2**: 198-200.

Nakano, M., Hoshino, Y. and Mii, M.(1994) Regeneration of transgenic plants of grape vine (*Vitis vinifera* L.) via *Agrobacterium rhizogenes*- mediated transformation of embryo genic calli. *J. Expt. Bot.*, **45**:649-656.

Naumova, T.N. (1993) Apomixis in angiosperms: Nucellar and integumentary embryony. Boca Raton, FL:CRC Press.

Nehra, N.S., Chibbar, R.N., Kartha, K.K., Datla, R.S.S., Crosby, W.L. and Stushnoff, C. (1990b) *Agrobacterium* mediated transformation of strawberry calli and recovery of transgenic plants. *Pl. Cell Rep.*, **9**:10-13.

Nehra, N.S., Chibbar, R.N., Kartha, K.K., Datla, R.S.S., Crosby, W.L. and Stushnoff, C. (1990) genetic transformation of strawberry by *Agrobacterium tumefaciens* using leaf disk re-generation system. *Pl. Cell Rep.*, **9**:293-298.

Neuhaus, G. and Spangenburg, G. (1990) Plant transformation by microinjection techniques. *Physiol. Pl.*, **79**:213-217.

Nyman, M. and Wallin, A. (1992) Transient gene expression in strawberry (*Fragaria* x *ananassa* Duch.) protoplasts and the recovery of transgenic plants. *Pl. Cells Rep.*, **11**:105-108.

Nyrgren, A. (1954) Apomixis in the angiosperms II. *Bot Rev.*, **20**:577-649.

Oliveira, M.M., Barroso, J. and Pais, M.S.S. (1991) Direct gene transfer into *Actinidia deliciosa* protoplasts: analysis of transient expression of the CAT gene using TLC autoradiography and GC-MS based method. *Pl. Mol. Biol.*, **17**:235-242.

Orford, S.J., Scott, S.N. and Timmis, J.N. (1995) A hyper variable middle repetitive DNA sequence from citrus. *Theor. Appl. Genet.*, **91**:1248-1252.

Pang, S.Z. and Sanford, J.C. (1988) *Agrobacterium* –mediated gene transfer in papaya. *J. Amer. Soc. Hortl. Sci.*, **113**:287-297.

Paszkowski, J., Saul, M.W. and Potrykes, I. (1989) Plant gene vectors and genetic transformation:DNA mediated direct gene transfer to plants. In: Cell Culture and Somatic Cell Genetics of Plants (Vol 6). Molecular Biology of Plant Nuclear Genes (Eds. J.Schell and I. Vasil), Academic Press, San Diego, pp. 52-68.

Pattnaik, S.K. and Chand, P.K. (1997) Rapid clonal propagation of three mulberries, *Morus cathayana* Hemsl. *M. ihoukoiz* and *M. serrata* Roxb. through *in vitro* culture of apical shoot buds and nodal explants from mature trees. *Pl. Cell Rep.*, **16**:503-508.

Pena, L., Cervera, M., Juarez, J., Navaro, A., Pian, J.A. and Navaro, L. (1997) Genetic transfomation of lime (*Citrus aurantifolia* Swing.) factors affecting transfomation and regeneration. *Pl. Cell Rep.*, **16**:731-737.

Raghaven, V. (1986) Embryogensis in angiosperms Cambridge: Cambridge University Press.

Reinicke, O. S. H. (1930). The relation of seed formation to fruit development in pear. *South Afr. J. Sci.*, **27**: 303-309.

Ronning, C.M., Schnell, R.J. and Gazit, S. (1995) Using RAPD markers to identify *Anona* cultivars. *J. Amer. Soc. Hort. Sci.*, **120**:726-729.

Rowland, L.J. and Levi, A. (1994) RAPD based genetic linkage map of blueberry derived from a cross between diploid species (*Vaccinium danowi* and *V. elliottii*). *Theor. Appl. Genet.* **87**:863-868.

Sagi, L., Panis, B., Remy, S., Schoofs, H., Desmet, K., Swennen, R. and Lammue, B.P.A. (1995) Genetic transformation of banana and plantain (*Musa* spp.) via particle bombardment. *Bio/Tech.*, **15**:481-485.

Sagi, L., Remy, S., Bart, P., Szannen, R. and Volckaert, G. (1994) Transient gene expression in electroporated banana (*Musa* spp. Cv. Bluggoe , ABB group) protoplast isolated from regenerable embryogenic cell suspensions. *Pl. Cell Rep.*, **13**:262-266.

Sanford, J.C. (1990) Biolistic plant transformation. *Physiol. Pl.*, **79**:206:209.

Schnell, R.J., Ronning, G.M. and Knight, R.L. (jr)(1995) Identification of cultivars and validation of genetic relationships in *Mangifera indica* L. using RAPD markers. *Theor. Appl. Genet.*, **90**:269-274.

Scorza, R., Cordts, J.M., Ramming, D.W. and Emershad, R.L. (1995) Transformation of grape(*Vitis vinifera* L.) zygotic derived somatic embryos and regeneration of transgenic plants. *Pl. Cell Rep.*, **14**:589-592.

Scorza, R., Ravelonandro, M., Collahan, A.M., Coraltsfuchs, M., Dunez, J. and Gonsalves, D. (1994) Transgenic Plum (*Prunus domestica* L .) expressing the plum box virus coat protein gene. *Pl. Cell Rep.*, **14**:18-24.

Serres, R. and Stang E. (1992) Gene transfer using electric discharge particle bombardment and recovery of transformed cranberry plants. *J. Amer. Soc. Hort. Sci.*, **117**:174-180.

Singh, B. D. (1990). Plant Breeding, Kalyani Publishers, New Delhi.

Singh, B. P. and Rana, R. S. (1993). Promising fruit introductions. In: Advances in Horticulture Vol. I (Eds. K. L. Chadha and O. P. Pareek), Malhotra Publishing House, New Delhi, pp 43-66.

Smigocki, A.C. and Hammerschlag, F A.(1991) Regeneration of plants from peach embryo cells infected with a shooty mutant strain of *Agrobacterium. J. Amer. Soc. Hort. Sci.*, **116**:1092-1097.

Sondur, S.N., Manshardt., R.M. and Stiles, J.I. (1996) A genetic linkage map of papaya based on RAPD markers. *Theor. Appl. Genet.*, **93**:527-533.

Songstad, D.D., Somers, D.A. and Griesbach, R.J. (1995) Advances in alternative DNA delivery techniques. *Pl. Cell, Tiss. Org. Cult.*, **40**:1-15.

Stebbins, G. L. (1958). Longetivity, habitat and release of genetic variability in the higher plants. *CSH Symp. Quant. Biol.*, **23**: 365-378.

Su, C.Y. and Tsay, H.S. (1985) Anther culture of papaya (*Carica papaya* L.). *Basic Life Sci.*, **32**:357.

Subramanyam, M. D. and Iyer, C. P. A. (1993). Improvement of guava. In: Advances in Horticulture Vol. I (Eds. K. L. Chadha and O. P. Pareek), Malhotra Publishing House, New Delhi, pp 337-347.

Tartarin, S. (1996) RAPD markers linked to the Vf gene for scab resistance in apple. *Theor. Appl. Genet.*, **92**:803-810.

Thomas, M.R., Matsumoto, S., Chain, P. and Scott, N.S. (1993) Repetitive DNA of grape vine classes present and sequences suitable for cultivar identification. *Theor. Appl. Genet.*, **86**:173-180.

Todorovic, R.R., Misic, P.D., Petrovic, M.D. and Mirkoric, M.A. (1992) Anther culture of peach cultivars, 'Cresthaven' and Visna. *Acta Hort.*, **300**:331-333.

Tsay, H.S. and Su, C.Y. (1985) Anther culture of papaya (*Carica papaya* L.). *Pl. Cell Rep.*, **4**: 28-30.

Tsay, H.S., Hsu, J.Y., Yang, T.P. and Yang, C.R. (1984) Anther culture of passion fruit (*Passiflora edulis*). *J. Agric. Res. China*, **33**:126-131.

Uematsu, C., Myrase, M., Ichikawa, H. and Imamura J. (1991) *Agrobacterium* –mediated transformation and regeneration of kiwifruit. *Pl.Cell Rep.*, **10**:286-290.

Vardi, A., Bleichman, S. and Aviv, D. (1990) Genetic transformation of *Citrus* protoplast and regeneration of transgenic plants. *Pl Sci.*, **69**:199-201.

Vardy, E. L. D., Gumzi, A. and Hewitt, J. (1989). Genetics of parthenocarpy in tomato under low temperature regime: I line R P 75/59, *Euphytica*, **41**: 1-8.

Vashishtha, B.B., Shukla, Anil Kumar and Shukla, Arun Kumar: (2004) Approaches for improvement of fruit crops. In: Advances in arid horticulture Vol. I (Eds. P.L. Saroj, B.B. Vashishtha and D.G. Dhandar), International Book Distributing Co., Lucknow, pp.159-176.

Walker, M.A. and Liu, L. (1995) The use of isozymes to identify 60 grape vine rootstocks (*Vitis* species). *Amer. J. Enol. Viticult.*, **46**:299-305.

Xianping, Q., Jaing, L. and Lamiknra, O. (1996) Genetic diversity in Muscadine and American bunch grapes based on RAPD analysis. *J. Amer. Soc. Hort. Sci.*, **121**:1020-1023.

Xiaojian, Y., Brown, S.K., Scorza, R., Cordts, J. and Sanford, J.C. (1994) Genetic transformation of peach tissue by particle bombardment. *J. Amer. Soc. Hort. Sci.*, **119**:367-373.

Yang J.S., Yu, T.A., Cheng, Y.H. and Yeh, S.D. (1996) Transgenic papaya plants from *Agrobacterium*-mediated transformation of petioles of *in vitro* propagated multishoots. *Pl. Cell Rep.*, **15**:459-464.

Yao, J.L., Cohen, D., Atkinson, R., Richardson, K., and Morris, B. (1995) Regeneration of transgenic plants from the commercial apple cultivar Royal Gala. *Pl. Cell Rep.*, **14**: 407-412.

Yao, J.L., Wu, J.H., Gleave, A.P. and Morris, B.A.M. (1996) Transformation of citrus embryogenic cells using particle bombardment and production of transgenic embryos. *Pl. Sci.*, **113**:175-83.

Yeung, E.C., Thorpe, T.A. and Jensen, C.J. (1981) *In vitro* fertilization and embryo culture. In: Plant tissue culture methods and application in agriculture (Ed. T.A. Thorpe), Newyork Academic Press, pp 253-271.

Zheng, X., Nishioka, M., Kawamura, A., Ukeela, H., Sawamura, M. and Fusuose, H. (1996) Cluster analysis by measurement of peroxidase and esterase for citrus flavedo. *Bioscience*, **60**;390-395.

2

Breeding for Biotic Stress Resistance

A plant is said to be healthy or normal when it carries out its physiological functions to the best of its genetic potential. These normal functions include division, differentiation, development, absorption of water and mineral from soil and translocation of these throughout the plants, photosynthesis and translocation of photosynthetic product to areas of utilization or storage, the metabolism of synthesized compounds, reproduction and storage of food supplies.

A plant becomes diseased when it is disturbed by pathogen under certain environmental conditions which interfere with one or more of its essential functions. Diseased plant refers to any disturbance brought about by living organism under environmental factors, which interfere with normal function of plant or in other words when any organ and part of plant is not doing their work properly and when either the growth or reproduction is not going forward in natural or regular manner.

2.1. Advantages of resistant breeding

i) It is very cheap- farmers can use resistant varieties without incurring any extra expenditure.

ii) It is a safe measure- fungicides and other pesticides leave some residual effect.

iii) It is more effective as compared to other measures of disease and pest control.

iv) In case of air borne diseases, it is impossible to cover long area with any other means of disease control.

2.2. Pathogenicity

Ability of the pathogen to attack a host is synonymous to virulence. Inheritance of resistance is a process only when a host with resistance gene is available.

Virulence of pathogen is under genetic control and more than one gene are generally involved in control of virulence at one locus. There are two alleles to explain the pathogenicity of apple scab. The apple scab disease caused by *Venturia inequalis*. The apple variety Wealthy is resistant to some strain of this fungus and susceptible to others, if cross between a virulence and virulence strain produces two types of ascospores, virulent and avirulent in 1: 1 ratio, indicating a single gene control with two alleles and if we include another variety i.e. Mcintosh different in level of resistance, the inheritance of pathogenicity can be confirmed. Some strains of fungus are virulent towards variety Wealthy but a virulence towards Mcintosh. When two such strains are crossed resulting ascospores are of 4 types-

- Virulent on both the varieties
- Virulent on Wealthy variety
- Virulent on Mcintosh variety
- Avirulent on both the varieties

Depending upon different types of the biotic agent involved in producing biotic stresses, breeding methods can be grouped into three-

- Breeding for insect resistance.
- Breeding for disease resistance.
- Breeding for resistance to parasitic weeds.

2.3. Breeding for insect resistance

2.3.1 Concept

The insects are usually specialized in their ability to attack the host or part of the host. An insect is capable of damaging or attacking every species of the host. The plant resistance includes those characters which enable a plant to avoid, tolerate or recover from the attack of insect under conditions that would cause greater injury to other plant of the same species.

Resistance is heritable characteristics possessed by the plant which influence the ultimate degree of damage done by the insect. In other words plant resistance is defined as being the collective heritable characteristics by which a plant species raise or individual may reduce the probability of successful utilization of that plant or a host by an insect species, race, biotype or individuals. The degree of resistance is a relative term which is measured by using susceptible cultivar of same plant species as check. The degree of resistance among specific host plants may vary between two extremes i.e. immunity and high susceptibility. Any degree of host reaction less than immunity is resistance.

2.3.2 Scales

2.3.2.1 Immunity

An immune species/cultivar is one which is never consumed or injured by insect under known conditions.

2.3.2.2 High resistance

A variety with high resistance is one which possesses quality resulting in small damage by specific insect under a given set of conditions.

2.3.2.3 Low resistance

A low level of resistance indicates the possession of quality which cause a variety to show less damage or less infestation by an insect than the average for the crop under consideration.

2.3.2.4 Susceptibility

A susceptible variety is that which shows average or more than average damage by insect.

2.3.2.5 High susceptibility

A variety show high susceptibility when much more than average damage is done by insect under consideration.

Painter, an eminent scientist proposed the term functional resistance in which he reported that there is certain phenomenon which are related to the resistance but not based on heritable traits. Under this he proposed the term pseudo resistance. The pseudo resistance is not the resultant of genetic character inherited in the plants but it is the result of some temporary shift in the environmental conditions favourable to the otherwise susceptible host plant. Variety showing pseudo resistance are important in economic entomology but it should be distinguished from cultivar that shows resistance through a wider range of a resistance i.e. with inherited mechanism. The functional resistance can be classified as-

2.3.2.5.1 Host evasion

Insect pest attack not only right host but also attack or contact the host at appropriate stage of development. The phonologies of host and insect must synchronize. Host evasion takes place when plant growth pattern is modified so as to bring an synchronism of insect host phonologies i.e. early maturity is a good example of host evasion where crop matures before severe damage.

2.3.2.5.2 Induced resistance

A term induced resistance may be used for temporary increase in resistance resulting from some condition of plant or environment such as amount of water and soil fertility. It is reported that aphid is highly sensitive to high level of nitrogen in the plant but respond negatively to the potash even in the presence of high nitrogen.

2.3.2.5.3 Escape

It is reported that even under very heavy infestation, susceptible plant will occasionally escape so that only studies of their progenies will establish their true relationship. Escape refers to the lack of infestation or injury to the host plant because of such transitory circumstances as incomplete infestation thus finding an infested plant in a susceptible population does not necessary that it is essential.

2.4. Type of genetic resistance

Host plant resistance may be due to a series of interaction between insect and plant which influence selection of plant as host and effect of plant in insect survival and multiplication.

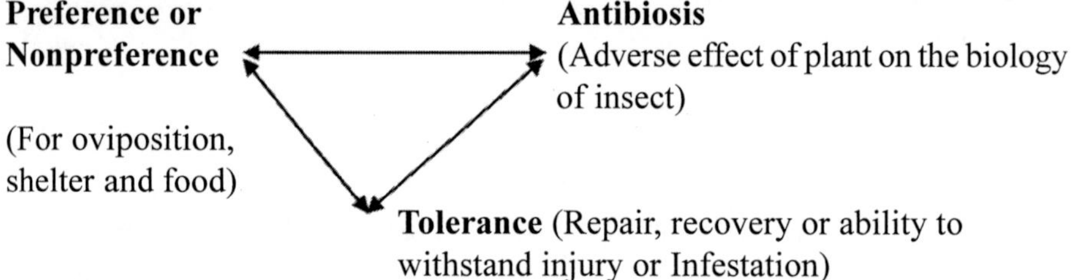

There are 3 interrelation of characteristics one or more of which is frequently present in resistant variety.

2.4.1 Non preference

Non preference is the insect response to plant that lack the characteristics to serve as host resulting from positive response or total avoidance during such search for food, oviposition, shelter or for the combination of these three. Kogan and Ortman proposed the term antigenesis for non preference in 1978. This antigenesis is parallel to antibiosis referred to the idea that the plant is avoided as a bad host.

Important factors which govern preference and non preference phenomenon are as under -

- Insect behaviour as a basis for preference.
- Preference differences as response to colour or intensity of light.
- Factor conditioning preference.
- Preference difference as response to mechanical stimuli from physical structure and surface of plant.
- Preference difference as response to chemical stimuli.
- Possibility of preference as a resistance mechanism.

2.4.2 Antibiosis

It is tendency to prevent injury or destroy insect life. Painter proposed antibiosis for those adverse effect on the insect life history which results when the insect uses a resistant variety of the host plant/species for their food. Following adverse physiological metabolic activities effect of temporary or permanent nature which occur when a insect feeds on a resistant plant, these are-

(a) Deleterious effect of specific chemicals including toxin.

(b) Food material present but for some reasons not available to insect.

(c) Lack of specific food material in part of the plant eaten by the insect.

(d) Presence of material so repellent, that insect will not eat on such resistant plant and virtually starve to death.

2.4.3 Tolerance

Tolerance is mechanism of resistance in which a plant shows an ability to grow and reproduce itself or repair injury. Resistant plant may be tolerant surviving under level of infestation that would kill susceptible plant. Tolerance differs from non-preference and antibiosis in its mechanism. Non preference and antibiosis require an active insect response or lack of response, however, tolerance response requires thorough understanding of the ways in which plant may repair the damage. Tolerance is more subjected to variation in environmental conditions than non preference and antibiosis. The age and size of the plant and size of insect population strongly influence the degree of tolerance, general vigour of plant is affected by its tolerance by insect attack.

2.4.4 Genetics of resistance

Plant can be evaluated in the field if there is high natural insect population or it can be evaluated in green house with field collected or laboratory reared insect. However, the field culture of insect may have variability for virulence. So, it is better to take laboratory reared insect for screening. Resistance may be determined on plant injury or symptom of insect attack or on the reaction of insect to plant insect damage in the form of defoliated plants. The pest host interactions are scored on a scale based on the reaction of the plant to the insect or sometimes reaction of insect to the plant. The reactions of parent F_1, F_2 and F_3 progenies are used to study the dominant or recessive or incomplete dominant and quantitative and qualitative nature of inheritance.

2.5. Problems associated to breeding for resistance to insect

An intimate knowledge of biology and feeding habit of insect is essential. The insect variability is an important problem and this must be closely monitored. Biotype difference within insect species may cause behavioural difference. The insects reared in laboratory may have different behaviour as compared to field selected population of insect. Natural variability in growth and other parameters of insect development must be carefully determined and accounted for interest for resistance. Method must be designed to examine large number of plants for insect infestation / damage in very short period of time. Seedling screening technique for mass selection should be developed and correlation of seedling resistance and adult plant resistance must be considered carefully.

It was hoped that chemical control measures will effectively control, or eliminate the insect pest but the experience with pesticides has shown that such a hope was entirely misplaced. Extensive pesticide application, (i) increases the cost of production of crops, (ii) reduces the population of natural enemies (predators and parasites) of insect pests, (iii) leads to the development of pesticide-resistance race of insects, and (iv) pollutes the environment. (The most tragic and horrifying example is the Bhopal gas tragedy of December, 1984).

Now a days, most of the entomologists speak of "pest management" in place of "pest control". The pest management involves, in addition to pesticide application, several divergent measures to minimize the losses due to insect pests.

"Insect resistance is the property of a variety of a host crop due to which it is attacked by an insect pest to a significantly lower degree than other varieties of the same host. From a practical point of view, an insect resistant variety produces relatively larger yield of good quality than susceptible varieties with the same level of initial infestation (insect attack) under comparable environmental conditions."

2.6. Mechanism of insect resistance

Mechanism of insect resistance is grouped into four categories i.e. (i) non preference, (ii) antibiosis, (iii) tolerance and (iv) avoidance. Further, the insect resistance may exhibit one, more often, two or more as these mechanisms. In some cases of resistance, non-preference, antibiosis and tolerance are simultaneously involved. Such combinations of resistance mechanism are highly effective and are not easily overcome by the concerned insect pest.

(i) Non preference

Host varieties exhibiting this type of resistance are unattractive or unsuitable for colonization, oviposition or both by an insect pest. These types of resistance is also termed as non-acceptance and antixenosis. e.g. Aphid resistance in Raspberry.

(ii) Antibiosis

It refers to an adverse effect of feeding on a resistant host plant on the development and or reproduction of the insect pest. In severe cases, it may even lead to the death of the insect pest. The insecticidal compounds are found across the complex spectrum of secondary metabolites produced in different plant species (Table 1). Some of these compounds act as antifeedants, by preventing the insect from recognizing the plant tissue as a suitable food source, while other kills the insect either directly or by interfering with normal developmental processes (toxins and antimetabolites)

Table 1: Plant secondary metabolites exhibiting insecticidal activity.

Compound	Plant source	References
Nonprotein antimetabolites		
Alkaloids		
2,5 dihydroxymethyl 3,4dihydroxypyrralidine (DMDP)	Lonchocarpus	Evans *et al.* (1985)
Castanospermine	Castanospermum australe	Nash *et al.* (1986)
Non protein amino acid		
P- animophenylalanine		Birch *et al.* (1986)
Pyrethroids Terpenoids	Compositeae	Mann (1987)
Protease inhibitors	Solanaceae Liguminoceae	Gatehouse *et al.* (1979), Hilder *et al.* (1987), Johnson *et al.* (1990)

(iii) Tolerance

An insect tolerant variety is attacked by the insect pest to the same degree as a susceptible variety. But the same level of infestation, a tolerant variety produces a large yield than a susceptible variety.

(iv) Avoidance

It is often as, or even more, effective as true resistance in protecting a crop from the pest damage. The pest avoidance is the same as the disease escapes and as such it is not a case of true insect resistance.

2.7 Breeding for disease resistance

2.7.1 Mechanism of disease resistance

The various mechanism of disease resistance are (i) mechanical (ii) hypersensitivity, and (iii) nutritional.

(i) Mechanical

Certain mechanical or anatomical features of the host may prevent infection.

(ii) Hypersensitivity

In large number of cases, immune reaction is due to the hypersensitivity reaction of the host. Immediately after infection, several host cells surrounding the point of infection die. This leads to the death of the pathogen or at least prevents its spore production. It is believed that phytoalexins are responsible for hypersensitive reaction. Phytoalexins are specific polyphenolic or terpenoid chemicals and are produced by the host in response to the infection by a pathogen.

(iii) Nutritional

The reduction in growth and spore production is generally supposed to be due to an unfavourable physiological condition within the host. Most likely, a resistant host does not fulfill the nutritional requirements of the pathogen and thereby limit its growth and reproduction.

2.7.2 Terminology

2.7.2.1 Disease escape

It refers to the freedom of susceptible host varieties from a disease due purely to the environmental factors. Disease escape occurs primarily by avoiding contact, but unfavourable weather conditions may prevent infection. Disease escape may be a result of the environmental factors, such as early varieties, use of resistant rootstock, balanced application of NPK, control of disease carrier etc.

2.7.2.2 Disease resistance

Host varieties are classified as susceptible or resistant according to their response to the pathogen. The various reactions of the hosts to the various pathogens may be grouped into the susceptible, immune, resistance and tolerance.

2.7.2.3 Susceptible reaction

In the case of susceptible reaction, disease development is profuse and is presumably not checked by the genotype of host. In practice the susceptible reaction is classifiable in relative terms only, that is in relation to that reaction of other host varieties available and the prevailing environment.

2.7.2.4 Immune reaction

When host does not show the symptoms of a disease, it is known as immune reaction. Immunity may result from the prevention of pathogen to reach the appropriate parts of the host.

2.7.2.5 Resistance

Resistance denotes a less disease development than the susceptible variety and is a relative attribute. Infection and establishment do take place, but growth of the pathogen in the host tissue is restricted. In case of resistance, disease symptoms do develop and the rate of reproduction is never zero i.e. $r > 0$, but it is sufficiently lower than one.

2.7.2.6 Tolerance

Tolerance implies that the host is attacked by the pathogen in the same manner as the susceptible variety, but there is little or no loss in biomass production or yield. Generally, tolerance is difficult to measures because it is confounded with partial resistance and disease escape.

2.7.3 Vertical and horizontal resistance

2.7.3.1 Vertical resistance

It is also known as race specific, pathotype specific or simply specific resistance. Vertical resistance is generally determined by major genes and is characterised by pathotype specificity. It denotes that the host carrying a gene for vertical resistance is attacked by only that pathotype which is virulent towards that resistance gene, to all other pathotypes the host will be resistant. Thus, a virulent pathotype will produce an immune response, i.e. $r=0$ or close to zero. But the virulent pathotype will lead to the susceptible reaction, that is $r=1$ clearly, immune or susceptible response in the case of vertical resistance depends on the presence of virulent pathotype. When the virulent pathotype becomes frequent, epidemics are common in the case of vertical resistance. A set of 14 rules governing the value of the vertical resistance in agriculture has been drawn up by Robinson (1971) -

(1) Vertical resistance is unlikely to be valuable in perennial crop that is difficult to breed, because replacement of variety (to neutralize new pathogenic races) is difficult with such crops.

(2) Vertical resistance is likely to be more valuable against a simple interest disease than against a compound interest disease, because the infection rate is slower in the former than in the latter.

(3) Vertical resistance is unlikely to be valuable against a pathogen with a higher vertical mortality, since frequent mutability of the pathogen jeoparadizes the variety with vertical resistance

(4) Vertical resistance is unlikely to be valuable when the host population is genetically uniform and is grown in large acreages of a single cultivar

(5) Vertical resistance is most likely to be valued if the stabilizing pressure can be exploited. The stronger gene for vertical resistance in host, the stronger is the pressure of stabilizing selection operating in favour of pathogenic races without unnecessary virulence.

(6) Against facultative parasites one strong gene is adequate for the exploitation of stabilizing pressure, against obligate parasites at least two strong genes are necessary.

(7) Crop and plant patterns vertical resistance in space are valuable chiefly against compound interest disease. The crop pattern (such as crop rotation) or plant patterns (e.g. multilines), both with strong vertical genes, tend to contain the infection rate (spread of disease) , hence are useful in controlling the compound interest disease.

(8) Crop pattern of vertical resistance in time (between two seasons/epidemics) are valuable chiefly against simple interest diseases, because replacement of one vertical host population with another involves only one crop in one field.

(9) Vertical resistance is likely to be less valuable against disease transmitted by the propagating material of the host.

(10) Vertical resistance will break down more quickly if the protection it confers is incomplete.

(11) Vertical resistance is likely to be more valuable if there is a closed season (winter, or long tropical dry season unfavorable for pathogen population).

(12) Vertical resistance is likely to be more valuable when legislative control is possible

(13) Vertical resistance is likely to be more valuable when it is reinforced with useful levels of horizontal resistance for epidemiological reason.

(14) The breakdown of a complex vertical resistance may result in the less disease resistance probably because of an inverse relationship between vertical and horizontal pathogenicity.

2.7.3.2 Horizontal resistance

It has number of synonyms, e.g. race non-specific, pathotype non-specific and partial or general resistance. Horizontal resistance is generally controlled by polygenes, that is many genes with small effects and it is pathotype nonspecific, that is why it is also known as "general resistance". In the case of horizontal resistance, reproduction rate of the pathogen is not zero, but it is less than one, i.e. $r > 0$ but < 1. Horizontal resistance therefore, does not prevent the development of symptoms of the disease but it slows down the rate of spread of the disease in the population. The horizontal resistance is evenly spread against all races of pathogen. It is more complex type comprising many components. By experience such resistance is expected to last longer. It is unlikely to be overcome by new races of pathogen. When such resistance are present host plant do not react differently to races of the parasite and such kind of balance is achieved minimizing the domination of certain races.

2.7.3.2.1 Hypothesis on the horizontal resistance

Vander Plank assumed that both polygenic and oligogenic uniform resistance is assumed to be governed by genes that are not special resistance genes but are simple genes that occur ordinarily on healthy plants and regulate ordinary process of host. According to this hypothesis there is a basic genetic difference between uniform and differential resistance.

Differential reactions i.e. vertical resistance is contributed by specialized gene, whereas, horizontal resistance is the resulted from presence of non-specialized genes. The use of horizontal resistance may guarantee to host parasite balance, minimizing development of new races of parasite and horizontal resistance is affected by many factors. Environment in addition to it is difficult to incorporate in to a host variety. Characteristics of horizontal resistance is that it does not break down and, in this way, it is permanent resistance. In case of horizontal resistance, differential interaction between pathodem and pathotype is absent that means there is a constant ranking of pathodem and pathotype.

2.7.3.2.2 Evidences for horizontal resistance-

2.7.3.2.2.1 Constant ranking

Use of pathodem and pathotype indicates that when significant differential interaction is absent, resistance could be of horizontal type, if it is present the resistance could be vertical.

2.7.3.2.2.2 Historical

If the resistance of old cultivated variety has not broken in several decades of scientific history, it is likely to be horizontal resistance.

2.7.3.2.2.3 Geographical evidence

If a series of pathodems are tested in a number of different regions and their resistance ranking does not change, they are likely to possess horizontal resistance.

2.7.3.2.2.4 Epidemiological evidence

The vertical resistance delays the onset of epidemic and horizontal resistance shows down epidemic after it has started.

2.7.3.2.3 Components of horizontal resistance

2.7.3.2.3.1 Tolerance

This is the capacity of a susceptible host to bear severe disease attack and yet give satisfactory result to the growers. There is no inherited resistance in this variety and in technical sense they are equavalent to susceptible variety. The nature of tolerance is not clearly known. It may be in the form of compensating growth.

2.7.3.2.3.2 Slow development of disease

Disease appears early but development of disease is slow.

2.7.3.2.3.3 Morphological

Plant characters i.e. thick cuticle, waxy bloom, small stomata, large number of leaf hairs and upright leaf are some characters which protect from the disease infection.

2.7.3.2.3.4 Functional

In some cases, stomata do not open in the early morning when the chance of spore germination is more. In this way disease infection is minimized significantly.

2.7.3.2.3.5 Hypersensitive reaction

When spore comes on leaf tissue it germinates and form mycelium but before the infection starts or spreads to other place on the leaf, the surrounding tissue dies. Thus, nutrition supply to pathogen is stopped. In case of susceptible host this activity takes place very slowly but in case of resistant variety it happens very quickly.

2.7.3.2.3.6 Escape mechanism

Early maturity is sometimes refer to as type of escape mechanism by which early vigour permit varieties to escape serious damage from a disease. Thus, early maturing varieties are usually less damaged from disease than late maturing variety.

2.8 Breeding methods for biotic stress resistance

2.8.1 Introduction

An introduced variety resistant to the concerned insect pest and diseases may be released for cultivation if it performs well in the new environment and is agronomically desirable. Thus, it is the quickest and perhaps, the earliest method of developing a biotic stress resistant variety. e.g. introduction of *Phylloxera vertifoliae* resistant grape rootstock from USA to France. Sometimes, the introduced variety may not perform well in the new environment and it may be susceptible to the biotypes of the concerned pest prevalent in the area or to a new insect pests and/or disease of the area.

2.8.2 Selection

Biotic stress resistant variants may be found in an existing variety of a crop. In such a case, selection for insect and disease resistance is practiced to isolate biotic stress resistant variety.

2.8.3 Hybridization

When the desired biotic stress resistance is present in an agronomically inferior variety of the crop or in a related wild species, hybridization is only course of action for the breeder e.g. breeding for fruit fly resistant variety in ber (Vashishtha *et al.* 1997). However, breeding in ber is difficult due to incompatibility, low fruit set etc. (Table 2). The performance of fruit characters of parent, hybrid and BC_1 germination is given in table 3.

Table 2. Reciprocal crosses, mature fruits, obtained and germination of ber hybrids.

Parents	Number of flowers pollinated	Number of mature fruits harvested	Number of seed germinated
Seb x Tikadi	2544	25	12
Tikadi x Seb	2050	-nil-	—
Gola x Tikadi	2000	-nil-	—
Tikadi x Gola	2100	-nil-	—

2.8.4 Mutation

Generally, it has not been used to produce a successful biotic stress resistant crop. The reason for this are difficulty in screening of suitable mutations, the failure of such mutagenesis to generate positive changes to the genome and large number of progeny that must be handled.

2.8.5 Source of resistance for different diseases

Seventeen cultivars of ber were screened against powdery mildew at Hisar, Rahuri, S. K. Nagar and Jobner (Table 4) and found that cultivar Banarasi Karaka, Kaithali, Umran etc were more susceptible to this disease than Sanaur –5 and Kathaphal. Similarly, cultivars Safed Rohtak, Sanaur-1, Seo Bahadurgarhia were resistant to *Isariopsis* leaf spot disease and cultivar Villaiti was resistant to *Alternaria* leaf spot (Table 5 and 6) at Hisar and Rahuri center (Pareek and Vishal Nath, 1996).

In an another study at Bikaner, Jodhpur and Hisar, it was observed that cultivar Khadrawy was resistant to *Graphiola phoenicis* leaf spot, but at other hand, Zahidi, Zaglool, Shamran etc cultivars of date palm were susceptible to this disease (Table 7). Cultivar Hayani and Khadrawy were resistant to black scorch disease (Table 7). Other cultivars like Abdul Rehman, Barhee, Halawy, Medjool, Muscut etc were susceptible (Pareek and Vishal Nath, 1996).

Table 3: Fruit characters of parents, hybrid and BC_1, generation in Ber (*Z. mauritiana*)*

Parent/Hybrids	wt. of fruit (g)	Size(cm)	Shape	Skin colour	Pulp colour	Cavity	Pulp/stone ratio	TSS °Brix	Resistance to fruit fly
Seb	25	3.6x3.3	Round	Yellowish green	White	Stem end	14.4	16.8	Susceptible
Tikadi	4.1	2.1x1.8	Ovate	Dark red	Cream	Stem end	7.6	26	Resistant
F_1 hybrid	8.25	3.0x2.5	Oval	Yellow	Cream	Stem end	13.6	22	Resistant
BC_1 (Seb x F_1)	17.9	4.1x3.5	Ovate	Golden yellow	Cream	Styler end or both	14.60	20	Resistant

* Source: Vashishitha *et al.* (2004)

Table 4: Reaction of some commercial ber varieties to powdery mildew

Varieties	Hisar (1976-81)	Rahuri (1985-87)	S.K. Nagar (1989-93)	Jobner (1985-93)
Banarasi Karaka	S	HS	S	—
Glori	—	MS	—	—
Gola	R	S	HS	MS
Illaichi Jhajjar	R	HS	MS	R
Jogia	—	—	MS	HS
Kaithali	HS	HS	S	MS
Kathaphal	R	—	—	—
Manukhi	—	MS	MS	—
Mehrun	—	S	MS	—
Mundia Murhera	S	—	S	MS
Safed Rohtak	R	HS	MS	—
Sanaur-5	R	—	—	—
Seb	—	—	S	R
Seo	R	—	S	—
Sukhavani	—	—	MS	—
Umran	S	HS	HS	MS
Vikas	—	—	MS	—

Table 5: Reaction of some ber varieties to *Isariopsis* leaf spot disease

Variety	Hisar (1976-81)	Rahuri (1987-89)
Bagwadi	—	FREE
Banarsi Kadaka	—	R
Dharkhi-1	—	R
Dharkhi-2	—	R
Glori	—	FREE
Gola	—	MS
Guli	—	FREE
Jhajjar Selection	R	MS
Jhajjar Special	R	MS
Kaithali	—	MS
Kharki-1	—	R
Kharki-2	—	FREE
Mundia Murhera	R	—
Safed Rohtak	R	R
Sanaur-1	R	R
Sanaur-6	—	FREE
Seedless	—	FREE
Seo Bahadurgarhia	R	FREE
Umran	—	MS
Villaiti	—	FREE

Table 6: Reaction of some ber varieties to *Alternaria* leaf spot.

Variety	Hisar (1976-81)	Rahuri (1987-89)	S.K. Nagar (1989-91)	Jobner (1990-93)
Banarsi Kadaka	—	S	S	—
Dandan	R	MS	S	—
Gola	—	—	S	S
Gola Gurgaon-1	R	MS	HS	—
Govindgarh Special	R	MS	—	—
Illaichi	R	MS	S	MS
Jhajjar Selection	R	MS	S	—
Jhajjar Special	R	S	HS	—
Kaithali	—	MS	HS	MS
Mundia Murhera	—	—	S	S
Popular Gola	R	MS	S	—
Safed Selected	R	MS	—	—
Safed Rohtak	R	MS	S	—
Seb	—	—	HS	S
Seo Bahadurgarhia	R	S	S	—
Umran	—	S	HS	MS
Villaiti	R	R	—	—
ZG-3	R	MS	—	—

Table 7: Reaction of some date palm cultivars to *Graphiola phoenicis* leaf spot and black scorch disease at varous locations

Variety	Graphiola disease			Black scorch disease
	Bikaner (1983-85)	Jodhpur (1989-91)	Hisar (1981-88)	Jodhpur (1989-91)
Abdul Rehman	R	MS	—	MS
Barhee	—	R	HS	MS
Halawy	MS	S	HS	HS
Hayani	R	R	S	R
Khadrawy	R	R	R	R
Khalas	MS	—	—	—
Medjool	R	S	S	S
Muscat	R	MS	—	MS
Sayer	—	HS	—	MS
Shamran	S	S	S	HS
Zaglool	MS	HS	HS	MS
Zahidi	MS	S	S	MS

2.8.6 Production of disease resistant plant by unconventional breeding

2.8.6.1 Basic technique in plant cell culture

a) Callus and suspension culture
b) Haploid culture from pollen
c) Protoplast isolation and culture
d) Embryogenesis in cell culture
e) Selection of mutation from pathotoxin resistant cells and clones
f) Regeneration within heterogeneous materials
g) Regeneration of plants from somaclonal / protoclonal variation
h) Resistant plant through fusion of protoplast
i) Disease resistance through uptake of foreign genetic material

2.8.6.2 Genetic engineering or Recombinant DNA technology

There is scope of genetic engineering in fruit crops for the development of transgenic varieties resistant to biotic stresses. This technology involves the isolation of gene of desired character. Insertion of this isolated gene in a suitable vector (making it a recombinant vector). Insertion of the recombinant vector into a suitable host (organism/cell) known as transformation. Selection of the transformed host. Multiplication followed by expression of the introduced gene into the host.

2.9 Breeding for resistance to parasitc weeds

Some weeds parasitise on number of economically important crop species particularly in vegetable crops and often cause a substantial reduction in their yield e.g. orobanche in solanaceous vegetables. A convenient and relatively cheaper way of reducing the losses in crop yields caused by such parasitic weeds is to develop resistant or tolerant varieties.

2.10 References

Brich, A.N.E., Fellows, L.E., Evans, S.V. and Docherty, S.V. (1986) Paraaminopheny lalanine Invigna: possible taxonomic and ecological significance as a seed defence against brunched. *Phytochemistry*, **25**: 2745-2751.

Evans, S.V., Gatehouse, A.M.R. and Fellows, L.E. (1985) Detrimental effect of 2,5 dihydroxy methyl, 3-4, dihydroxy pyrrolidine in some tropical legeme seeds on leaves of the bruchid (*Callosobruchus maculates*). *Entomology: Experimental and Applied*, **37**: 257-261.

Gatehouse, A.M.R, Gatehouse, J.A., Dobie, P., Kilminstor, A.M. and Borther, D. (1979) Bio-chemical basis of insect resitance in *Vigna unguiculata*. *Journal of the Science of Food and Agriculture* **30**: 949-958.

Hilder, V.A., Gatehouse, A.M.R., Sheerman, S.E., Barker, R.F. and Boulter, D. (1987) A noval mechanism of insect resistanc engineered in to tobaco. *Nature,* **330**: 160-163.

Johnson, R., Narvaez, J., An, G. and Ryan, C.A. (1990) Expression of potato proteinase inhibitors I and II in transgenic tobaco plant: Effect on natural defence against manduca sixta larvae. *Proceeding of the National Academy of Science*, USA **86**: 9871-9875.

Mann, J. (1987) Secondary metabolism, 2nd Edn, Oxford University Press, Oxford.

Nash, R.J., Fenton, K.A., Gatehouse, A.M.R. and Bell, E.A. (1986) Effects of the plant alkaloid castanospermine as on antimetabolites of storage pests. *Entomology: Experimental and Applied,* **42**: 71-77.

Pareek, O.P. and Vishal Nath (1996) Coordinated Fruit Research in Indian Arid Zone – A Two Decade Profile. NRC for Arid Horticulture, Bikaner.

Robinson, R. A. (1971) Vertical resistance. *Rev. Pl.Path.*, **50**: 233-239.

Vashishtha, B.B., Singh, M.P. and Singh, G.K. (1997) Breeding for the resistance to fruit fly (*Capomyiana vesuviana*) to protect environment hazards. Symposium on Recent Advances in Management of Arid Ecosystem, *Abstracts*: 121, March 3-5.

Vashishtha, B. B., Shukla, Anil Kumar and Shukla, Arun Kumar (2004) Breeding for biotic stress in arid fruits. In: Advances in arid horticulture Vol. I (Eds. P.L.Saroj, B.B.Vashishtha and D.G.Dhandar), International Book Distributing Co. Lucknow, pp.177-184.

3

Plant Growth Regulators in Fruit Breeding

Plant growth regulators are organic substances other than nutrient, which is required in minute quantity to modify the plant growth processes. PGR has very important role in fruit breeding which is used in every facet of crop improvement e.g. multiplication of hybrids/selection through asexual means or by micro propagation, breaking of seed dormancy, growth control, regulation of flowering, fruiting and induction of sterility in reproductive organs etc (Dhatt, 1985). Details of the various uses of plant growth regulators are given as under.

3.1. Multiplication of hybrids/selections

3.2. Root initiation

Plant growth regulator particularly auxin has important role in root initiation of cutting and layering. Synthetic auxins like IBA and NAA are very effective in root initiation.

Sl.no.	Name of the fruit crop	Method of propagation	Name of the PGR and its concentration
1.	Guava	Cutting Air layering Stooling	IBA 5000 ppm + Benzoic acid 200 ppm IBA 5000 ppm IBA 5000 ppm
2.	Custard apple	Stooling	IBA 20000 ppm
3.	Grape	Cutting	IBA 3000 ppm
4.	Litchi	Layering	Pre-treatment of stock plants with etherel 250 ppm + treatment with IBA 5000 ppm + 200 ppm hydroxy benzoic acid
5.	Sapota	Air layering	mixture/IBA and NAA 10,000 ppm
6.	Lime and Lemon	Stem cutting	IBASO - 100 ppm
7.	Pomegranate	Cutting	IBA 1500-2000 ppm Air layering - 10,000 ppm
8.	Phalsaa	Cutting	IBA 1500 ppm

3.3. Seed germination/overcoming seed dormancy

Good germination of hybrid seed is the prime requisite in fruit breeding programme to raise a large population for evaluation. Seeds of some fruits require a specific period of after ripening, means keeping the seeds at low temperature for certain period for proper germination and growth of seedling is termed as stratification. Attempts have been made to use growth regulators for complete or partial substitution of stratification. Gibberellic acid has been found to be the best chemical for inducing germination e.g. Padmanabhaiah and Satyanarayana (1969) and Chadha (1965) recommended the use of 100 to 500 ppm GA_3 in grape. Further, Kachru *et al.* (1972), Uppal *et al.* (1972) and Randhawa and Negi (1964) observed that GA_3 not only substituted stratification for 40 days but also stimulated the growth of seedlings in grape. Pandey and Singh (1988) found that seeds extracted from immature berries (5 to 6 weeks after anthesis) were capable of germinating with 1500 to 2500 ppm GA_3 and 1500 to 2000 ppm Kinetin treatments.

3.4. Substitution for chilling requirement

Dormancy of the resting buds is controlled by environmental factors affecting levels of phytohormones. Breaking of bud dormancy in fruit trees, use of GA_3 has been reported by many workers (Donoho and Walker, 1957, Chauhan *et al.*, 1961). In peach, only leaf buds are GA_3 inducible and flower buds were unaffected or had a very week response (Hatch and Walker, 1969). The dormant pear buds showed very week response to GA_3 (Brown *et al.*, 1960), however, apple buds did not show any response (Walker, 1970). Further, Biggs *et al.* (1974) and Mc Eachern and Storey (1974) reported that the GA_3 and thiourea sprays break the dormancy of leaf as well as flower buds in peach, pecan and grapes.

3.5. Induction of male sterility

Male sterile line can be used to do away with the tedious process of emasculation. Pollen sterility can also be induced by chemicals which do not cause ovule sterility. Iyer and Randhawa (1965 and1966) reported that aqueous solutions of maleic hydrazide (MH) at 500 to750 ppm, Tri-iodo-benzoic acid (TIBA) at 400 to 500 ppm and 1,2 Dichloro-iso-butyrate (FW-450) at 0.30% applied twice to 13-15 days old inflorescence of grape, induced pollen sterility. Similar reports were made by Singh (1967), Dhillon and Singh (1970) and Randhawa (1971).

3.6. Shortening of breeding cycle in fruit crops

The efforts to shorten the breeding cycle in order to enhance the breeding efficiency are an integrated approach. It would facilitate a quicker evaluation of genotypes. Further, "fruiting cutting" (Mullins, 1966) and conversion of tendril in to inflorescence by use of plant growth regulators can help in this regard (Srinivasan and Mullins, 1976).

3.7. Influence on reproductive growth

Regulation of flowering through chemical means has been achieved in some fruit crops. Since, juvenile phase in some of the fruit plants ranges from 3 to 12 years in seedling progenies. Therefore, the early assessment of hybrid/ selection is not possible. To obtain early flowering, it seems desirable to shorten both the juvenile period and the transition period. To speed up the growth of seedling it should be grown in green house for 2-5 months and screened for resistance to disease and insects during this period. Then promising type should be shifted to the field for the induction of flowering through growth regulators. There are some examples that growth regulators like paclobutrazol in mango and SADH (Bruinsma, 1966, Dennis, 1968) TIBA (Dayton, 1966) CEPA (Kender, 1971 and 1974) have been reported to induce flowering in apple and pear seedlings. Application of paclobutrazol promoted flowering in mango, citrus and cherry. Application of NAA and 2,4-D were reported beneficial in pineapple, papaya, and litchi. Application of ethylene promoted flowering in pineapple.

3.8. Control of fruit drop

To prevent the dropping of fruits from the tree, PGR has important role. The important fruit crops in which fruit drop can be minimise by the use of plant growth regulators are as under

S.No.	Name of the fruit crops	Name of PGR and their concentration
1	**Mango**	
	Alphonso	NAA 25 ppm 2,4-D 25 ppm
	Neelam	2,4-D 30 ppm
	Dashehari	2,4-D 20 ppm
	Langra	2,4-D 20 ppm
	Chausa	2,4-D 20 ppm
	Bombay green	2,4-D 20 ppm
2	**Citrus**	
	Naval orange	2,4-D 20 ppm
	Valencia orange	2,4-D 4 - 40 ppm
	Pineapple orange	2,4 ,5-T 10-23 ppm
	Mosambi	2,4, S -T 30 ppm
	Jaffa	2,4 -D 20 ppm
	Sathgudi	2,4 -D 5-10 ppm
3	**Cashewnut**	**2,4-D 10 ppm NAA 10 ppm**

3.9. Making distant crosses

In distance crosses, some times embryo gets degenerated. This problem can be over come by the use of embryo rescue technique. In this technique for the development of embryo and organogenesis again PGR has an important role.

3.10. References

Biggs, R.H., Buchanan, D.W. and Arnold, C.E. (1974) Response of several *Prunus persica* cultivars to chemical formulation to terminate bud dormancy. *Hort. Sci*., **9** (3): 17.

Brown, D.S., Griggs, W.H. and Iwakiri, B.T. (1960) The influence of gibberellin on resting pear buds. *Proc. Amer. Soc. Hort. Sci*., **76**: 52-58.

Bruinsma, J. (1966) On dersook naar de toepassing van chemische regulatinen in de plantenteelt wageningen. Centrem Voor Plantentalt. Onderzock, *Jaarverslag*, **1965**:19-21.

Chadha, K.L. (1965) Studies on phenotypic variability and seed germination in some grape cultivars. Ph.D. Thesis, IARI, New Delhi.

Chauhan, K.S., Biggs, R.H. and Sites, J.W. (1961) Influence of gibberellic acid, naphthalene acetic acid, indole-3- acetic acid and maleic hydrazide on peach bud dormancy. *Proc. Fla. Sta Hort. Soc*., **74**: 374-378.

Datt, A.S. (1985) Role of growth regulator in fruit breeding. In: Fruit breeding in India (Ed. G.S. Nijjar), Oxford &IBH Publishing Co. New Delhi, pp 9-14.

Dayton, D.F. (1966) Chemical induction of flowering in triploid apple progenies. *Hort. Sci*., **1**: 92-93.

Dennis, F.G. (1968) Growth and flowering responses of apple and pear seedlings to growth retardants and scoring. *Proc. Amer. Soc. Hort. Sci*., **93**: 53-61.

Dhillon, B.S. and Singh, D. (1970) Induction of male sterility in grapes. *Indian J. Hort*., **27**. (1/ 2) : 80-85.

Donoho, C.W. (Jr.) and Walker, D.R. (1957) Effect of gibberellic acid on breaking dormancy of rest period in Elberta peach. *Science*, **126**:1178-1179.

Hatch, A.H. and Walker, D.R. (1969) Resting intensity of dormant peach and apricot leaf buds as influenced by temperature, cold hardiness and respiration. *J. Amer. Soc. Hort. Sci*., **94**: 304-307.

Iyer, C.P.A. and Randhawa, G.S. (1965) Chemical induction of pollen sterility in grapes. *Curr. Sci*., **34**: 411-412.

Iyer, C.P.A. and Randhawa, G.S. (1966) Induction of pollen sterility in grapes (*Vitis vinifera*). *Vitis*, **5** : 433-445.

Kachru, R.B., Singh, R.N. and Yadav, I.S. (1972) Physiological studies on dormancy in grape seeds (*Vitis vinifera* var. Black Muscat). II On the effect of exogenous application of growth substances, low chilling temperature and subjection of the seeds to running water. *Vitis*, **11**(4): 289-295.

Kender, W.J. (1971) The induction of early fruiting in juvenile apple tree by 2-chloroethyl phosphonic acid. *Hort. Sci*., **6** : 283.

Kender, W.J. (1974) Ethephon induced flowering in apple seedling. *Ibid*, **9** : 444-445.

Mc Eachern, G.R. and Storey, J.B. (1974) The influence of photoperiod, gibberellic acid thio-urea and uran on the emergence of peach, grape and pecan buds from dormancy. *Hort. Sci*., **9** (3) : 40.

Mullins, M.G. (1966) Test plant for investigation of the physiology in *Vitis vinifera* L. *Nature* **239**:419-420.

Padmanabhaiah, D.R. and Satyanarayana, G. (1969) Studies on the methods of accelerating germination in grape seed. *Andhra Agric. J*., **16**(2): 33-37.

Pandey, S.N. and Singh, R. (1988) Germination of seeds extracted from immature berries of grapes. *Indian J. Hort.*, **45**(1/2): 56-60.

Randhawa, G.S. (1971) Induction of pollen sterility in different crops use of plant growth regulators and gibberellins in Horticulture. ICAR Bull. No. 34, p. 62.

Randhawa, G.S. and Negi, S.S. (1964) Preliminary studies on seed germination and subsequent seedling growth in grape. *Indian J. Hort.*, **21**(3/4): 186-196.

Singh, D. (1967) Induction of male sterility with plant growth regulators in Anab-e-Shahi grapes (*Vitis vinifera*). M.Sc. Thesis, PAU, Ludhiana.

Srinivasan, C. and Mullins, M.G. (1976) Reproductive anatomy of the grape vine (*Vitis vinifera* L.) origin and development of the anlagen and its derivatives. *Ann. Bot.*, **40**: 1079-1084.

Uppal, D.K., Jawanda, G.S. and Sharma, J.N. (1972) Studies on the grape seed dormancy. Third International Symposium on Subtropical and Tropical Horticulture (Abst.) Banglore, P.17.

Walker, D.R. (1970) Growth substances in dormant fruit buds and seeds. *Hort. Sci.*, **5**: 414-417.

4

Breeding of Mango

Botanical name : *Mangifera indica* L.

Family : Anacardiaceae

Chromosome number: 2n=2x=40

Scientific classification

Kingdom	:	Plantae- Plants
Division	:	Magnoliophyta–Flowering plants
Class	:	Magnoliopsida – Dicotyledons
Subclass	:	Rosidae
Order	:	Sapindales
Family	:	Anacardiaceae
Genus	:	*Mangifera*
Species	:	*indica L.*

Mango is one of the choicest fruits of India, grown over an area of 1.23 million hectares in the country. Mango occupies the same position in India as occupied by the apple in temperate and grape in subtropical areas. In India mango is acclaimed as king of fruits. The name *Mangifera* was given for the first time by Bontius in 1658, when he referred to this plant as arbor *Mangifera* (The tree producing mango). Linnaeus also referred to it as *Mangifera arbor* in 1747, prior to changing the name to its present form (*Mangifera indica*) in 1753. Mango is good source of vitamin A and C apart from the usual content of minerals and other vitamins. Mango is also considered to have some medicinal properties. Ripe fruits of mango is fattening, diuretic and laxative. Smoke of burning is supposed to cure hiccups and some throat troubles and the kernal is effective against diarrhoea and asthma. Besides table purpose, fruits of mango can be used for the preparation of pickles, preserves, jam, *amchur* (mango powder), mango leather (*ampapad*) and mango fool (mango+milk+sugar) (Singh, 1992).

4.1. Centre of diversity

The Indo-Burma region is supposed to be known as centre of origin of mango. Geographical distribution, phylogenetic trend, pollen morphology, chromosome number, cytogenetical and breeding behaviour indicates highest concentration of species of *Mangifera* in Malayan Peninsula followed by Sunda Island and the Eastern Peninsula, comprising of Burma, Thailand and Indo-China (Mukherjee, 1949, Mukherjee, 1958).

Mangifera indica is economically most important species. It is closely related to *M. longipes* Grift. and *M. sylvatica* Roxb. In India many wild types occur almost throughout the tropical and subtropical hilly forests and ravines upto 1000 M above mean sea level and in the Andamans. Important zones of variability of wild *M. indica* are given below: (Yadav and Rajan, 1993)

(a) Humid subtropical region (Manipur, Tripura, Mizoram and South Assam)

(b) i. Chhottanagpur Plateau and part of Madhya Pradesh, Bihar and Orissa.
ii. Rajmahal hills (Santhal Pargana)

(c) South Madhya Pradesh (tribal area), adjoining Orissa and Andhra Pradesh

(d) Dhar Plateau of Madhya Pradesh and adjoining South Rajasthan and Gujarat.

(e) Andaman and Nicobar Islands.

Broadly, the *Mangifera indica* germplasm can be classified into (a) Seedling races: (wild and cultivated types) which is most common in Tropical Asia and almost all the commercial cultivars of mango have arisen as a result of selection from these seedlings. (b) Horticultural races: (Vegetatively propagated types) which include both mono and polyembryonic cultivars. Monoembryonic cultivars are the rule in India, South America, Africa and Florida. However, polyembryonic cultivars predominate in South East Asia, Central America , Haiti, Hawai, Australia and South Africa. In Indonesia, Philippines and Holland the major commercial cultivars are of polyembryonic type, while in Malaysia and Singapore a mixture of mono and polyembryonic cultivars are grown.

4.2. Germplasm resources

India is home of mango germplasms where more than thousand varieties are existing, which are widely distributed in different agroecological zones (Yadav and Rajan, 1993, Singh, 1992). Central Institute for Subtropical Horticulture, Lucknow has largest collection of mango (633 accessions in the national repository) and they have greater genetic variability with respect to fruit shape, skin colour, stone size, period and time of maturity, pulp thickness, colour, bearing habit, yield and quality parameters (Anon., 2002). Further, IIHR, Bangalore, IARI, Pusa New Delhi, Sabour (Bihar), Sangareddy (Andhra Pradesh) etc. are also maintaining the germplasm of mango. In India, majority of varieties are monoembryonic whereas in most tropics region polyembryonic types are predominant.

4.3. Problems in mango breeding

- Seedling plants take 5-10 years for bearing i.e. long juvenile phase.
- Presence of single seed in a fruit.
- Heterozygosity and cross pollination makes it difficult to predict the quality of hybrid fruits.
- Complex nature of panicle and flower.
- Excessive fruit drop.
- Large area of land is required for raising hybrid seedlings.
- Lack of knowledge on inheritance pattern.
- Lack of pre-selection procedure for isolation of superior hybrid seedlings.
- Problem of polyembryony which makes identification of hybrids defficult.

4.4. Objectives

- Dwarf growth habit.
- Precocious and prolific bearer.
- Regular bearer.
- Attractive and good quality fruits.
- Resistance to biotic and abiotic stresses.
- High productivity.
- Good keeping and processing quality.

4.5. Botany

Botanically, mango belongs to the family Anacardiaceae to which cashewnut (*Anacardium occidentale* L.), pistachionut (*Pistacia vera* L.), Indian hogplum or amra (*Spondias pinnata*) and Chironji (*Buchanania latifolia*) also belongs. It is an erect, branched, evergreen tree, living for 100 years or more with dense, dome shape canopy, long tap root up to 6 m in depth. Leaves are spirally arranged, glabrous exstipulate, produced in flushes, young leaves usually reddish in colour later on turned to dark shining green and remain on tree for a year or more. Petiole 1-10 cm long, some what flattened on upper surface, lamina is 8-40 x 2-10 cm size narrowly elliptic or lanceolate, somewhat leathery, apex acuminate, base tapering, margin is usually undulate, midrib is prominent and upto 30 pairs of lateral veins. Stomata are found on both the surfaces but maximum on lower surface. Almost all the commercial cultivars of mango are related to a single species *Mangifera indica*. However, a few commercial cultivars of South East Asia belong to the other edible species such as *M altissima* , *M. caesia, M. cochinchinensis, M. foetida, M. griftithi, M. langinifera, M. longipes, M. macrocarpa, M. odorata, M. pajang, M. pentandra, M. sylvatica and M. zeylanica.* There are different reports regarding the number of species in Genus *Mangifera,* Singh (1969) reported 62 Species whereas Mukherjee (1949) reported 41 species but later on he reported that only 39 species are existing (Mukherjee, 1985). There are five species of *Mangifera* reported from the India e.g. *M. andamanica, M. indica, M. khasiana, M. sylvatica* and *M. comptosperma* (Mukherjee, 1985).

Further, Mukherjee (1985) reported the following morphological characters for the identification of *Mangifera* species.

- Type of disc present e.g. swollen, lobed or narrow.
- Pentamerous or tetramerous or of both types of flower on the same inflorescence.
- Number of stamens and their fertility level.
- Presence of ridges on petals.
- Nature of inflorescence ramification and pubescence.
- Shape, size, texture and venation of leaves.
- Fruit size and shape.

4.6. Floral biology

Flowering in mango is preceded by the differentiation of the flower bud in the shoots. Period of differentiation is reported to be October-December depending upon the climatic conditions (Sturrock, 1934, Sen and Mallik, 1941, Singh, 1958). In Baramasi strain of mango, the critical time of differentiation is twice in a year i.e. May - June and September - October. Apart from the inherent character of the variety, the time of flowering in different regions is mainly governed by the local climate, for example the time of flowering in South India is December but in North India it is February-March.

Mango inflorescence also arise from axillary buds quite frequently. Panicle is much branched having several hundred tiny flowers (1000-6000), which may be male and hermaphrodite (andromonoecious). The percentage of perfect flowers varies from 0.74 to 70% depending upon the varieties (e.g. 0.74% Rumani, 30.6% Dashehari, 42.9% Chausa, 69.8% Langra). Sex ratio and pollen abnormalities are the main problems in pollination. Occurrence of few abnormal flowers, e.g. double hermaphrodite flowers with ten sepals and ten petals, two stamens and two ovaries were observed by Burns and Prayag (1921). However, Randhawa and Damodaran (1961) reported other abnormalities like double staminate flower, flowers with extra petals between lobes of disc and staminoides, fusion of two hermaphrodite flowers etc. The flowering period of mango is usually of short duration of 2-3 weeks. The low temperature may extend and high temperature may shorten the flowering period of mango.

4.7. Pollination

Pollination is essential for fruit set in mango. Randhawa and Damodaran (1961) observed that pollination in mango is mainly entomophilous. It was found that under natural pollination, about 40-60% perfect flowers failed to receive any pollen grains (Mukherjee, 1951). Insects responsible for effective pollinations are *Musca domestica* (common house fly), *Melipona* species (a small dipterous fly) and *Syrphidae* species (a kind of hover fly). Sporophytic type of incompatibility is reported in mango. Singh *et al.* (1962) reported the presence of self incompatibility in Dashehari. Later on Sharma and Singh (1970) reported that Dashehari, Langra, Chausa and Bombay Green are self unfruitful, similar for Dashehari, Langra and Chausa was reported by Ram and Sirohi (1979). Ram (1976) reported that Bombay Green and Dashehari appears to be the best pollinizer for Dashehari and Chausa, respectively. Singh and Chadha (1979) also corroborated that Bombay Green is the best pollinizer for Dashehari. Neelum is partially self compatible whereas Mallika is self incompatible (Raja Sekaran, 1984).

4.8. Inheritance pattern

- Dwarfness of the plant is governed by the recessive gene because of the fact that upright growth habit of the tree is dominant over spreading habit and spreading growth habit is dominant over dwarfness (Majumder *et al.*, 1972).
- Fruit size is governed by polygenes (Additive genes).
- Biennial bearing character of a variety is dominant over regular bearing (Sharma and Majumder, 1985).

- Red colour appears to be dominant over green colour and influenced by duplicate gene (Sharma and Majumder, 1985)
- Bunch bearing habit of mango is a dominant character.

4.9. Pre selection criteria for dwarfness (Majumder *et al.*, 1969)

- High bark percentage
- More number of stomata per unit area
- Smaller size of xylem vessels

4.10. Breeding methods and achievements

4.10.1 Introduction

For incorporation of good colour to boost export of fresh fruits number of mango verities were introduced from the different countries for use as donor parent (Singh and Rana, 1993). Tommy, Zilete, Haden, Sensation, Julie, coloured variety of mango were introduced from Miami, Florida (USA), PI 24927, M 4336 (Carbao) from USA and EC 201556 (Carbao) from Philippines were introduced for regular bearing. Cultivar Amolie and Sweet were introduced from Belgium and Thailand, respectively.

4.10.2 Selection

Almost all the present commercial varieties of mango in the world were developed from open pollinated seedling selection e.g. Dashehari, Langra, S.B. Chausa, Rataul, Swarnrekha etc.

The evolution of Florida varieties which are the leading mango cultivars of the world is interesting. In 1889 introductions were made from India of which Mulgoba became popular. Cultivar Haden was a seedling of Mulgoba, subsequently many promising seedlings were selected which became popular cultivars. Tommy Atkins from Haden, Keitt from Mulgoba, Dyke and Palmer from unknown origin, Irwin from Lippins, Golden Nuggets and Brooks from Sandersha, Sensation from unknown origin etc are promising seedling selections.

4.10.3 Clonal selection

Clonal selection have also resulted in identification of few elite clones. Dasheshari-51 from Dashehari a regular bearer (CISH Lucknow), Subash chance seedling from Zardalu (BAC, Sabour), Red blush strain of Alphanso (Vengurla) heavy yielder strains of Langra and Himsagar (Kalyani W.B.), resistant to bacterial black spot in clones of Kensington , superior clones of

Rumani and Neelam (Tamil Nadu) and Regular bearer Cardoz Mankhurad from Goa Mankurad and Niranjan clonal selection from Parbhani and Pusa Surya from Eldon at IARI, New Delhi.

4.10.4 Hybridization

In recent years, emphasis has also been laid on evolving varieties tolerant to mango malformation. Variety Bhadauran, tolerant to this disorder, is being utilized in crossing with Neelam and Dashehari. (Singh *et al.*, 1985). A beginning was made in this direction as back as 1911 when Burns and Prayag initiated planned hybridization work on mango at Pune (Burns and Prayag, 1921). Since then the work was started at a number of research stations in the country, especially at Sabour (Sen *et al.*, 1946) and at Kodur (Bhujanga Rao and Rangacharlu, 1958) during early forties. The work at Sabour yielded two promising hybrids named Mahmud Bahar and Probha Shankar from the parental combinations of Bombai x Kalapady. Hybrid Mahmud Bahar was found to be a regular bearer for four years whereas Probha Shankar, was not a regular bearer. Further, the mango improvement work was initiated at Saharanpur in 1951 and also in Punjab in 1950 to develop regular bearing varieties. Later on in India nearly 20 inter-varietal hybrids of mango have been released for cultivation from IARI, New Delhi, CISH, Lucknow, IIHR, Bangalore, FRS, Sangareddy, FRS, Periakulam, AES, Paria (Gujarat), FRS, Vengurla etc.

4.10.4.1 Interspecific hybridization

Interspecific hybridization did not receive more attention but it can be useful tool to transfer the useful gene in cultivated varieties. Because of fact that all the *Mangifera* species have the same chrosomes number (2n=40) therefore, they can inter cross easily (Mukherjee, 1963).

4.10.4.2 Improved Hybridization technique

a. Single day pollination of limited number of flowers in a panicle. Here the main emphasis was given on utilizing large number of panicles and crossing what ever few flowers opened on the panicle during single day. Bagging with perforated polythene bags of 24" x 123 size of 100 gauge, instead of pollinating more number of flowers in a panicle for a number of days and bagging with muslin cloth (Mukherjee *et al.*, 1961).

b. Caging technique: The discovery of self incompatibility in some of the popular cultivars at IARI, New Delhi led to further improvement in the technique of hybridization it is known as caging technique (Sharma and Singh, 1970, Singh *et al.*, 1962). In this technique grafted plants of parent varieties are enclosed in an insect proof cage and pollination is affected through freshly reared houseflies.

c. Marker gene: The purple colour of new leaves and panicle and beak character of fruit helps in identifying the hybrid seedlings in the nursery (Sharma and Majumder *et al.*, 1985).

4.10.4.3 Promising hybrid of mango

IARI, New Delhi -Mallika, Amrapali, Pusa Arunima,

IIHR, Bangalore - Arka Anmol, Arka Puneet, Arka Aruna, Arka Nilkiran

RFRS, Vengurla - Ratna, Sindhu, Konkan Ruchi

CISH, Lucknow - CISH-M1, Ambika

FRS, Sangareddy - Au Rumani, Manjeera

BAC, Sabour - Safari, Jawahar,

AES, Paria - Neeleshan Gujarat, Neeleswari, Neelphanso

FRS, Periyakulam: PKM-1, PKM-2

In Israel, a new cultivar, Naomi, has been released which has smooth skin and red pigmentation. In Australia, a hybrid of Sensation x Kengsington has shown promising results. In Israel rootstock breeding is also in progress and a polyembryonic rootstock 13/1 has been released that is tolerant to salinity.

4.10.5 Mutation

Naturally occurring useful mutants like Rosica from the Peruvian variety Rasado de Ica has been isolated. Similarly, Davis Haden is mutant of Haden. However, no induced mutant is known to be released.

4.10.6 Polyploidy

There is much scope of this method, however, till date there is no report on this line. Variety Vellaicollamban cultivar of mango is tetraploid in nature (2n=4x=80).

4.10.7 Heterosis

Iyer and Subramanyam (1984) observed large fruits in the some progenies of Alphonso x Banganpalli. Transgressive segregation of this character was also observed. The population with bigger fruits was large among hybrid progenies obtained with Banganpalli as one of the parents. This effect may be due to an accumulation of dominant allele each having additive effects and masking the effect of deleterious recessive allele.

4.10.8 Biotechnological tool

Repetitive and proliferative somatic embryogenesis has been achieved using nucellar explants in monoembryonic varieties. High frequency somatic embryogenesis from nucellus tissue of monoembryonic mango 'Arka Anmol' has been achieved. Media composition for conversion of embryogenic callus cells to embryos, their development, maturation and germination leading to the formation of complete plantlets has been worked out. Suspension cultures have been raised from nucellar callus in mango hybrid 'Amrapali' to maximize production of somatic embryos. Direct plantlet regeneration was induced from higher embryos and somatic embryogenesis was induced in immature zygotic embryos of certain mango varieties (Ghosh, 1997).

4.11. Important wild species of mango and their uses

S. No.	Name of the species	Potential uses
1.	*M. laurina, M. gedebe , M. griftith*	Root stock in inundated soils.
2.	*M. pentandra*	Five stamens species and hence good pollinizer as breeding parent.
3.	*M. minor*	Resistant to anthracnose
4.	*M. odorata, M. foetida M. caesia, M. castruri, M. griftithi and M. torgenda*	For good quality
5.	*M. zeylanica*	Salinity tolerance
6.	*M. pajang*	Can be peeled like Banana

4.12. Important new cultivars of mango (Pandey *et al.*, 2002)

Alfazli (Alphonso x Fazali)

Developed at Bihar Agricultural College, Sabour in 1981. Tree is tall and regular bearer, fruit is medium to large size (228.70 to 460g) with 79% pulp, moderate T.S.S. content (16.20 to 18.5° Brix) with acidity of 0.24% , keeping quality is good. Fruit characteristics is similar to Fazali, but the maturity of fruits is earlier. It is late maturing cultivar and free from malformation and fruit fly.

Amrapali (Dashehari x Neelum)

This cultivar was released in 1979 from IARI , New Delhi. Plant is small, medium spreading, suitable for high density plantation. Fruit is small to medium (130g), late maturing, ovate-oblong, ventral shoulder slightly more prominent than dorsal. Peel thick, light greenish apricot yellow, pulp fibereless, plentiful (74.8%) colour deep orange, excellent fruit quality. T.S.S. 22.8° Brix, total sugar 17.20 %, acidity 0.12%, vitamin C 35 mg/100 g of pulp, high vitamin A content with total carotenoid content 16830 mg/100 g of pulp and keeping quality is very good. This cultivar is suitable for table purpose and nectar making.

Arka Anmol (Alphonso x Janardan Pasand)

Tree is medium vigorous and regular and prolific in bearing, fruit weight is 250 g with uniform yellow peel, pulp colour is orange. TSS is 20° Brix, keeping quality is excellent. This cultivar is free from spongy tissue disorder and suitable for export.

Arka Aruna (Banganpalli x Alphonso)

Tree is dwarf, regular and precocious bearer, it is late maturing cultivar and fruit is large (500-750 g), pulp content is high (73.3-78.5%) and free from spongy tissue. TSS of the fruit is 22° Brix.

Arka Neelkiran (Alphonso x Neelum)

This is semi-vigorous and regular bearer. Average weight of fruit is about 340 g, pulp content is 68%, pulp colour orange, TSS-21^0 Brix. This cultivar is also free from the spongy tissue disorder.

Arka Puneet (Alophonso x Banganpalli)

Tree is vigorous with regular and prolific bearing. This is mid season cultivar. Average fruit weight is 284 g, pulp content is 75%, with orange yellow colour and free from fibre. TSS is 21° Brix, the flavour of fruit is just like Alphonso. This cultivar is also free from the spongy tissue and fruit fly.

AU Rumani (Neelum x Mulgoa)

This variety was evolved at Fruit Research Station, Kodur (Andhra Pradesh). The tree is medium vigorous and heavy yielder, fruit is large (450 g), peel is smooth and yellow, pulp moderately firm, fibreless, and good in flavour.

Dashehari –51

It is a superior clonal selection of Dashehari from the orchard of CISH, Lucknow. This was released during 1998. This clone produces good crop every year without 'off bearing rhythm. This has higher productivity by 38.8% over normal Dashehari even in off year. Dashehari-51 produces an average yield of 43.4 kg per tree against very poor or no crop on normal Dashehari tree.

Eldon

This cultivar is an introduction from Brazil, which showed promising for table use at IARI, New Delhi. Fruit is medium to large i.e. 220 to 300g broadly ovate, ventral shoulder slightly rising and then rounded, dorsal shoulder ending in a moderate curve, beak perceptible, sinus slight, peel thick, golden yellow with attractive red coloration on ventral shoulder, pulp firm, fiberless, sweet, TSS-17.25% highly juicy, pleasantly flavoured, stone weight 30 g and shelf life excellent, at room temperature (11-14 days). It is suitable for long distance transportation.

Jawahar (Gulabkhas x Mahmudbahar)

This variety was evolved at Sabour in 1989. It was named on the occasion of Jawahar centenary. Tree is semi-dwarf, precocious and regular bearer. It is a mid season variety. Fruit is medium in size (215g) with 80% pulp. T.S.S. 22.5% and medium in acidity (0.19%).

Konkan Ruchi (Neelum x Alphonso)

This variety was evolved at Regional Fruit Research Station, Vengurla, Maharashtra and released in 1999. This is a regular bearing mango variety developed for pickle making. The tree is heavy yielder (70 kg/tree), large fruit size (430g) and pulp recovery is about 78%.

Mallika (Neelum x Dashehari)

This variety was released in 1971 by IARI New Delhi. The tree is semi vigorous, heavy and fairly regular bearer. This variety performs exceedingly well under Bangalore conditions. Fully ripe fruit has attractive apricot yellow peel. Fruit medium to large (300 to 500g) in size and pulp recovery is 75%. The taste and flavour of fruits are excellent with high TSS (24-20° Brix), good keeping quality (15 days). It is suitable for table and canning purposes.

Manjeera (Rumani x Neelum)

This variety was evolved and released at Fruit Research Station, Sangareddy (Andhra Pradesh). It is a dwarf and regular bearing variety suitable for high density planting at the distance of 4.5 x 4.5 m (about 500 tree/ha). Fruit is very attractive and large (300-350 g). It looks like Rumani. The peel is light yellow, pulp is firm, fibreless and sweet. Average yield varies from 120-150 kg/tree at 10 year age. Fruits come to maturity in the first fortnight of June.

Menaka

It is selected as chance-seedling of Gulabkhas at Sabour and released in 1991. Due to its attractive colour, name Menaka was given. Fruit is medium (300g) with very attractive red blush on ¾ based side of fruits on yellow back ground. It is late in maturity. Fruits have firm pulp, 20%TSS and 0.14% acidity. Keeping quality is good.

Nileshan Gujarat (Neelam x Baneshan st.1)

It was developed at Agricultural Experiment Station, Paria (Gujarat) and released in 1986. Tree is dome shaped, moderate in growth and bearing with sub erect to spreading branches. Fruit is medium (318g) and suitable for table use. It has 16.26° Brix TSS, total sugar 11.51%, acidity 0.12% and vitamin C content 16.09 mg/100g pulp). Keeping quality is good and average fruit yield is 48 kg./ tree at 12 year age.

Neeleshwari (Neelum x Dashehari st.1)

This variety was evolved at Agricultural Experiment Station, Paria (Gujarat) and released in 1986. Tree is dwarf, moderate bearer, late maturity, it takes 120-140 days flowering to fruit maturity. Fruit medium in size (228 g), peel apricot colour at ripening and smooth, pulp adhering to peel, texture moderately firm, non fibrous, fruit quality is very good, TSS 20.53%, total sugar 10.78% , acidity 0.11%, vitamin C content 36.04 mg/100 g of pulp. It is suitable for table purpose. Fruit yield is about 35kg /tree form 12 years old tree.

Neelphanso (Neelum x Alphonso st.1)

It was evolved at Agricultural Experiment Station, Paria (Gujarat) and released in 1986. Tree is moderately vigorous, moderately regular bearer with late ripening tendency (120-140 days after flowering i.e. 2nd week of July), fruit is medium in size (196g), thick peel with apricot colour at ripening. The pulp texture is firm and non fibrous, fruit quality is excellent, TSS 19.0%, total sugar 13.13%, acidity 0.07% and vitamin C content 22.80 mg/100g pulp. It is suitable for table and juice purpose. Average fruit yield is about 51 kg/ tree from 12 years old tree.

Niranjan

It was selected at Himayatbagh, the private garden of His Highness, Nizam of Hyderabad in Aurangabad District of Maharashtra. It is a regular, off season bearer. It flowers in the month of June-July and fruit are harvested in the month of October. Fruit is medium (200 g) similar to Langra, peel yellow-green in colour, pulp yellow with less fibre, TSS 16.5%, acidity 0.8% with pleasant flavour.

PKM-1 (Chinnasurarnarekha x Neelum)

This variety was evolved at FRS, Periyakulam (Tamil Nadu) and released in 1980. This is a regular bearer and high yielding cultivar.

This hybrid was originally evolved at Fruits Research Station Kodus, Andhra Pradesh. A few grafted of this hybrid were taken and planted af Fruit Research

Station, Periakulam, Tamil Nadu of after bi furcation of erst while Madras State into Andhra Pradesh and Madras States. The hybrid pestoremed well in Periakulam and it was relased by Tamil Nadu Agricultual Universit.y

PKM-2.(Neelum x Mulgoa)

This is a regular bearer with excellent fruit quality.

Ratna (Neelum x Alphonso)

This variety was evolved at the Regional Fruit Research Station, Vengurla (Maharashtra) and released in 1981. Tree is very dwarf in North India with weak and spreading branches, in which some tissues make drooping habit in weaker case. Newly emerged leaves are dark red colour and leaves are small to medium in size. This is a regular bearing variety. The fruit maturity is late (fruit mature in 3rd week of July) in North India. Fruits are medium to large (315g), peel is deep yellow, pulp recovery 79% and orange in colour. The fruit is free from spongy tissue. The TSS is 23.8^{o} Brix, acidity 0.25% and fruit taste is sweet.

Sindhu (Ratna x Alphonso)

Sindhu is the first seedless parthenocarpic variety of mango evolved at Regional Fruit Research Station, Vengurla (Maharashtra) and released in 1992. Tree is semi-vigorous, having regular bearing habit, fruit is medium in size (215 g) with attractive red peel, deep orange pulp, pulp to stone ratio is high (26:1), fibreless and free from spongy tissue. Taste of fruit is pleasant with a better sugar acid ratio than Ratna. TSS above 22° Brix and acidity medium 0.248%. The stone of fruit is very thin 6g.

Safari (Gulabkhas x Bombai)

This cultivar was evolved at BAC, Sabour and released in 1989. This is an improvement of Bombai with red blush on the base. Fruit is medium (153.0g) with 74% pulp, 22.5% TSS and 0.804% acidity.

Subhash

This is a chance-seedling selection of Zardalu made at BAC, Sabour and released in 1991 and named on the occasion of Subhash centenary. Fruit is almost similar to Langra and medium in size (10x7.7 cm) with attractive yellow fruit colour of Zardalu, TSS 24.0% and acidity 0.294%.

Sunder Langra (Langra x Sunder Pasand)

This cultivar was evolved at Bihar Agricultural College, Sabour and released in 1981. This is mid season cultivar. Tree is semi-vigrous and spreading with regular and moderate bearing habit, The fruit is medium in size (240 g), with TSS-18.5 to 22° Brix and acidity 0.19 to 0.26%.

Arka Udaya (H12: Amrapali x Arka Anmol)

This is released by ICAR-IIHR, Bengaluru. Fruits are deep yellow in colour and Fruit weight is 230- 250g, TSS:23-24°Brix with 68-70% pulp. Yield is 190 q/ha at 5x5 m plant spacing.

Arka Neelachal Kesri

It is developed by CHES-IIHR, Bhubaneswar. It is suitable for 650-700 growing degree days under Eastern Coastal region. Yield is 50-60 kg/plant (>10 years old). Maturity: 85-90days, fruit weight is 200-230g. Peel colour: Crimson red, TSS : 18 °Brix, Pulp : 68.70%, Sugar/acid ratio :26-28 (sweet) and Fibre : medium.

Pusa Lalima (Dashehari X Sensation)

It is released through Institute Variety Release Committee on 2012. Suitable for Plains of North and Central India. Yield : 50-60 kg per trees. Regular bearer. semi-vigorous and are suitable for closer planting. Fruits are attractive in shape and having red peel and orange pulp. The fruit (209 g) having attractive red peel colour and higher pulp content (70.1%) Pulp with medium total soluble solids (19.7%), vitamin C (34.7 mg/100 g pulp) and high ß-carotene content (13,028 µg/100 g pulp). It has good flavour with shelf life (5 to 6 days) at room temperature. It is suitable for domestic market as well as international market.

Pusa Shreshth (Amaravalli X Sensation)

It is released by IARI, New Delhi in 2012 which is suitable to Plains of North, West and South India. Yield : 20-22 kg per tree.It is unique hybrid having regularity in bearing, attractive elongated shape, red peel and orange pulp. Plants are semi-vigorous and suitable for closer planting (6m x 6m). Fruit medium sized (228g) with attractive red peel colour and higher pulp content (71.9%). The total soluble solids are 20.3%, vitamin C (40.3mg/100g pulp) and ß-carotene content (10,964 ug/100g pulp). It has pleasant flavour with self life (7 to 8 days) at room temperature. It is suitable for domestic market as well as international market.

Pusa Surya

Yield is 12-15 kg per tree (10th year). Trees semi-dwarf and suitable for closer planting (6m x 6m). Fruit ripens by mid-July in northern India, medium to large (260 to 290 g) with attractive apricot-yellow peel colour with pink-blush.

Pulp with medium TSS (19.0 °Brix) with long shelf-life (10 to 12 days) at room temperature after ripening. Suitable for both domestic and international markets.

Pusa Pratibha (Amrapali X Sensation)

Reguler bearer, atractive fruit shape, bright red feel and orange colour feed. Plants are seed vigour and about 278 plant may be accomodation in hectare (6m x 6m). It has good sugar: acid belend and above are uniformity in fruit six. It was released by IARI in 2012.

Pusa Arunima

It is suitable for al over India. Yield : 15-20 kg per tree (10th year). Regular bearer, semi-vigorous and suitable for closer planting (6 m x 6 m). Fruits medium to large in size (230 to 250 g) with attractive red peel and medium TSS (19.5 °Brix). It is suitable for both domestic and international markets, with long shelf-life (10 to 12 days) at room temperature after ripening.

Pusa Peetamber (Amrapali x Lal Sundari)

It is suitable to Plains of North, West and South India. Yield : 20-25 kg per tree. It is regular bearer, semi-vigorous and suitable for closer planting (278 plants/ha). Fruits are having attractive ablong shape, bright yellow peel. It is moderately resistant to mango malformation and major insect pests of mango. The fruits weight about 213 g with attractive yellow peel colour with higher juicy pulp (73.6%). It has medium total soluble solids (18.8%), rich in vitamin C (39.8 mg/100 g pulp) and β-carotene content (11,737 μg/100g pulp). It has appealing flavour with shelf life.

Ambika – 2000

It is released by ICAR-CISH, Lucknow. A cross between Amrapali and Janardhan Pasand, fruits oblong oval, colour bright yellow with dark red blush, pulp dark yellow, firm with scanty fibre and weighs about 300-350g. TSS is 21°B. It is a regular bearing variety and late in maturity. The hybrid has potential for both internal and export markets because of its attractive fruit colour. Yields about 80 kg/plant by about 10 years of planting under improved cultural practices.

Arunika - 2008

It is released by ICAR-CISH, Lucknow. A cross between Amrapali and Vanraj, the tree produces dwarf and compact canopy. Fruits are smooth, orange yellow with red blush, weigh about 190-210g, medium sized, ovate oblique, pulp orange yellow, firm with scanty fibre, pulp about 65%, TSS 24.6oB. The hybrid has potential for both internal and export markets because of its attractive fruit colour. Average fruit yield is about 69 kg/plant at 8 years of planting. It has demonstrated wider adaptability and is performing well under northern plains. A regular bearer, has red peel color.

Arka Suprabhath (H-14)

It is a new double cross hybrid between Amrapali (Dashehari × Neelam) × Arka Anmol (Alphonso × Janardhan Pasand).

Chhattisgarh Nandiraj

This is a clonal selection of local mango from Nayapara, Jagdalpur. It has attractive shape, appearance and pleasant aroma. The proposed strain has an average productivity of 280 kg/tree and average fruit size 200-250 g with acid blend. It is very much comparable to Deshari and having good keeping quality and suitable for export (TSS is 21.30 Brix and total sugar is 17.87%).

CISH-M-2

A cross between Dashehari and Chausa. Fruits of this hybrid are also medium with an average weight 220 g & oblong in shape. Skin is smooth, tough and yellowish green when ripe. Flesh is firm, Dark yellow in colour with scanty fibre. TSS 23 °Brix. This hybrid has commercial potential beacause of its sooty mould free fruits surface even after exposure to heavy rains. The fruits are similar to those of Dashehari, a leading variety of North India, but mature 15 days after Dashehari.

Hybrid-1084

A cross between Amrapali & Janardhan Pasand was found promising with attractive dark red colour on the peel and better shelf life. These hybrids have regular bearing character.

Hybrid-949

This is a cross between amrapali and Vanraj with firm flesh, attractive dark red colour on the peel and better shelf life. This hybrid has regular bearing character and late in season.

Sabour Mango-1

It has regular bearing habit. The fruits matures during 14th June to 18th June. Average fruit weight is 454.4g and pulp content is 82.73 per cent.

Manjeera, KMH-1, Neeleshan, Neeluddin, Neelgoa, Swarna, Jehangir, AU Rumani are some of the mango varieties which was released by Dr. YSRHU (APHU), Andhra Pradesh.

4.12.1 Important Commercial cultivars of mango for different regions:

S.No	Name of the state	Commercial cultivars
1	Northern region (U.P. and Punjab)	Dashehari: Mid-season cultivar, good for canning and cold storage. Susceptible to malformation Langra: Mid-season cultivar, poor keeping quality, alternate bearer. Excellent sugar/ acid blend and pleasant flavour.Bombay green: (Syn. Malda in U.P. and Sehroli in Delhi) Early cultivar, very susceptible to malformation, best pollinizer for Dashehari.Samarbehisht Chausa: (Syn-khajri) Sweetest cultivar, late in season, highly susceptible to malformation, biennial in habit.Others: Gopal Bhog and Zardalu-early season.Gulabkhas, Fajrizafrani, Khasul-Khas, Krishan Bhog, mid season Taimuriya-Late season.
2.	Eastern region (Bihar, W. Bengal)	Bombai: (Syn Malda in Bengal) Early cultivar poor keeping quality, alternate bearer.Himsagar: Mid season cultivar, most popular cultivar of West Bengal, biennial bearer.Fazali: (Syn Malda), Very late cultivar, keeping quality is moderate.Sukul: Late cultivar of Bihar, keeping quality is good.Others: Gulabkhas, Langra, Krishan Bhog etc.
3.	Western region	Alphonso: (Syn. Hafus, Appus, Badami, Gundu, Khader, Patnan, Jathi). Believed to be originated in Goa. Export cultivar, susceptible to spongy tissue. Pairi: (Syn. peter, yerra Goa, Nadusalai,Grape, Peter Pasand, Gohabunder, Raspuri). Alternate bearer, keeping quality fair.Fernandin: Important late cultivar of West coast of India, fruit quality is good.Kesar: Taste is very good and sugar/acid blend is excellent. Famous cultivar of Gujarat.Rajapur: Another famous cultivar of Gujarat good fruit quality, biennial bearer.Other: Malkurad, and Jamadar.
4.	South region	Bangalora: (Syns-Totapari, Collector, Sundersha, Gitti Mukku, and Thevadiyamuthi) Keeping quality is good, preferred by processing industries. Neelum: (Syn Kayaladdu). Late cultivar regular bearer, good donor parent for regular bearing. Rumani: Very important cultivar due to its good keeping quality. Banganpalli: (Syn Baneshan, Chapta, Chaptai and Safeda in North India). Mid season cultivar, fruit quality and keeping quality is good. Others: Mulgoa, Swarnrekha, Raspuri (Pairi)

4.12.2 Evaluation of mango germplasm for various traits

S.no.	Characteristics	Varieties
1.	Dwarfness	Creeping, Ladavio, Rumani
2.	High yield	Totapari, Vashi Badami
3.	High sex ratio	Langra, Dashehari, Neelum, Amrapali
4.	Large fruit size	Banganapalli, Kensington, Fazali, Langra, Mulgoa, Mallika
5.	Red peel colour	Janardan Pasand, Sinduria, Dori, Suvarna Rekha, Vanraj, Tomy Atkins, Sensation, Irwin.

6.	High pulp content	Langra, Dashehari, Pairi, Arka Aruna, Sindhu, Amrapali, Sunderja, Mallika, Fazli, Chausa Vanraj.
7.	High TSS content	Sahib Pasand, Doodth Peda, Dashehari, Langra, Arka Puneet, Mallika, Lazzat Baksh, Amrapali, Sakkarchina, Gopal Bhog.
8.	Keeping quality	Alphonso, Totapari, Sunderja, Sashejhari, Banganpalli, Kesar, Rataul.
9.	Early maturing	Lazzat Baksh, Guruvarm, Gaurjee, Zardalu, Chandrakaran, Panakalu.
10.	Processing	Alphonso, Rani Pasand, Sabib Pasand, Banganpalli, Totapari, Hyder Sahib, Pedda Rasam, Alif Laila, Probha Shankar, Kesar, Arka Anmol, Arka Puneet.
	Canned Slice	Dashehari, Baneshan, Langra, Bombay Green, Chausa
	Pickles and chutney	Amlette, Chandrakaran , Karanjio, Ramkela, Kharbooja, Ashwina
11.	Tolerant to malformation	Elaichi, Bhadauran

4.13. Future thrust

- Although commendable works have been done towards mango breeding but much has to be done in the direction of resistant breeding for biotic and abiotic stress. There is urgent need to evolve disease and pest resistant variety by exploiting potential of wild species.
- Mutation breeding another thrust area for mango improvement
- Application of biotechnological tools
- Post harvest technology
- Development of market oriented coloured variety with good keeping quality.

4.14. References

Anonymous, (2002) DARE Annual Report. ICAR New Delhi.

Bhujanga Rao, C. and Rangacharlu, V.S. (1958) Breeding new mango varieties in Sounth India. *Indian J. Hort.,* **15**:173-83.

Burns, W. and Prayag, S.H. (1921) The Book of the mango *Bull.* 103, Bombay Dpt. Agric.

Ghosh S.P. (1997) Research and development in horticulture: Indian fruit industry. *Indian Hort.*, **42** (3):4-9.

Iyer, C.P.A. and Subramanyam, M. P. (1984) Report Fruit Research Workshop on Subtropical and Temperate fruits, Dapoli, pp 1-11.

Majumder P.K., Sharma, D.K., Singh, R.N. and Mukherjee, S. K. (1972) Preliminary studies on inheritance in *Mangifera indica* L. *Acta. Hort.*, **24**:120-125.

Majumder, P.K., Mukherjee, S.K. and Chakladar, B.P. (1969) Mango rootstock research in India. Paper presented at ICAR Workshop on Fruit Research, Ludhiana.

Mukherjee, S. K., Majumder, P.K. and Chatterjee, S.S. (1961). An improved technique of mango hybridization. *Indian J. Hort.*, **18**: 302-304.

Mukherjee, S.K. (1949) A monograph on the genus *Mangifera* L. *Lloydia*, **12**: 73-136.

Mukherjee, S.K. (1951) Pollen analysis in *Mangifera* in relation to fruit set and taxonomy, *J. Indian Bot. Soc.*, **30**:49-55.

Mukherjee, S.K. (1958) The origin of mango. *Indian J. Hort.,* **18**: 302-304.

Mukherjee, S.K. (1963) Cytology and breeding of mango, *Punjab Hort. J.*, **3**: 107-115.

Mukherjee, S.K. (1985) Systematic and ecogeographic studies on crop genepool:1. *Mangifera* L. International Board for Plant Genetic Resources, Rome, Italy p.p. 86.

Pandey, S.N., Pandey, D., Chandra, R. and Singh, R. (2002) Fruit varieties released by centres of AICRP on subtropical fruits. AICRP (STF) Technical bulletin No. 2, CISH, Lucknow.

Raja Sekaran, R.T. (1984) Compatibility studies in mango (*Mangifera indica* L) M.Sc. thesis, P.G. School, IARI, New Delhi.

Ram, S. 1976 Search of suitable pollinizers for mango cultivars. *Acta. Hort.*, **57**:253-263.

Ram, S. and Sirohi, S.C. (1979) Study on selection of pollinizers for commercial mango cultivars. Paper presented in Mango Workers Meeting. Panaji Goa (Unpublished).

Randhawa, G.S. and Damodaran, V.K. (1961) Studies on floral biology and sex ratio in mango (*Mangifera indica* L.) Var. Chausa, Dashehari, and Krishan Bhog. III Anthesis, dehiscence, receptivity of stigma, pollination, fruit set and fruit development. *Indian J. Hort.*, **18** :51-64.

Sen, P.K. and Mallik, P.C. (1941) The time of differentiation of the flower bud of the mango. *Indian J. Agric Sci.*, **11**:74.

Sen, P.K., Mallik, P.C. and Ganguly, S.R. (1946) Hybridization of the mango. *Indian J. Hort.*, **4** : 4-15.

Sharma, D.K. and Majumder, P.K. (1985) Studies on the inheritance of mango (*Mangifera indica* L.) Second Int. Symp. Mangoes Bangalore, pp.10.

Sharma, D.K. and Singh, R.N. (1970) Investigation onself incompatibility in mango (*Mangifera indica* L.) *Hort. Res.,* **10**:108-118.

Singh, B.P. and Rana, R.S. (1993) Promising fruit introductions. In: Advances in Horticulture Vol I (Eds. K.L. Chadha and O.P. Pareek), Malhotra Publishing House, New Delhi, pp.43-66.

Singh, Harmail and Chadha, K.L. (1979) Studies on the selection of some suitable pollinizers for Dashehari cultivars of mango. Paper presented at mango workers meeting Panaji, Goa(Unpublished).

Singh, L.B. (1969) In: Outlines of Perennial crop breeding in the tropics (Eds. F.P. Ferwerda and F. Wit), Wageningen, pp. 309-327.

Singh, R. (1992) Fruits. National Book Trust, India New Delhi, pp 18-48.

Singh, R.N. (1958) Studies in the differentiation and development of fruit buds in mango (*Mangifera indica* L.) II morphological and histological changes. *Hort. Ad. Saharanpur*, **2** : 37.

Singh, R.N., Majumder, P.K. and Sharma, D.K. (1962) Self incompatibility in mango (*Mangifera indica* L.) var. Dashehari. *Curr. Sci.*, **31**:209.

Singh, R.N., Majumder, P.K. and Sharma, D.K. (1985) Improvement of mango through hybridization. In: Fruit Breeding in India (Ed. G.S. Nijjar), Oxford and IBH Publishing Co. New Delhi, pp. 25-30.

Sturrock T.T. (1934) Flower bud formation in the mango. In: A bibliography of the mango (*Mangifera indica* L.) (Ed. M.J. (Jr) Souls,) Univ. Fla. Res. Prob., Florida mango forum University of Maiami.

Yadav, I.S. and Rajan, S. (1993) Genetic resources of *Mangifera.* In: Advances in Horticulture Vol. I (Eds. K. L. Chadha and O.P. Pareek), Malhotra Publishing House, New Delhi, pp 77-93.

5

Breeding of Banana

Botanical name: *Musa paradisiaca* L.

Family: Musaceae

Chromosome number: 2n=3x=33

Scientific classification

Kingdom	:	Plantae- Plants
Division	:	Magnoliophyta–Flowering plants
Class	:	Liliopsida – Monocotyledons
Subclass	:	Zingiberidae
Order	:	Zingiberales
Family	:	Musaceae – Banana family
Genus	:	*Musa*
Species	:	*paradisiaca L.*

Banana is one of the oldest fruits and second largest growing fruit crop in the world. It is also known as the Adam's Fig and Apple of Paradise. Unlike other fruits banana is an herbaceous fruit plant. Banana breeding started in Trinidad, West Indies in 1922 and in Jamaica in 1924 (Shepherd, 1994). The driving force for this breeding programme was to develop improved Fusarium wilt (Fusarium oxysporum F. sp. Cubense) resistant banana for export trade (Buddenhagen, 1990). In 1960 both the programmes were combined under the Jamaica Banana Board. United Fruit Company also started a small breeding programme in Panama in 1920s. In India hybridization work was started at Central Banana Research Station, Aduthurai, Tamil Nadu in 1949. Important banana growing states are Maharashtra, Karnataka, Kerala, Tamil Nadu, Andhra Pradesh, Orissa, Bihar, West Bengal and Assam. Now a days in some districts of Uttar Pradesh, Harichal cultivar of banana is cultivated at commercial scale. In South India banana is extensively used in all auspicious occasions such as wedding, festival and worship. Banana is a good table fruit, besides, the cultivar Nendran is used for cooking, it is also used for preparation of halwa, sweet and chips etc. Main bud or heart of banana bunches is also used as vegetable in India, Thailand, Philippines, Malaysia and Indonesia. Musa textiles is known for strong fibre quality.

5.1. Centre of diversity

Edible banana is native to old world especially South East Asia (Simmonds,1962). Malayan area seems to be the primary centre of origin of cultivated banana (*M. acuminata*). Probably *M. acuminata* spread in India and Burma where *M. balbisiana* is native species. The natural hybridization in these two species may resulted many hybrid progenies (AAB, ABB etc).

5.2. Germplasm resources

At botanical garden, Howrah, seeds of few banana species were collected from Chittagong and Madras (Roxburg, 1832). More number of genotypes of banana are also maintained at Central Banana Research Station, Aduthurai (Nayer, 1957). At present 303 named varieties of banana have been maintained in germplasm collection at various centres (Singh,1989, Singh, 1990, Singh and Chadha, 1993). According to Singh, (1989) collection of Musa genotypes in India is as under:

Related species and genomic group	Number
Musa wild species	06
Musa AA group	08
Musa AAA group	48
Musa AAAA group	01
Musa BB group	01
Musa AB group	12
Musa AAB group	84
Musa ABB group	78
Musa ABBB group	03
Unidentified	52

Further, different institutes where field gene banks of banana is being maintained are as under :

S.no.	Name of the Institute/University	Number of collections
1.	Indian Institute of Horticulture Research, Bangalore	250
2.	Tamil Nadu Agricultural University, Coimbatore	243
3.	Banana Research Station, Kannara (Kerala Agricultural University)	257
4.	Central Horticulture Experiment Station, Ranchi	109
5.	Fruit Research Station, Kahikuchi (Assam Agricultural University)	95
6.	Banana Research Station, Hajipur (Rajendra Prasad Central Agricultural University)	115
7.	Navsari Agricultural University, Gandevi	74
8.	Banana Research Station, Kovvur (Dr. YSR Horticultural University)	51
9.	National Research Centre on Banana, Trichy	597
10.	National Bureau of Plant Genetic Resources, Regional Station, Trichur	412
11.	Dr. Rajendra Prasad Central Agricultural University, Bihar	81

5.3. Objectives

- To develop dwarf stature banana suitable for high density planting and to prevent damage through high wind velocity
- Production of good quality fruits
- Resistant to biotic and abiotic stresses i.e. nematode, panama wilt, bunchy top, sigatoka, moko disease and weevil etc
- To develop varieties with wider agro-ecological adaptability
- Development of male fertile parthenocarpic diploids with resistant to major disease and pests
- Developing longer finger size
- Suitability for export
- Good keeping quality

5.4. Botany

Banana belongs to family Musaceae and genus Musa. Banana is monocotylednous plant. It consists of pseudostem, rhizome, leaf blades and inflorescence. Rhizome is the real stem on which large number of buds or eyes develop. Initiation of the inflorescence primordium of the banana bunch takes place in the heart of the pseudostem. Ensete is another genera of this family probably originated in Asia. Genus Ensete has 6-7 species of which E. ventricosa is reported to be grown in Ethiopia as a food crop. Genus Musa has about 50 species and this genus is divided in to five sections:

a. ***Emusa:*** Includes about 13-15 species of edible and wild banana. The chromosome number 2n=22 in wild species and most of the cultivated varieties are having 2n=33 (2n=44 rarely) e.g. *M. acuminata, M. balbisiana, M. basjoo* etc.

b. ***Rhodochlamys:*** Mostly diploid, spread from India to Indonesia. Five to seven species are kept in this group. Parthenocarpy is absent in this group e.g. *M. ornata, M. velutina*

c. ***Callimusa:*** This is of ornamental value and x = 10 and 2n = 20. It is are found in Indo-China, Malaya and Borneo. Parthenocarpy is absent in this type. It includes about 5-6 species e.g. *M. coccinea.*

d. ***Australimusa:*** Like Callimusa it has 2n = 20. Species of this group is common in Queensland to Philippines. Important species of this group are M. textiles abaca or manilahemp, *M. maclayi* etc.

e. ***Incertae sedis:*** It includes *M. ingens* (x = 7, 2n= 14) of New Guinea which grows to a height of over 10m. This is largest known herbs. Another species in this group is *M. beccarii* (x = 9, 2n = 18) from North Borneo.

The most important Musa cultivars are the almost sterile triploid (2n = 3x = 33) and also tetraploid and diploid banana cultivars have also local importance in Asia. All banana and plantain land races are farmers selection from intra and interspecific hybridization of two different species, *M. acuminata* Colta, donor of the A genome and *M. balbisiana* Colla, donor of the B genome. Simmonds and Shepherd (1955) workers of Trinidad reported scoring technique to indicate the relative contribution of the two wild species for the constitution of a given cultivar. Fifteen distinguishing characters between *Musa acuminata* and *Musa balbisiana* were identified by them. Score one was given for each character to which cultivar agreed with *Musa acuminata* and score five was given for each character to which agreed with *Musa balbisiana.* Intermediate expression of the characters was assigned score of 2, 3, or 4 depending to their intensity.

Table 1: Taxonomic scoring of banana based on distinguishing features

Sl.no.	Characters	*Musa acuminata*	*Musa balbisiana*
1	Pseudostem colour	More or less heavily marked with black or brown blotches	Blotches slight or absent
2	Petiolar canal	Margin erect or spreading with scarious wings below, not clasping pseudostem	Margins not winged below, clasping pseudostem

Contd.

3	Peduncle	Usually downy or hairy	Glabrous
4	Pedicel	Short	Long
5	Ovules	Two regular rows in each locule	Four irregular rows in each locule
6	Bract shoulder	Usually high (ratio: 0.28)	Usually low (Ratio: 0.30)
7	Bract curling	Bracts roll	Bracts lift but do not roll
8	Bract shape	Lanceolate or narrowly ovate tapering sharply from the shoulder	Broadly ovate, not tapering sharply
9	Bract apex	Acute	Obtuse
10	Bract colour	Red dull purple or yellow inside pink, dull purple	Distinctive, brownish purple outside, bright crimson inside
11	Colour fading	Inside bract colour fades to yellow towards base	Inside bract colour continues to base
12	Bract scars	Prominent	Scarcely prominent
13	Free tepal of male flower	Variabily corrugated below tip	Rarely corrugated
14	Male flower colour	Creamy white	Variably flushed with pink
15	Stigma colour	Orange or rich yellow	Cream, pale yellow or pale pink

According to this scoring technique, the score range from 15(15 x 1) for *Musa acuminata* to 75(15 x 5) for *Musa balbisiana*. Cultivar would have a larger score if it were derived from *Musa balbisiana* and smaller if it were derived from *Musa acuminata*. Depending upon the contribution of these two parents to the constitution of the progeny and combining their chromosomal status, the naturally occurring edible banana can be grouped in to six category i.e. two diploid, three triploid and one tetraploid. This classification has been used widely to classify a range of edible banana (Shephered, 1990). Silayoi and Chomchalow (1987) classified 137 accessions in the Thai banana gene bank on scoring basis. They observed some deficiencies in the original classification and modified it. The main difference between two classification was the introduction of pure balbisiana clone in the Thai grouping which did not appear in the Simmond and Shepherd's classification.

Table 2: Taxonomic classification of edible banana (Simmonds and Shepherd, 1955)

Genome Constitution	Ploidy level	Score and nomenclature
AA	2x	16-23 Matti
AAA	3x	15-21 i) Gros Michel ii) Cavendish
AAAA	4x	15-20 Bodles Altafort (Synthetic hybrid of West Indies)
AB	2x	46-49 Ney Poovan
AAB	3x	26-46 Champa
ABB	3x	59-63 Kancha Kela
ABBB	4x	63-69 Klue Teparod

Table 3: Classification of banana according to Silayoi and Chomchalow (1987)

Genome Constitution	Ploidy level	Score
AA/AAA	2x/3x	15-25
AAB	3x	26-46
AB	2x	-
ABB	3x	59-63
ABBB	4x	67-69
BB/BBB	2x/3x	70-75

Recently, Principal Coordinate and clustering techniques have supported the conventional intragenomic classification which placed both species in section Emusa of the genus Musa (Simmonds and Weatherup, 1991). Moreover, this analysis suggested that the geographically wide spread species, *M. balbisiana* is not very viable and with no obvious close relatives. Similarly, four (malacccensis, microcarpa, burmannica, siamaea) of five sub species of *M. acuminata* showed relative homogeneity which the sub species banksii was more distant phonetically.

Numerical taxonomy, based on genomic contributions and ploidy (Simmonds and Weatherup, 1990), confirmed that Musa land races should be designated as AAA dessert banana (e.g. Cavendish for export trade), AAA high land cooking and beer bananas of East Africa, AAB plantains, AAB dessert banana of Brazil and India and ABB cooking bananas of Asia. The best edible banana and todays major export banana cultivars belongs to the AAA group. Further, the plantain land races have been defined in four sub-groups based on their inflorescence morphology i.e. French, French horn, False horn and True horn. The distinguishing features of different types of plantains are as under -

(1) **True horn plantain:** Incomplete inflorescence, hermaphrodite and male flowers are absent. Inflorescence axis terminates in a tail or a deformed glomeruls. Examples Gabon3, One Hand Planty etc.

(2) **False horn plantain:** Inflorescence is incomplete, hands consisting of large fingers followed by few hermaphrodite flowers. Examples Agbaba, Esang, Corne etc.

(3) **French horn plantain:** In this group inflorescence is again incomplete, hand consisting large fingers and many hermaphrodite flowers. Examples Boofo, Koi etc.

(4) **French plantain:** Complete inflorescence, at maturity hands having many fruits, inflorescence axis covered with persisting hermaphrodite and male flowers. Male bud is large and persistant. Example: Nendran.

5.5. Floral biology

Inflorescence develop from the heart of pseudostem, type of inflorescence is spadix. The interesting feature of banana inflorescence is the production of a series of different types of flowers i.e. female, hermaphrodite and male in the same floral stalk. Chailakhyan, (1961) reported the dual factor hypothesis which is fit for the flowering of banana. According to this hypothesis one gibberellin – like substance acts upon the growth and elongation of the main stem and the other, anthesin acts as flowering hormone to produce flower. Most common type of inflorescence consists of pistillate flowers at basal portion which develops into the fruits with deciduous staminate flowers e.g. Poovan, Monthan and Kaali, sometimes male bud (heart) continues to produce staminate flowers till the fruits ripen but in some groups such as Kaali (Nadan) heart withers and dries up long before the maturity of the bunch. Second group of inflorescences does not consist male bud. The whole inflorescence bears pistillate flowers and hence all the flowers develop into fruits e.g. Thatilla Kunnan, Ayirank Rasthali and Moongil. In the third type of inflorescence, basal flowers develop into fruits followed by persistent male flowers consisting of green rudimentary ovaries with persistent perianth and bracts e.g. Dwarf Cavendish, Nendran, Kullan etc. In the fourth types of inflorescence basal portions having female flowers, developed into fruits followed by persistent male flowers, which is again followed by deciduous male flowers. The bracts of persistent male flowers are deciduous e.g. Rasthali and Chakkarakeli. The ovary of flower is trilocular, with axile placentation. Although the edible banana is highly parthonocarpic where fruits develop without pollination but in seeded species/variety pollination is mainly done by bees and birds which visit the flowers for nectar secreted at the tips of the ovary and collected at the base of the perianth (Rao, 1984).

5.6. Inheritance pattern

Edibility in cultivated bananas is a result of combination of seedlessness and parthenocarpy and is not associated with polyploidy since edible diploids (AA and AB) do exist. In addition, the edible bananas are mostly sterile from both male and female sexes with some exceptions. Parthenocarpy does not appear in Musa balbisiana.

Inheritance of fruit parthenocarpy in diploid bananas was the only subject of genetic research on an important morphological descripter until the early 1990's (Ortiz, 1995). Occurrence of albino seedlings, i.e. complete lack of chlorophyll in any plant tissue in diploid plantain banana hybrid provided a means to re-investigate the almost unapproachable Musa genome. Ortiz and Vuylsteke, (1994a) established that albinism in Musa species is under the genetic control of at least two independent recessive alleles with complimentary gene action.

This was the first report on Musa genetics almost 40 years after the report on the inheritance of fruit parthenocarpy. The seed sterility in banana is due to cytogenetic factors. Further, the edibility in banana may be due to parthenocarpy and seed sterility.

Although, the major recessive genes control apical dominance, the dominant gene is responsible for parthenocarpy (Pi) whereas recessive gene (bsi) is responsible for black sigatoka resistance. Polyploidy level significantly attracted the quantitative variation observed in plantain banana, tetraploid, triploid and diploid hybrids (Craenen and Ortiz, 1995, Ortiz and Vuylsteke, 1995, Vandenhout et al., 1995). Further, fruit size of high yielding plantain banana hybrids may be enhanced by the epistasis, both (additive x additive) and (additive x intralocus interaction), of the bsi and Pi loci (Ortiz and Vuylsteke, 1996). Host–plant response to weevil in Musa is controlled by gene (s) exhibiting partial dominance towards the resistant parent and modified genes with additive and dosage effect for susceptibility in plantain parent. The dominant ad alleles improve the suckering (Ortiz and Vuylsteke, 1994 b) and dwarfism in banana is controlled by a single recessive gene (dw). Further, bunch orientation appears to be an oligogenic trait regulated by the epistatic effects of at least three dominant loci.

5.7. Breeding methods and achievements

5.7.1 Introduction

Introduction of some cultivars of banana was made for resistance to biotic stresses e.g. Lady Finger (EC 160160) resistant to bunchy top virus introduced from Australia and is being evaluated at IIHR, Bangalore and TNAU, Coimbatore. Further, cultivars Nain MS (EC 27237) from France and Velery from West Indies were introduced for utilization in improvement programme (Singh and Rana, 1993).

5.7.2 Hybridization

In India breeding work was started at Central Banana Research Station, Aduthurai (Tamil.Nadu) in 1949 (Sathiamoorthy and Balamohan, 1993). Afterwards breeding programme was also initiated at TNAU, Coimbatore and Kerala Agricultural University, Trichur.

5.7.2.1 Interdiploid hybridization (2 x 2)

Diploids are the main source of genetic variability in breeding new commercial hybrids. Since 1971, inter diploid hybridization were done to synthesize new diploid forms at TNAU, Coimbatore (Sathiamoorthy and Balamohan, 1993).

Initially, breeding was started with the objective to incorporate the resistance against wilt in Gros Michel (as female parent) by using wild *M. acuminata* diploid as male parent. The tetraploid (progeny of hybrid) was very tall but it was resistant to wilt. The dwarfness was incorporated by using a dwarf mutant Highgate as a substitute for tall Gros Michel (Ray, 1999). Later work utilized Lowgate which was smaller than Highgate and gave progeny the same height as Grand Naine. Since Lowgate gave fewer seeds than Highgate and its bunch qualities were also inferior to Highgate. Matti (AA) as adopted cultivar commercially grown in Southern India exhibits a strong resistance to sigatoka but susceptibility to nematodes. Even Matti is male sterile but it sets seeds after pollination. This cultivar can be utilized as female parent in diploid breeding (Sathiamoorthy and Balamohan, 1993). Among the wild species *M. acuminata* spp. *burmannica* and *M. acuminata* spp. *malaccensis* have shown resistance to both races 1 and 2 of the panama disease fungus (Vakili, 1965), Sigatoka leaf spot (Rowe, 1984, Rowe and Richardson, 1975) and nematodes (Rajendran, 1984). Other diploid clones involved in diploid male parent synthesis are Anaikomban (AA) resistant to nematode and Fusarium wilt but susceptible to Sigatoka leaf spot, Namarai susceptible to Sigatoka and nematode but tolerant to panama wilt, Pisanglilin (AA) and Tangot (AA) resistant to panama wilt and nematodes (Simmonds, 1962). Thc pollination is carricd out bctween 7.00 to 10.00 AM, undehisced anthers of male flowers are collected and twisted gently to force them to dehisce. Using a soft stable hair brush, the pollen grains are taken out and smeared gently over the stigmatic surface of the female flowers opened on the day of pollination. The pollinated flowers are to be covered with soft cloth bag (Sathiamoorthi and Balamohan, 1993). Most of the seeds (74.9%) are found in 1/3 part of the distal end of fruits, about 20.9% in 1/3 mid portion and rest at (4.2%) in the proximal 1/3 (Sathiamoorthi and Balamohan, 1993).

5.7.2.2 Triploid breeding

The crossing of diploid and tetraploid results in the production of a triploid having three sets of chromosomes, one from one parent and two from another parent. Because of fact that tetraploids tend to produce a higher proportion of abortive gametes than diploids it is better to use them as male parents. Production of pollen grains is so great that numerous fertile one will still be present inspite of the very large majority being non-functional. The presence of sterile female gametes on the other hand (if tetraploids are used as females) would result in drastic reduction in number of seed. Natural AAA triploid arose from the AA cultivars by chromosomes restitution at meiosis as an evolutionary course (Simmonds, 1962). Several synthetic (man made) AAA triploids have been bred out using High gate / Valery as female parents and wild/edible/ or improved acuminata diploids as the male. A good number of such triploids are now available for field testing or multilocation screening for resistance to specific disease or pests. For developing AAB triploids, AABB tetraploids are crossed with AA diploids. However, there has been an indication that an AABB plant would not necessarily produce exactly AB spores, but they would be the best possible approximation. The better approach for developing AABB tetraploids will assuredly be ABB x AA. This will permit a two stage use of developed AA germplasm in producing tetraploids and subsequently in producing triploids. At Coimbatore, cultivars Poovan (AAB), Rasthali (AAB) Peyan (ABB), Thote (ABB), Peykunnan (ABB), Rajavazhai (ABB), and Ney Vennan (ABB) were used as female parents for developing commercial triploids of bispecific origin (Sathiamoorthy and Balamohan, 1993).

5.7.2.3 Tetraploid breeding

Tetraploids are bred by crossing a triploid female (High Gate AAA) with the improved diploid male (AA or AB). Seed fertile tetraploids are analyzed for potential usefulness in (4x) x (2x) crosses for synthesizing triploid hybrids. Diploids can be selected for longer green life, firm pulp, stronger pedicels and disease or pests resistance. A number of good AAAA clones exist but as yet remain virtually unexploited. They were bred while breeding better export quality AAA clones. Similarly, AABB tetraploids have been developed to breed AAB/ABB types. Important tetraploids for commercial adoption are e.g. Gold Finger (FHIA-01 AAAB), FHIA-02 (AAAA), FHIA-03(AABB), FHIA-17, FHIA-21, BITA-1, BITA-2.

5.7.3 Mutation

Several natural sports of well-established commercial clones have been recognized for e.g. Highgate (AAA) is a semidwarf mutant of Gros Michel (AAA), Motta Poovan (AAB) is a sport of Poovan (AAB). Ayiranka Rasthali is a sport of Rasthali (or Silk), Barhari Malbhog is a sport of Malbhog, Krishna Vazhai is a natural mutant of Virupakshi (or Pome), and Sombrani Monthan (ABB) is a mutant of Monthan (ABB). Recently a new banana mutant Novaria has been derived in Malaysia from the gamma irradiation of the clone Grande Naine (Mak et al., 1996). This is early flowering with outstanding bunch yield. The effect of irradiation has also been studied upon rhizomes of cultivars Gros Michel (Broertjes and Vanharten, 1978), at INIVIT Cuba with the induced mutation of ABB cooking banana, Buoro CEMSA: triploid with FWR grown by Cuban Farmers (Rodriguez Nodals et. al., 1992). At QDPI Australia DPM: putative FMR triploid mutants (M1V3) from AAA Cavendish banana is in preliminary yield trials (Smith et. al., 1993) and at TBRI, Taiwan Tai Chiao1and GCTCV triploid with FWR is a result of some clonal variation of AAA Cavendish banana (Hawang 1991, Hwang and KO, 1988, 1989). The early flowering Fatom 1 was developed with induced mutation of AAA Cavendish banana at TBRI, Taiwan (Tan et. al., 1993).

5.7.4 Biotechnology

Plant tissue culture and molecular biology techniques are applied to enhance the handling and improvement of banana. Important application of a cell biology are micropropagation for rapid multiplication and germplasm exchange, embryo culture / rescue for in vitro seed germination, cryopreservation of germplasm and genome manipulation through genetic engineering using cell suspensions or protoplast culture. Although, Vuylsteke et al. (1996) repeated that somaclonal variation through micropropagation is of limited use in plantain breeding it has been successfully applied in Taiwan for the development of improved Cavendish banana cultivars with resistance to fusarium wilt and acceptable fruit quality (Hwang 1991, Hwang and Ko, 1989). The recent advances in gene transfer methods, Sagi et al. (1995), from KUL Belgium reported that the transgenic triploid cooking banana showing transient expression of GUS marker gene in pot growing in the green house from DNA particle bombardment on ABB cooking banana. In another work May et al. (1995) from Texas A&M, USA reported transgenic triploid banana expressing Kanamycin resistant and GUS marker genes through Agrobactorium transformation or AAA Cavendish banana in test tube in laboratory. The molecular markers are providing tools for phylogenetic investigations and cultivar identification, basic genetic research, marker assisted selection and diagnostics in pathogen identification.

5.8. Source of resistance

Name of the clone/cultivars	Name of the biotic and abiotic stress
Musa balbisiana	Drought
Calcutta-4	Black sigatoka
Pisang Lilin	Panama wilt (race1)
SH3142 (Diploid hybrid)	Race of Fusarium
Musa acuminata sp *malaccensis*, *Musa acuminata* sp *burmannica*	Race 1 and Race 2 of Fusarium
SH699	Bacterial wilt race 2, Moko disease
Pisang Jari Buaya (PJB)	Burrowing nematodes
Tangat, Anaikomban	Nematodes

5.9. Important hybrids of banana

H-1 (Agniswar x Pisang lilin)

Short cropping cycle, resistant to leaf spot, Fusarium wilt and burrowing nematodes. Developed by Kerala Agricultural university Kannara. It has early ratooning ability. Average weight of bunch is 14 to 16 Kg.

H-2 (Vannan x Pisang Lilin)

It is tolerant to leaf spot and nematodes. Developed by KAU, Kannara. Average weight of bunch is 15 to 20 Kg.

Co-1 {Kadali x (Kallar Laden (AAB) x Musa balbisiana clone sawai (BB))}

It is promising pome hybrid developed by Tamil Nadu Agricultural University Coimbatore. It retains typical apple flavour of Virupakshi even when grown in plain.

FHIA –1 (SH3142 x Dwarf Prata)

Belongs to AAAB group, fruits have apple flavour, highly resistant to black sigatoka, Fusarium wilt (race1 and race 2) and burrowing nematodes. It is developed at FHIA, Honduras.

FHIA-2 (Williams x SH 3393)

It belongs to AAAA group, It is highly resistant to sigatoka but susceptible to Fusarium wilt.

FHIA-3 (SH3386 x SH3320)

It belongs to AABB group. It is drought tolerant hybrid. It is dual purpose hybrid mostly cultivated for culinary purpose. It is also resistant to Sigatoka disease.

FHIA-17 (Highgate x SH3362)

Good quality attributes and resistant to black sigatoka.

FHIA-21 (AVP67 x SH3142)

Resistant to black sigatoka and produces larger bunch.

BITA-1 (Bluggoe x Calcutta 4)

Developed at IITA, Nigeria. It is tetraploid and resistant to black sigatoka and Fusarium wilt.

BITA-2 (Pisang Awak x *M. balbisiana*)

It is also tetraploid and resistant to black sigatoka and Fusarium wilt. It was developed at IITA, Nigeria.

5.9.1 Important varieties of Banana

Udhayam

This is a single plant selection belonging to Pisang Awak subgroup. The average bunch weight is 37 kgs and has a potential to yield upto 45 kgs under good management practices. It gives 30-35 per cent higher yield than local Karpuravalli. Crop duration is 14 - 15 months and yield remains stable even in ratoons. Exhibits field tolerance to Sigatoka leaf spot and nematodes. The cylindrical bunch and well-spaced hands make it suitable for long distance transportation and less prone to fruit damage during transit. It has high sugar content with 31o B with harmonious blend of acidity and sweetness and hence suitable for use in Fig industry. Udhayam performs well in Kerala, Karnataka Andhra Pradesh Bihar, West Bengal and North eastern India.

Bangrier

This is a single plant selection from the local land race, Bangrier, having genomic group ABB and Bluggoe subgroup. Being starchy in nature, it is mainly used for cooking purposes. Selected for its higher yield of 28-30 kgs which is 20% over local Monthan and stable yield over a period of 4 years (main crop + 2 ratoon crops). Good fruit quality for cooking with mealy texture. Fruits are elongated with pointed tip. This reduced the wastage of fruits which is more in Monthan.

Saba

Saba is an exotic introduction from Philippines and belongs to ABB genomic group and Bontha subgroup. Weight of the bunch ranged from 26- 29 kg, and the duration is 360 – 380 days. Fruits are dark green, flattened with blunt tip. Fruit pulp is starchy. It is growing well both in plains and higher altitudes. Saba is more suitable for marginal cultivation and saline sodic soils with pH ranging from 8.8 to 9.0. Among the ABB types, this is more tolerant to drought. This can be used for both culinary and dessert purpose.

Namwa Khom

It is also one of the exotic introductions from Philippines. It belongs to ABB genomic group and Pisang Awak sub group. Plant is dwarf statured with 2-2.4m height with about 90 cm circumference at the base and therefore suitable for high density planting and needs no propping. Leaves are almost erected in position. Petiole bases are tightly clasped on the pseudostem. Among the Pisang Awak sub group, it exhibits the shortest duration of 12 months and hence suitable for annual cropping system. Average yield is 20 kgs, with a potential to yield upto 25 kg. Bunch is acceptable size and the fruits are marketable quality with sugary taste and 29 °Brix.

Popoulou

Popoulou (AAB, French plantain) bananas are indigenous to the Pacific region. Under Kerala conditions, it performed well and the yield was found to be stable for the third year (2014-2015). It was similar to Nendran with respect to crop duration (290 days) and superior to Nendran with respect to yield (17.5 kgs and 75 fruits per bunch as against 11.5 kgs and 54 fruits per bunch). This had a short duration of 310-320 days with shorter maturation time of 50-55 days, yielding 14-15 kgs, 6-7 hands with 10-12 loosely packed fruits and 60-65 fruits per bunch. Fruits are very bold fruits with 15-16 cm circumference. Individual fruit weight is nothing less than 150g. Pulp of the ripened fruit is orange yellow and mild taste. However, the unripe mature fruits are highly suitable for making banana snacks like bajji, chips making etc. though the pulp looks orangish it is not as sweet as plantain. Fruits are eaten raw as well as cooked.

Manoranjitham variant

This belongs to AAA genomic group and has unique features like good aromatic and fragrant pulp. There are 13-14 fruits in a hand. Fruits are dark in green at maturity and yellowish green at ripening. Average yield ranged from 13-15 kg. Crop duration is 13-14 months. Tolerant to leaf spot disease. Plant height ranged from 4.9-5.2 m with a pseudostem girth of 110 cm. The bunch weight almost 51 kgs (more than 300% increase in yield) with 12-14 hands and 17-20 fingers per hand as against 13- 15 kg in local Manoranjitham. The yield was almost stable even during the third ratoon (46 kgs).

Bontha

This is the foremost cooking variety of the State. The fruits are long, slightly curved with prominent ridges and blunt apex. The rind is thick and green with whitish pulp. The male bud is also used for culinary purpose. Suitable for entire State. It is resistant of Leaf spot but is susceptible to Panama disease. Each bunch weighs 15 kgs with 5-6 hand and 70-80 fruits. The duration is 13 months and the spacing is 2×2m.

BRS-1: It is released in 1998. It is cross between Agniswar X Pisanglilin

BRS-2: It is released in 1998. It is cross between Vannan x Pisanglilin.

Godavari Bontha (KBS-5)

It is a culinary variety and comparatively high yielder than Kovvur Bontha with 8-9 hands and 90-100 fingers per bunch. It can be grown as pure crop and also as inter crop in coconut Orchards. The bunch weight is 23-24 kg. It is tolerant to thrips and aphids and moderately resistant to leaf spot diseases.

Kovvur Bontha (Cooking type)

It is a cooking variety and a sport with heavier bunches of bigger size and superior quality fruits. Fruits are larger, stout with less prominent ridges and bottle neck apex. The rind is thick and green with whitish pulp. The plants are tall sturdy and sucker freely, waxy bloom is present over petioles and young regions, susceptible to rhizome rot disease. Tolerant to Sigatoka leaf spot, good cooking quality and it is the only cooking banana variety commercially cultivated in Andhra Pradesh Fruit bunches are heavy, pendent and not compact, suitable for the entire State. Each bunch weighs 16-18 kgs with 6-7 hands and 75-90 fruits. The duration is 13 months and the spacing is 2 x 2m.

5.10. Important cultivars of banana with genome

Rose (AA), Pisang Lilin (AA), Gros Michel (AAA), Dwarf Cavendish (AAA), Giant Cavendish (AAA), Robusta (AAA), IC2 (AAAA), Bodles Altraffort (AAAA), Ney Poovan (AB), Kunnan (AB), Poovan (AAB), Silk (AAB), Pome (Virupakshi, Sirumalai AAB), Nendran (AAB), Monthan (ABB), Klue Teparod (ABBB), Atan (AAAB), Gold Finger (AAAB).

5.11. References

Broertjes, C. and Vanharten, A.M. (1978) Application of mutation breeding methods in the improvement of vegetatively propagated crops. Elsevier Pub. Co., New York, pp 316.

Buddenhagen, I.W. (1990) Banana breeding and Fusarium wilt. In: Fusarium wilt of Banana. APS Press, St., Paul, Minnesota, pp 107-113.

Chailakhyan, M. Kh. (1961) *Canad. J. Bot.*, **39:** 1817-1841.

Craenen, K. and Ortiz, R. (1995) Effect of the black sigatoka resistance gene bsi and ploidy level in the fruit and bunch traits of plantain banana hybrids. *Euphytica*, **87:** 97-101.

Hwang, S.C. (1991) Somaclonal resistance in Cavendish bananas to Fusarium wilt. *Plant Protection Bulletin* (Taiwan), **33:** 124-132.

Hwang, S.C. and Ko, W.H. (1988) Mutants of Cavendish banana resistant to races 4 of *Fusarium oxysporum* f. sp. Cubense. *Plant Protection Bulletin* (Taiwan), **30**: 386-392.

Hwang, S.C. and Ko, W.H. (1989) Improvement of fruit quantity of Cavendish banana mutants resistant to race 4 of Fusarium oxysporum f. sp. Cubense. *Plant Protection Bulletin* (Taiwan), **31:** 131-132.

Mak, C., Ho, Y.W. and Tan, Y.P. (1996) Novaria – A new banana mutant induced by gamma irradiation. *Infomusa,* **5(1):** 35-36.

May, G.D., Afza, R., Mason, H.S., Wiecko, A., Novak, F.J. and Arntzen, C.J. (1995) Generation of transgenic banana (Musa acuminata) plants via Agrobacteriium mediated transformation. *Biotechnology*, **13:** 486-492.

Nayar, P.N. (1957) The Nendran, The malabar plantain. *Madras Agric. J.*, **29:** 470-473.

Ortiz, R. (1995) Musa genetics. In: Banana and plantains. Chapman & Hall, U.K, pp 84-109.

Ortiz, R. and Vuylsteke, D. (1994 b) Genetic analysis of apical dominance and improvement of suckering behaviour in plantain. *J. Amer. Soc. Hort. Sci.*, **119:**1050-1053.

Ortiz, R. and Vuylsteke, D. (1994a) Inheritance of albinism in banana and plantain. *J. Amer. Sucher. Sci.*, **119:** 1050-1053.

Ortiz, R. and Vuylsteke, D. (1995) Effect of parthenocarpy gene P1 and ploidy in bunch and fruit traits of plantain and banana hybrids. *Heredity*, **75**: 460-465.

Ortiz, R. and Vuylsteke, D. (1996) Recent Advances in Musa genetics breeding and biotechnology. *Plant Breeding Abs,* **66(10):** 1355-1363.

Rajendran, G. (1984) Studies on Helicotylenchus multicinctus (Cobb) Golden, 1956 with special reference to banana. Ph.D. Thesis, TNAU, Coimbatore.

Rao, V.N.M. (1984) Banana. Published by ICAR, Krishi Anusandhan Bhawan, New Delhi.

Ray, P.K. (1999) Breeding and improvement of tropical fruits In: Tropical Horticulture Vol. I (Eds. T.K. Bose, S.K. Mitra, A.A. Farooqi and M.K. Sadhu), Naya Prokash Calcutta, pp 114-124.

Rodriguez Nodals, A., Ventura Martin, J.C., Rodriguez – Rodriguez, R., Lopez – Torres, J., Pino–Algora, J. and Roman–Martinez, M.I. (1992) Banana genetic improvement programme at INIVIT in Cuba : A progress report. *Infomusa*, **1:**3-5.

Rowe, P.R. (1984) Breeding bananas and plantains. *Plant Breeding Rev.*, **2:** 135-155.

Rowe, P.R. and Richardson, D.L. (1975) Breeding bananas for disease resistance, quality and yield. Tropical Agril. Research Service (SIASTA), La Lima Honduras. Tech. Bull. No. 2.

Roxburg, W. (1832) *Flora India*. **1:** 663-669.

Sagi, L., Panis, B., Remy, S., Schoofs, H., De Smet, K., Swennen, R. and Cammve, B.P.A. (1995) Genetic transformation of banana and plantain (Musa sp.) via particle bombardment. *Biotechnology*, **13**: 481-485.

Sathiamoorthy, S. and Balamohan, T.N. (1993) Improvement of banana. In: Advances in Horticulture Vol. I (Eds. K.L. Chandha and O.P. Pareek), Malhotra Publishing House New Delhi, pp 303-335.

Shepherd, K (1994). History and methods of banana breeding. *Infomusa,* **3(1):** 10-11.

Shepherd, K. (1990) Observations on the Simmonds – Shepherd system of genomic classification of banana cultivars. In: Musa conservation and Documentation, INIBAP, Montpellier (FRA) pp 47.

Silayoi, B. and Chomchalow, N. (1987) Cytotaxonomic and morphological studies of the banana cultivars In: Banana and Plantain breeding strategies (Eds. G.J. Parsley and E.A. Delanghe). *ACIAR Proc* **21:** 157-160.

Simmonds, N.W. (1962) The evoluation of the bananas. Longmans green & Co. Ltd., London.

Simmonds, N.W. and Shepherd, K. (1955) The taxonomy and origin of the cultivated bananas. *J. Linn. Soc.*, **55**: 302-312.

Simmonds, N.W. and Weatherup, S.T.C. (1990) Numerical taxonomy of the cultivated banana. *Tropical Agri.,* **67:** 90-92.

Simmonds, N.W. and Weatherup, S.T.C. (1991) Numerical taxonomy of the wild banana. *New Phyto.,* **115:** 567-571.

Singh, B.P. and Rana, R.S. (1993) Promising fruit introductions. In: Advances in Horticulture Vol .I (Eds. K.L. Chandha and O.P. Pareek), Malhotra Publishing House, New Delhi, pp. 43-66.

Singh, H.P. (1989) Genetic resources of banana in India. INIBAP/IBPGR Workshop Musa Germplasm conservation, Belgium, pp.14.

Singh, H.P. (1990) Country report on banana and plantain in India. Proc. Regional Consultation on Banana and Plantain (Ed R.V. Valmayor), Banana and Plantain R&D in Asia and Pacific, Manila, INIBAP, pp.161-185.

Singh, H.P. and Chadha, K.L. (1993) Genetic resources of banana. In: Advances in Horticulture Vol. I (Eds. K.L. Chadha and O.P. Pareek), Malhotra Publishing House New Delhi, pp. 123-142.

Smith, M.K., Hamill, S.D., Langdon, P.W. and Peggi, K.G. (1993) Mutation breeding programme produces a plant with potential Fusarium wilt (race 4) resistant Cavendish variety. *Mutation Breeding Newsletter*, **40:**4-5.

Tan, Y.P., Ho, Y.W., Mak, C. (1993) Fatom – 1: an early flowering mutant derived from mutation induction of Grand Nain, a cavendish banana. *Mutation Breeding Newsletter*, 40:5-6.

Vakili, N.G. (1965) Fusarium wilt resistance in seedlings and mature plants of Musa species. *Phytopathology*, **55:**135-140.

Vandenhout, H., Ortiz, R., Vuylsteke, D., Swennen, R. and Bai, K.V. (1995) Effect of ploidy on stomatal and other quantitative traits in plantain and banana hybrids. *Euphytica*, **83:** 117-122.

Vuylsteke, D.R., Swennen, R.L. and De Langhe, E.A. (1996) Field performance of somaclonal variants of plantain (Musa spp. AAB group). *J. Ame. Soc. Hort. Sci.*, **121:** 42-46.

6

Breeding of Papaya

Botanical name: *Carica papaya* L.

Family: Caricaceae

Chromosome number: 2n=2x = 18

Scientific classification

Kingdom	:	Plantae- Plants
Division	:	Magnoliophyta–Flowering plants
Class	:	Magnoliopsida – Dicotyledons
Subclass	:	Dilleniidae
Order	:	Violales
Family	:	Caricaceae – Papaya family
Genus	:	*Carica*
Species	:	*papaya L.*.

Papaya is an ideal fruit crop for growing in kitchen garden, backyards of home as well as orchards, especially those places nearer to cities or big town. It is also grown as a filler plant in orchard. The ripe fruits of papaya are consumed throughout the tropics and sub tropics. Fruits are also used in preparation of jam, soft drinks, ice cream flavour, crystalized fruits and syrups. Unripe fruits are commonly used as vegetables. Papain is prepared from the latex of immature fruits. It is proteolytic enzyme used for tendering of the meat, preparation of chewing gum and cosmetics etc. The ripe fresh fruit is rich source of vitamin-A (2500 I.U.), carbohydrates and minerals (Singh, 1969). Papaya on account of producing fruit in a short period after planting has attracted the attention of fruit growers to be cultivated on a large scale in the country. It is known as wholesome fruit which is valued for its nutritional and medicinal properties. Due to its nutritional, industrial and export demand, papaya is recognized the most potential fruit crop for its economic cultivation. The variable sex forms, propagational problems, susceptibility to frost and water logging, fungal and viral diseases are well identified problems in its cultivation. Economically the production life of papaya is 3 years and every time new plantation has to be raised. Papaya being transplanted at closer spacing is highly suitable for high density planting. Now high yielding variety with better fruit quality are needed to boost the production under varied agro edaphological situation.

6.1. Center of diversity

Papaya is native to Tropical America (Badillo, 1971). The South America and Costa Rica are the micro center of origin of papaya. It is a close relative of *Carica peltata.* In India it was introduced in early part of the 16th century from Philippines through Malaysia. It was widely spread in different part of the country particularly tropical and sub-tropical zones. India is largest producer of papaya in world. It is also cultivated in Brazil, Mexico, Australia, Hawai, Malaysia, Tiwan, Peru, Florida, Gold coast, South Africa and Bangladesh. In India it is widely cultivated in Uttar Pradesh, Bihar, Karnataka, Jharkhand and Madhya Pradesh etc.

6.2. Germplasm resources

Abundant germplasm is available in Madhya Pradesh, Bihar, Jharkhand, Utter Pradesh and Tamil Nadu. The variation with respect to plant stature, sex types, fruit shape and size, seed content is also observed in papaya. Presently, plenty of germplasm is being maintained at IIHR, Bangalore, IARI, New Delhi, Reg. Station, Pusa, Bihar, CHES, Ranchi, CHES, Bhubeneshwar and CISH, Lucknow for further characterization and evaluation.

6.3. Objectives

- To develop varieties of dwarf stature, smaller petiole length and early bearing.
- To evolve varieties with high yield and good quality fruits.
- To develop varieties with low seed cavity and more pulp thickness.
- To breed varieties having good keeping quality and suitable for export.
- Breeding for high latex yield for papain production.
- To develop varieties resistant against biotic and a biotic stress (virus, frost, water logging etc.)

6.4. Botany

Papaya belongs to family Caricaceae and genus Carica having about 40 species (Ram, 1993). It is a dioecious plant but gynodioecious cultivars are also available. The stem is hollow and soft wooded. It is usually unbranched while young but at later stage upright shoots develop at its terminal growth due to obstruction. The leaves are palm like with long stalks. Flowers are fragrant axillary panicles. The fruit is fleshy berry. The shape of fruits from pistillate flowers is ovoid-oblong to nearly spherical and from hermaphrodite flowers it is pyriform, cylindrical or grooved.

6.5. Floral biology

Dioecious papaya produces male and female flowers separately on different plants while gynodioecious cultivars produces both female and hermaphrodite flowers on same plant. Inflorescence of Washington papaya emerged out 45 – 48 days after planting. Male flower appears in the axil of the 24th leaf and that of female in 18th to 20th leaf. Female and male flowers developed with in 32 and 42 days respectively after bud initiation. The period from bud initiation to anthesis was shorter for male than female flower bud (Dhaliwal et al., 1991). Stamen development occurred prior to ovary development in the hermaphrodite flower and stamen differentiation was observed 56-59 days before anthesis. Anther dehiscence was completed with in 18-36 hours before the flowers opened and the stigma becomes receptive a day before the flowers opened remaining receptive for 6 days (Khuspe and Ugale, 1977).

Carica papaya, *C. monoica*, *C. cauliflora* and *C. goudotiana* flower throughout the year in India. Generally, 21days and 36 days are required from bud emergence to anthesis in *C. quercifolia* and *C. monoica*, respectively. The peak anthesis was observed between 5.00-6.00AM in all the species except in the pistillate flowers of *C. cauliflora* and staminate flowers of *C. goudotiana*. The receptivity of stigma was found on the day of anthesis in most of the species (Subramanyam and Iyer, 1986).

6.6. Types of sex and sex inheritance

Papaya is considered as a dioecious species. However, Ram (1993) reported that papaya is polygamous plant nad has many sex forms i.e. pistillate, staminate, andromonoecious, and hermaphrodite etc. Basically, there are three types of sex in papaya i.e., staminate, hermaphrodite and pistillate out of which only pistillate form is stable whereas hermaphrodite and staminate forms show sex reversal in different environmental conditions. Following eight categories of flowers have been reported in papaya by Storey (1958).

- Staminate produced by male plants.
- Teratological staminate produced by sex reversing male plants.
- Reduced elongata produced by hermaphrodite plants.
- Elongata produced by hermaphrodite plants.
- Carpelloid elongata produced by hermaphrodite plants.
- Pentendria produced by hermaphrodite plants.
- Carpelloid pentendria produced by hermaphrodite plants.
- Pistillate produced by female plants.

Further, male plants can produce all types of flowers during sex reversal while hermaphrodite plants can produce all types except staminate and teratological staminate whereas female plant can produce only pistillate flowers. There are no clear-cut picture about the genetic basis of sex. However, there are some reports that sex in papaya is controlled by single gene with three alleles (Hofmeyr, 1938, Storey, 1941). Male and hermaphrodite are heterozygous for sex and it is homozygous for female. Later on, Storey (1953) reported that sex in papaya is determined by the complex gene which is closely linked in differential segments occupying identical region on sex chromosomes. Storey (1953) also reported that there are two independent sets of factors which modify the sex expression in males and hermaphrodites under certain environmental conditions. The scheme for sex determination in papaya as adopted by Storey and Jones (1941) using the Mendalian symbol proposed by Hofmeyer (1938), is as under – (Cited from Ram, 1993)

M1 dominant factor for maleness.

M2 dominant factor for hermaphrodite

m recessive factor for femaleness

The genetic constitution of all the three sexes are as follows -

M1m male plants having staminate flowers

M2m hermaphrodite plants having bisexual flowers

mm female plant having pistillate flowers

Further, Ram et al. (1983,1985) suggested the following symbols for sex reversing and pure male.

M1RRm sex reversing male

M1Rrm sex reversing male

M1rrm pure male (do not go under sex change)

Ram et al. (1983) also reported that when sex reversing male was selfed, it gave the progeny of sex reversing male and pure male in the ratio of 3:1 and when it was crossed with pure male the segregation ratio was 1:1 which shows the operative multiple allelism. According to Samson (1980) if female (mm) is crossed with male (M1m) the segregation in progeny will be 1:1 for male and female plants. If hermaphrodites are selfed or crossed the ratio of female and hermaphrodite is 1:2, whereas if hermaphrodite is crossed with a male the ratio of female: hermaphrodite: male is 1:1:1.

Table 1: Sex inheritance in different cross combinations is as follows (Ram, 1993)

S. No.	Cross or Self	Female	Hermaphrodite	Male	Non-viable
1.	mm x M1m	1mm	-	1M1m	-
2.	mm x M2m	1mm	1 M2m	-	-
3.	M2m x M2m	1mm	2 M2m	-	1 M2M2
4.	M1m x M1m	1mm	-	2 M1m	1 M1M1
5.	M2m x M1m	1mm	1 M2m	1 M1m	1 M2M1

The plants shown as non-viable failed to appear because of lethality in early stage of development of M_1M_1, M_1M_2 and M_2M_2. A significant deviation of male and hermaphrodite from expected ratio of female to male and female to hermaphrodite (1:1), hermaphrodite to hermaphrodite and hermaphrodite self (2:1) and hermaphrodite to male and male to hermaphrodite (1:1:1) was observed. It shows that there was some unknown factor operative in reducing the number of (m) gametes or the viability of (mm) zygote in early stage of development. The sex in papaya has been reported to change due to injury (Iorns, 1928) and interaction of certain genetic and environmental factors (Higgins and Halt, 1914, Storey, 1953).

6.7. Pollination

To ensure good fruit set in the dioecious cultivars of papaya (e.g. Pusa Giant, Pusa Dwarf, Pusa Nanha, CO 1, CO 2, CO 5 and CO 6 etc.) minimum 10% male plants should be planted in the orchard. However, the pollinizers are not required in gynodioecious varieties i.e. Pusa Majesty, Pusa Delicious, Solo, Sunrise Solo, Coorg Honey Dew etc. Further, insects are the major pollinating agents in papaya.

6.8. Breeding methods and achievements

6.8.1 Introduction

Commercial cultivars such as PI 41436, Soniyimma, Malinchly, Maru Ank LG (Nigeria), Sunrise, Wilder, Sunset, T88/211, Kapohosolo, HCAR 27, 7836,

HCAR 16, Sunrise HCAR-29, Maradol HCAR-19 and California (USA) have been introduced (Singh and Rana, 1993). The cultivar California is dioecious in nature with high degree of tolerance to papaya ring spot virus. Besides cultivars, several Carica sp. like *C. goudotiana, C. cauliflora, C. sphaerocarpa, C. querciflora, C. monoica, C. pubescens* and *C. stipulata* are having resistance to various diseases, pests and frost etc have been introduced from USA and Venezuela (Singh and Rana, 1993).

6.8.2 Inbreeding and selection

For varietal development in papaya selection and inbreeding methods are mainly adopted (Ram, 1982 a). Being a highly cross-pollinated crop, flowering and fruiting habit of this crop varies considerably. Systematic work for breeding of uniform varieties with high yield and good quality for wider adaptability was started at IARI Regional Station, Pusa, Bihar in 1966. As a result of inbreeding and selection for 8 generations during 1966-1982 (Ram, 1981, 1982a, 1982b, 1984a, 1984b) uniform lines Pusa Delicious, Pusa Majestic, Pusa Giant and Pusa Dwarf with desirable attributes were developed.

Due to continuous selection several seedlings races are found to produce fairly large proportion of progenies true to the parents from seed, and of these Washington of Maharashtra and Honey Dew enjoy some popularity in South India. At Coorg Aiyappa and Nanjappa (1959) selected Coorg Honey which is gynodioecious. Based on the two-season evaluation of papaya germplasm in Nainital, cultivars Barwani Red, Coorg Honey Dew, BR 1, 2, 3, 4, Coimbatore 1B and Washington have been found promising for commercial cultivation. Considerable genetic variability was found in many plants with respect to fruit characteristics like fruiting height, fruit weight, size, shape, TSS, taste, number of fruits per tree. Selection practiced from a mixed population of CO1 and others resulted in 12 promising selections which were superior to CO1 arising as a chance seedling at the Chettali Citrus Research Station. The Coorg Honey Dew payapa has not only spread throughout Coorg but also in whole country. Solo variety of papaya is true breeding cultivars in the world which is established variety of Hawaii. This is because hermaphrodite flowers of this cultivar are 100% bud pollinated and Solo papaya has become predominantly homozygous and highly inbred. Development of cultivar with high papain content was started at TNAU, Agriculture College and Research Institute, Coimbatore. As a result of intensive breeding programme, four cultivars i.e. CO1, CO2, CO5 and CO6 were developed through inbred selection (Rao, 1974, Sundarajan and Krishnan, 1984, Shanmugavelu et al., 1988) at Coimbatore.

Table 2: Important cultivars of papaya developed through selection are as under-

S. no.	Cultivars	Selections
1.	CO1	Selection from cv. Ranchi
2.	CO2	Pure line selection from local material
3.	CO5	Selection from variety Washington
4.	CO6	Selection from Pusa Majesty
5.	Pusa Delicious	Selection from cv. Ranchi
6.	Pusa Majestic	Selection from cv. Ranchi
7.	Pusa Giant	Selection from cv. Ranchi
8.	Pusa Dwarf	Selection from cv. Ranchi
9.	Coorg Honey Dew	Selection from Honey Dew
10.	Punjab Sweet	Developed at PAU Ludhiana by selection
11.	Pant 1, 2, 3	Developed at GBPUAT, Pantnagar through selection
12.	BR 30	Selection from Pusa Nanha

6.8.3 Induction of polyploidy

Hofmeyer (1945) reported polyploidy in papaya. They found that the quality of tetraploid fruit was better than the diploid and it was also compact with small seed cavity. But tetraploids were less fertile than diploid as indicated by comparative seed count. However, according to Singh (1955) there was complete sterility in both female and male tetraploids and expressed doubt about their commercial utilization. Further, Zerpa (1957) reported that colchicine induced tetraploid hermaphrodite plants, which were used as male parent in a cross with a female diploid produced a few seeds without endosperm, by embryo culture, two triploid plants were obtained which turned out to be hermaphrodite.

6.8.4 Hybridization

Hybridization is important tool to develop varieties by incorporating specific gene. Although some hybrid varieties have been developed by the intervarietal or intergeneric hybridization even though there is great scope for development of superior cultivars with better quality and yield. Hybridization was initiated through intervarietal crosses between CO1 x Coorg Honey Dew and CO1 x Washington. The reciprocal combinations were also tried. The progenies were evaluated over a number of generations through sib mating and hybrid derived CP80 was identified as a promising selection out of CO1 x Washington cross combination. This is characterized by a purple tinge shade on stem petiole and flower.

At TNAU, Coimbatore three varieties have been developed viz. CO3 (CO2 x Sunrise Solo), CO4 (CO1 x Washington) and CO7 (Coorg Honey Dew x CP85). However, at IIHR, Bangalore two hybrids IIHR-39 named as Surya (Sunrise

Solo x Pink Fleshed Sweet), and IIHR– 54 (Waimanalo x Pink Fleshed Sweet) were developed (Dinesh and Yadav 1998). Hybrid HPSC-3 (Tripura local x Honey dew) was developed by the ICAR Research complex, Tripura (Singh and Sharma, 1986).

Cultivar Cariflora was developed by crossing K2 x K3 line of papaya which is tolerant to PRSV (Conover et al., 1986).

Interspecific hybridization was also attempted in the genus Carica. The cross between the *Carica papaya* and *Carica cauliflora* did not form mature seed but immature embryos could be germinated and grown by employing embryo culture. F1 hybrids of *Carica papaya* x *Carica pubescens* and *Carica papaya* x *Carica quercifolia* were vegetatively vigorous. *Carica papaya* x *Carica cauliflora* F1 progenies are slow growing and those of *Carica papaya* x *Carica stipulata* developed apical necrosis before reaching maturity. Iyer and Subramanyam (1984) attempted interspecific hybridization, according to them F1 hybrids of *C. cauliflora* x *C. monoica* when crossed with *C. papaya* gave fertile hybrids, although the *C. cauliflora* and *C. monoica* are incompatible to *C. papaya* but hybrids of *C. cauliflora* x *C. monoica* is compatible.

6.8.5 Heterosis breeding

Heterosis is superiority of F1 over the parents evolved in cross. Dai (1960) reported heterosis in the cross between Philippines x Solo varieties. F1 hybrid tended to have reduced seed number and enhanced plant vigour. Heterosis up to 111.4% for yield and yield traits was obtained in Solo yellow x Washington whereas high heterosis for potential economic competitiveness was noticed in Thailand x Washington (Iyer and Subramanyam, 1981).

6.8.6 Mutation

Ram and Majumder (1981) developed a dwarf mutant line by treating papaya seed with 15K gamma rays. Initially, 3 dwarf plants were isolated from M2 population. Repeated sib mating among the dwarf plants helped in establishing a homozygous dwarf line (Ram, 1984c). Example Pusa Nanha.

6.8.7 Biotechnology

In vitro propagation and genetic engineering technique can serve as a potential tool to overcome major constraints in *Carica papaya.*

6.8.8 Embryo culture

In papaya incompatibility is mainly due to the failure of endosperm formation. The hybrid embryo resulting from interspecific cross of *C. papaya* and *C.*

cauliflora has been successfully rescued on White's medium in 30 days by Yung (1986).

6.8.9 Somatic embryogenesis

Callus derived from cotyledon and hypocotyl segments of papaya cv. Sunrise on MS medium was induced to form embryoids and shoots. The peroxidase isoenzyme patterns of leaf tissues from 5-6-month-old in vitro derived plantlets of hermaphrodite, pistillate and staminate plants were identical as those of mature plants and successfully distinguished the sex and cultivar of each plantlets. Protoplast isolated from suspension cultures of somatic embryos on culture in liquid KM8p-S medium. Fresh somatic embryos obtained from suspension cultures of papaya contained 90% water had poor regeneration capacity and could not be stored. After desiccation to 50-70% water, 82% of somatic embryos produced seedlings on MS medium.

6.8.10 Transgenic papaya

Transgenic papaya has been developed against papaya ring spot virus (PRSV) using coat protein mediated resistance in University of Hawaii by Dennis Gonsalves. The coat protein gene from PSRV was isolated, cloned and used for transforming papaya to provide resistance against the severe strain of the same virus.

The target cultivar used in transforming papaya were the Red fleshed, Sunset and the Yellow Fleshed Kapoho. Transformation with coat protein gene done using micro projectile bombardment technique using embryogenic tissues of payapa.

Two transgenic lines Sun UP from Sunset and UH Rambow from Kapoho were developed which have shown excellent resistance to PRSV. Sun UP, which is homozygous for C P (Coat Protein gene) was resistant to most isolate of PRSV, from other geographical location except Tiwan's YK isolate of PRSV. Rainbow which is hemizygous for CP was susceptible to PRSV, isolates from outside Hawaii but was resistant to the severe strain of Hawaiian PRSV (HA isolates).

6.9. Important varieties and cultivars

Coorg Honew Dew - It is gynodioecious selection from Honey Dew at Coorg.

CO1 – Plant is dwarf, fruit is round, selected from cultivar Ranchi.

CO2 - It is pure line selection from local type, suitable for papain production.

CO3 – It is hybrid between CO2 x Sunrise Solo, plant is vigorous, fruit is medium in size with good keeping quality.

CO4 – It is hybrid between CO1 x Washington, fruit is large, keeping quality is good, flesh colour is yellow.

CO5 - It is selection from Washington, it is also good for papain production.

CO6 – It is selection from Pusa Majesty, plant is dioecious and dwarf, suitable for papain production.

CO7 – It is gynodioecious hybrid between Coorg Honey Dew x CP85. It has red flesh.

Pusa Majestic – It is selection from cv. Ranchi. It is gynodioecious variety suitable for high papain.

Pusa Delicious - A gynodioecious selection from cv. Ranchi, it has good fruit quality.

Pusa Giant – A dioecious selection from cv. Ranchi, it is tolerant to strong wind.

Pusa Dwarf – It is selection from cv. Ranchi, plant is dwarf and dioecious in nature, fruit shape is oval.

Pusa Nanha – It is developed through muntation, plant is dwarf, suitable for high density planting, fruit quality is good.

Pant Selection-1 – Selection made at GBPUAT, fruit shape is oblong.

Punjab Sweet – It is selection made at PAU Ludhiana. It is frost tolerant and dioecious in nature.

Surya - It is gynodioecious hybrid (Sunrise Solo x Pink Flesh Sweet), developed at IIHR, Bangalore, it has high yield with good quality fruit. It was selected from F14 generation. Hence, seeds can be produced by bagging the hermaphrodite flowers or by crossing the female flowers with hermaphrodite flowers. The plant is gynodioecious in nature with no male plants. Fruits resemble Sunrise Solo in shape. The plants are shorter compared to Solo. Skin is smooth, becomes uniformly yellow in colour on ripening. Fruits are medium in size of about 600 to 800 g with a small fruit cavity. Pulp is about 3 & 3.5 cm thick, deep red in colour and sweet with a TSS of 13.5 & 15 Brix. It does not have the odd flavour. Keeping quality of fruits is good. Yield per plant is approximately 55 & 65 kg (60 & 65 t/acre).

HPSC-3 – It is hybrid between Tripura Local x Honey Dew. This cultivar is resistant to papaya mosaic potex virus.

Arka Prabhath - It is from the cross (Surya x Tainung-1) x Local Dwarf. It is gynodioecious in nature. The plants are semi-vigorous and bearing starts at a lower height (60-70 cm). Since the variety is gynodioecious, seed production is easy, as bagging of bisexual flowers ensures true to type plants. The pulp is firm (5.9 kg/cm2) and colour is deep pink. The fruit weight on an average is 900-1200 g, the TSS is 13-14 & Brix and yield per plant is 90-100 kg. The keeping quality is good.

Washington - It is a table purpose variety. Fruits are round to ovate, medium-large in size with few seeds. When ripe, skin attains a bright yellow colour. The average weight of fruit ranges from 1.5-2 kg. Male and female plants are separate.

IIHR39 and IIHR54 - Developed at IIHR, Bangalore. This variety bears medium sized sweet fruit with high TSS (14.5° Brix) and better shelf life.

Red Lady - Early, vigorous, productive, and tolerant to papaya ring spot virus. Plants begin to bear fruit at 60-80 cm height and have over 30 fruits per plant in each fruit-setting season. Fruits are short oblong on female plants and rather long-shaped on bisexual plants, weighing about 1.5-2 kg. Flesh is thick, red, with 13% sugar content, and aromatic.

Sinta- It is a gynodioecious hybrid of UPLB philipines released in 1995. It is yellow flesh firm and solid with small cavity index. It is moderately tolerant to viral problem. Fruit weight ranged from 0.8 by to 1.5kg.

Arka Surya: It is the sprng of sunrise solo x Pink Flesh Sweet Poor IIHR Bangalore. It was related prour F14 generation and released as gynodioecious hybrid. Fruits medium in size with a small fruit cavity. Pulp in deepred in colour and sweet. It does not have to add flavour.

Sunrise: Fruits are small, Pear shaped, red fleshed, very high sugar content and aromatic, high quality, verification about 400g.

Dioecious line	Pusa Giant, Pusa Dwarf, CO1, CO2, CO5, CO6
Gynodioecious line	Pusa Majesty, Pusa Delicious, Coorg Honey Dew, CO7, Surya
Hybridization	CO 3, CO 4, CO7, HPSC-3
Mutation breeding	Pusa Nanha

6.9.1 Important species of papaya

The genus Carica is having about 40 species but Carica papaya is the only species which produces edible fruits. Some Carica species are used in interspcific hybridization for resistance breeding e.g. *Carica cauliflora* is resistant to viruses, *Carica candamarcensis* and *Carica pentagona* are resistant to frost. Based on the crossbility and compatibility it has been observed that *Carica monoica, Carica cauliflora* and *Carica candamercensis* are easily crossable with each other and producing viable seeds. But interspecific hybridization of these species with Carica papaya do not produce mature seeds. However, by using embryo culture technique immature embryo can be developed into mature embryo. Further, cross between *Carica papaya* and *Carica goudotiana* always gave negative results. According to Capoor and Verma (1961) *Carica cauliflora* is resistant to mosaic virus. However, Horoviz and Jiminez (1967) reported that *Carica cauliflora, C. pubescens, C.stipulata* and *C. candicans* are resistant to ring spot virus.

6.10. References

Aiyappa, K.M. and Nanjappa, P.P. (1959) Coorg Honey, a new find in papaya. *Indian Hort.*, **3**:3.

Badillo, V.M. (1971) Monografia de la familia caricaceae, Editorial Nuestra, America C.A., Maracay, Venezuela, pp 111.

Capoor, S.P. and Verma, P.M. (1961) Immunity to papaya mosaic virus in the genus Carica. *Indian Phytopatho.*, **14**: 96-97.

Conover, R.A., Litz, R.E. and Malo, S,F. (1986) Cariflora – a papaya ring spot virus tolerant papaya for South Florida and the Caribbean. *Hort. Sci.,* **21(4)**: 1072.

Dai, B.B. (1960) Experiments on utilization of F1 papaya hybrids. Nung.Yeh Yenchin Agril. *Res. Taipai.* 8: 17-29.

Dhaliwal, G.S., Gill, M.I.S. and Kalra, S.K. (1991) J. Res. Punjab Agric. Univ., 28:354-356.

Dinesh, M.R. and Yadav. I.S. (1998) Surya : a promising papaya hybrid. *Indian Hort. 43(3)* : 21-33.

Higgins, J.E. and Holt, V.S. (1914) The papaya in Hawaii. *Hawaii Agri. Expt. Sta. Bull.*, **32.**

Hofymer, J. D. J. (1938) Genetical studies in *Carica papaya* L. l. Inheritance of sex and certain plant characteristics. ll Sex reversal and sex forms. Bull. Rep. Agri. for Sci. South Africa, pp187.

Hofymer, J. D. J. (1945) Further studies of tetraploidy in *Carica papaya* L. *South African J. Sci.*, **41**:225-230.

Horovitz, S. and Jiminez, H. (1967) Cruzamientos interespecificos e intergenericos en caricaceas y sus implicaciones fitotecnias. *Agron. Trop.*, **17**:353-359.

Irons, M.J. (1928) Observation on change of sex in *Carica papaya* L. *Science,* **92**:125-126.

Iyer, C.P.A. and Subramanyam, M.D. (1981) Exploitation of heterosis in papaya (*Carica papaya* L). National Symposium on Tropical and Subtropical fruit crops, Bangalore.

Iyer, C.P.A. and Subramanyam, M.D. (1984) Using of bridging species in interspecific hybridization in genus Carica. *Curr. Sci.*, **53(24)**:1300-1301.

Khuspe, S.S. and Ugale, S.D. (1977) *J. Maharashtra Agric. Univ.*, **2**:115-118.

Ram, M. (1981) Pusa 1-15, an outstanding papaya. *Indian Hort.*, **26**: 21-22.

Ram, M. (1982a) Papaya improvement through selection and breeding technique. Punjab J. Hort., 22:8-10.

Ram, M. (1982b) What boosts papaya production. *Indian Hort.*, **26**:8-10.

Ram, M. (1984a) Pusa Majesty for higher yield. *Farmers*, **4**:46-48.

Ram, M. (1984b) Promising varieties of papaya. Proc. National Seminar on papaya and papain production. Pp 37-39.

Ram, M. (1984c) Pusa Nanha – A dwarf variety of papaya. *Seeds Farms*, **19**:37-39.

Ram, M. (1993) Improvement of papaya. In: Advance in Horticulture Vol. I (Eds. K.L. Chadha and O.P. Pareek), Malhotra Publishing House New Delhi, pp 383-397.

Ram, M. and Majumder, P.K. (1981) Dwarf mutant of papaya (*Carica papaya* L) induced by gamma rays. *J. Nuclear Agric. Biol.,* 10: 72-74.

Ram, M., Majumder, P.K. and Singh, B.N. (1983) Genetics of sex reversing male in papaya. XV International Congress of Genetics, New Delhi, pp.596.

Ram, M., Majumder, P.K. and Singh, B.N. (1985) Genetics of sex reversing male in papaya. *Indian J. Hort.,* **42:** 63-65.

Rao, V.N.M. (1974) Papaya in India. Farm Information Unit, Directorate of Extension, Bull No. 9.

Samson, J.A. (1980) Tropical Fruits. Longman, Landon.

Shanmugavelu, K.G., Thambraj, S. and Abdul Khader, Md. J.B.M. (1988) New varieties of horticultural crops. TNAU, Coimbatore, pp 1-2.

Singh, B.P. and Rana, R.S. (1993) Promising fruit introductions. In: Advance in Horticulture Vol. I (Eds. K.L. Chadha and O.P. Pareek), Malhotra Publishing House New Delhi, pp 43-66.

Singh, I.P. and Sharma, C.K. (1986) HPSC-3: a high yielding new papaya hybrid for Tripura. *J. Hill Res.,* **9(1)**: 73-75.

Singh, R. (1969) Fruits. National Book Trust, New Delhi.

Singh, R.N. (1955) Further studies in colchicine induced polyploidy in papaya (*Carica papaya* L). *Indian J. Hort.,* 12:63-71.

Storey, W.B. (1941) The botany and sex relation of papaya. *Hawaii Agri. Expt. Sta. Bull.*, **87:** 5-22.

Storey, W.B. (1953) Genetics of papaya. *J. Hered.,* **44:**70-78.

Storey, W.B. (1958) Modification of sex expression in papaya. *Hort. Adv.,* **2:**49-60.

Storey, W.B. and Jones, W.W. (1941) Papaya production in Hawaii Island. *Hawaii Agri. Expt. Sta. Bull.,* **87**: 5-31.

Subramanyam, M.D. and Iyer, C.P.A. (1986) Flowering behaviour and floral biology in different species of Carica L. genus. *Haryana J. Hort. Sci.,* **15 (3-4)** :179-187.

Sundarajan, S. and Krishnan, B.M. (1984) Improved papaya varieties. Proc. National Seminar on Papaya and Papain production, pp.40-41.

Yung, J.S. (1986) Interspecific hybridization and immature embryo culture of Carica sp. J.College of Sci and Engg,National Chung Hsiang Univ. *Taichung, Taiwan*, **23:** 13-26.

Zepra, M.D.E. (1957) Triploids De *Carica papaya* L. *Agronomic Trop.*, **7:** 83-86.

7

Breeding of Citrus

Botanical Name: *Citrus* sp.

Family: Rutaceae

Sub family: Aurantiodae

Chromosome number: 2n=2x=18

Scientific classification

Kingdom	:	Plantae- Plants
Division	:	Magnoliophyta–Flowering plants
Class	:	Magnoliopsida – Dicotyledons
Subclass	:	Rosidae
Order	:	Sapindales
Family	:	Rutaceae – Rue family
Genus	:	*Citrus L.*

Citrus constitutes a major group of fruits comprising of mandarins, oranges, lemon, pummelo, grape fruits, tangelo, trifoliate orange, citron, citranges etc. Despite of inter-specific and inter-generic hybrids, Poncirus and Fortunella also belong to genus Citrus. During its long history, citrus has given the world numerous varieties both by open pollination, bud sports and of recently by controlled pollination and artificial induction of bud variation. Citrus fruit cultivation lies between latitude 40 0N also 40 0S where conditions are neither cold nor moist and dry. India is considered to be the home of several citrus species and they are found growing wild in some parts of the country. Many types of citrus still remain unexploited by man and such types are considered as semi wild. The productivity of citrus plant in our country is very low which is 7.65 tonnes/ha as acompared to Spain, Italy and Japan where it lies between 17-35 tonnes/ha. Sweet oranges cover an area of 42.4 thousand ha with an annual production of 510.8 thousand tonnes (Ghosh, 1985).

7.1. Centre of diversity

There are three major centres of diversity in India. The first in the North-East including Assam and adjoining areas. It includes papedas, pummelos and their hybrids, Citrus indica, citron, lemons and mandarins and other interesting types like jenera-tenga, soh synteng, a sour fruit similar to the sweet lime and soh siem, a sour mandarin. The second in South India, indigenous types include gajanimma or baduvapulli, kichili and some wild mandarin types. The third in North-West region at the foot of Himalayas where the hill lemon (galgal) and attani are common. The various types of mandarins, hybrids of pummelo, citron, lemons, karna-khatta and rough lemon are found all over the country. In general, the wild types are more common in the foot hill regions and in the North East, where the extreme hot dry climate is not found. Many of the progenitors of citrus fruits are believed to have originated in India (Bhattacharya and Datta, 1951). These include *Citrus ichangensis*, *C. latipes*, *C. limonia*, *C. karna*, *C. pennevesiculata*, *C. maderaspatana*, many of these are wild types. Presence of Sah-Niangriang, a wild sweet orange and a wild mandarin (C. indica) furnishes a strong evidence that Eastern India might be a centre of origin for many citrus fruits. Tanaka (1958) also believed that sweet oranges originated in India in addition to many other citrus types. Thus, North-Eastern and parts of North-Western India become a major centre of diversity and the best locations for collecting primitive germplasm of citrus (Singh, 1981).

7.2. Germplasm resources

Exotic collection of citrus germplasm was started in 1940. Kinnow mandarin was one of the collections which is now a leading cultivar in North Western India. Besides, other exotic collections were Valencia late, Washington Navel, Jaffa, Malta Blood red, Pineapple, Ruby orange, Satsuma, Dancy Tangerine, Climentime, Cleoptera wilking, Temple, Duncan, Marsh Seedless, Lisbon lemon, trifoliate orange, Inchange lemon. More than 650 accessions are being maintained at CHES, Chethali, IIHR, Bengalore, CHES, Ranchi, RFRS, Abohar, NRC on citrus, Nagpur, Horticultural Experiment Station, Bathinda, IARI, New Delhi, MPKV, Rahuri, Citrus Improvement Project, Tirupati, Citrus Experiment Station Katol, Nagpur, Horticultural Experiment Station, Periakulam, Citrus Experiment Station, Tinsukia, Assam etc. During 1988 as a result of systematic exploration by NBPGR in North Eastern region, C. indica and many endangered species were collected for conservation.

North-Eastern region is a hunting ground of biodiversity of Citrus species. The first systematic attempt to explore and collect the samples and conserve them for further, characterization (Bhattacharya and Dutta, 1956). Chakrawar et al. (1988) identified two promising clones of acid lime Vikram and Pramalini in Maharashtra. At Nagpur, Seedless Santra has been selected which has commercial potentiality (Chadha and Singh, 1990). At Periakulum, PKM-1 acid lime has been selected which gives high yield and quality fruits (Anon.,1989).

Attempt has been made during 1978 by NBPGR to preserve the *C. indica* which is progenitor of *C. reticulata* (Singh, 1981). For establishment of Gene Sanctuary, National Park, the natural genetic diversity of *C. indica* was observed in the forest of Garo hills in Meghalaya which exhibited plant characters varying from bush to climber with high frequency of distribution in dense forest and showing resistance to biotic stresses. Therefore, a gene sanctuary for *C. indica* was established in Tanga Range in Garo hills. Genetic material of citrus is conserved in field gene bank or repository.

7.3. Problems in citrus breeding

There are five major problems which hinder the success of citrus breeding.

1. **Time** - Cirtus being perennial in nature takes more time to come into bearing. However, this period can be reduced to a maximum of half by top working the seedling on a old tree.
2. **Polyembryony** - It is peculiar feature found in citrus in which seed consists more than one embryo. In addition to the zygotic embryo, one or more sometimes as many as fifteen additional embryos are developed from the nucellar tissue called nucellar embryos and found in the embryo sac. Most often, the zygotic seedling is crowded out by the vigorous nucellar seedlings. So even if a seedling comes out from a seed, one is not sure whether it is a zygotic seedling. For getting large number of hybrids, citrus breeder should select a seed parent known to be either monoembryonic or week polyembryonic. Unfortunately, all the economic citrus species are polyembryonic except *C. medica, C. latifolia* and *C. grandis* which restricts the choice of breeder and complicate the procedure required to attain the desirable objectives.
3. **Sterility** - Sterility is inability of gametic or sexual reproduction. Prevalence of high generative sterility is obviously a serious hindrance with use of a particular parent in hybridization. Complete pollen sterility is problematic, where proportion of nucellar embryos is very high. High level of sterility often leads to production of seedless fruits which is a serious hinderance to develop varieties.
4. **Self-incompatibility** - Self compatibility and cross incompatibility is common phenomenon which widely occurs in citrus. Most of the varieties of grape fruit (*C. grandis*) are found to be self-incompatible besides,

some varieties of lemon, sweet orange and mandarins exhibit cases of self incompatibility. Satsuma mandarin and Washington Navel orange are pollen sterile type and they produce parthenocarpic fruits. Self incompatibility in citrus plant is of gametophytic type governed with oppositional alleles. Cultivars of C. grandis are self incompatible, however, hybrid cultivars including Clementine, Orelando, Minneola, Sukega, Nova, Robinson, Fairchild. Orelando and Minneola are cross incompatible. Nova, Robinson are also suspected to be cross incompatible. Sweet orange varieties Washington Navel and Satsums mandarin are having sterile pollen, thereby they produce parthenocarpic fruit if cross pollination is not done through viable pollens.

5. **Long juvenility** - It is a major barrier in the progress of citrus breeding in India. General treatments to shorten the period or induce flowering have not been generally effective. It was reported that neither chemical treatment nor incorporation by genetic transfer has yet been effective in combating long juvenility in citrus.

7.4. Objectives

- To breed varieties suitable for different agro-ecological situations.
- To develop rootstock which should have resistance against biotic and abiotic stresses.
- To breed virus free nucellar strains for existing varieties.
- To breed varieties free from insect-pest and diseases.
- To develop varieties having less tendency of granulation.
- To breed varieties suitable for export.

7.5. Botany

Genus Citrus, Poncirus and Fortunella belong to family Rutaceae and sub family aurantoidae. The botanical descriptions of the important genera of the sub family aurantoidae are as under

7.5.1 Kumquats (Fortunella species)

Kumquats are evergreen densely branched, small spines, axillary or lacking, leaves unifoliate, blunt, thick, dark green at upper part and paler beneath, thickly dotted with glands on lower side, petioles normally winged, articulated or not articulated with the lamina, flowers white, smaller, solitory or in branches in leaf axils either on current season flush or previous season growth, pedicels bracted with one or two smaller bracts, 4-5 petals, stamens 16-20, with two ovules in each cell, stigma capitate, fruit small, oblong or globose, rind thick sweet and edible, juice vesicles small, seeds small, cotyledons green, polyembryonic, grown for ornamental purposes. These are 4 types of Fortunella

1	*F. margarita*	Nagami kumquat
2	*F. japonica*	Marumi kumquat
3	*F. crassifolia*	Meiwa kumquat
4	*F. hindsi*	Hongkong wild kumquat

7.5.2 Poncirus

Central and Northern China is native home of Poncirus trifoliata where it is widely found in wild or cultivated form. It is shrub or tree with deciduous trifoliate leaves, it flowers during summer, bud is protected from scales, flowers nearly sessile, borne on wood of previous season growth, petal narrowed and claw like at the base, opening wide and flat, stamens numerous, free, ovary 6-8 celled, fruits contain plenty of acrid oil mixed with the acid juice in the pulp vesicles, seeds are many, cotyledons white and polyembryonic in nature (Rajaput and HariBabu ,1985).

7.5.3 Citrus

Citrus species are native of Cochin China, Malaya Archipelago and adjoining parts of Asia. Plants are evergreen tree or shrubs, normally spiny, but sometimes absent, leaves thick, leathery, dotted with translucent oil cells unifoliate, petioles winged, articulated with lamina, bisexual flowers are axillary, solitary or in clusters or in small terminal cymes, white or whitish purple with pink tinge on outer surface of the petals, scented, calyx small, cupped, petal usually5, imbricated, thick dotted with oil glands, stamen are numerous (15-60), polyadelphous, united in bundles, ovary compound, 8-16 celled , style deciduous, fruit berry (hesperidium), globose, subglobose or oblong pointed, orange coloured, dotted with oil glands, seeds few to many, mono/ polyembryonic.

7.6. Floral biology

Study of floral biology of citrus and other related genera is an essential pre-requisite for initiating hybridization programme. Flowering in citrus takes place during February - April. In North India, sweet orange and mandarins bloom only once in March. However, it is reported that sweet oranges bloom twice in a year under Bihar conditions i.e. February-March and June –July. Inflorescence in Citrus species is of Cymose type. Generally, anthesis takes place in the morning hours between 9.00 A.M. to 12.00 noon. Flowers on shaded side of the tree have been observed to open later than those exposed to sunshine. Dehiscence of anthers takes place at different time viz., in lemon before anthesis and in sweet mango, sweet lime etc. after anthesis oozing of a gummy substance on the stigmatic surface is indication of stigma receptivity. In various citrus varieties receptivity starts 2-3 days earlier to anthesis and lost 4 days after anthesis with maximum receptivity on the day of anthesis. In citrus both staminate and hermaphrodite flowers are produced on the same plant.

7.7. Inheritance pattern

Species of Citrus, Poncirus, *Severinia buxifolia* have 2n=18 chromosome number. The allied genus Atlantia ceylinica is allopolyploid, 2n=4x=36 of *A. monophylla* and *A. guellaumini* (Agrawal, 1988). Cytomorphological studies show that species is probable hybrid of *A. ceylanica* (2n=36) and *S. buxifolia* (2n=18). Hybrid vigour and inbreeding depression have been observed in citrus. Narrow crosses within the genus produced weak offspring as in progenies of Ruby x Valencia and Eureka x Lisbon lemon. Wider crosses like mandarin x grape fruit produced vigorous progenies.

Hybridization between oranges and mandarins can yield progenies with rind and flesh colour vary from pale yellow to deep red orange. In most of the crosses, variability occurs with respect to plant vigour, growth habit and leaf characters considerably. Trifoliate leaf characters of Poncirus show essentially complete dominance over monofoliate condition of citrus. Purple colour in young leaves in lemon is governed by a dominant gene and nucellar embryo in citrus is controlled by one or two dominant gene. Crossing between polyembryonic cultivars and monoembryonic cultivar resulted both type of offspring in 1:1 ratio (Soost and Cameron, 1975). Small size of fruits, sadiness and paleness of colour appear to be dominant. However, characters like presence of thorns, pubescence and oil gland are dominant characters whereas reduction of nectaries and scaleness of flower plants are recessive (Soost, 1987).

7.8. Breeding methods and achievements

7.8.1 Introduction

It is usually attempted when existing germplasm is felt unsuitable or lack some desirable characters merely to enrich the existing germplasm.

- Mandarin-Santra from Aurangabad
- Mandarin - Kinnow, Cleoptera, Clementine-USA
- Sweet orange- Washington Navel, Valencia, Jaffa, Blood Malta, Tangerines,

- Grape fruit- Marsh seedless, Duncan, Foster, Thompson-USA
- Lemon- Lisbon, Eureka, Villafrance-USA

Kinnow mandarin was introduced in South India in 1958 and in Punjab during 1959 and has performed extremely well in Punjab and also adapted well in South India.

7.8.2 Selection

Selection plays a key role in isolating the superior types from a hybrid population obtained by hybridisation. Development or creation of variability in citrus was due to mutation, hybridization and bud sport etc. The process of such selection has gained considerable impact with the introduction of vegetative propagation. At present selection is employed in selecting superior types as per the objectives by screening from mass population of germplasm e.g. Kagzi lime, Vikram, Pramalini.

7.8.3 Hybridization

Citrus improvement through hybridization poses greater problem due to its long juvenile phase and polyembryonic nature which makes the assessment delayed. The crosses made using these as female, result in very few weak hybrids which are difficult to identify from nucellar seedlings. Electrophoretic techniques, separating the isozymes of parents and hybrids may be of great value in scion breeding programme as no morphological markers are available at present. Citrus is highly heterozygous in nature which leads to wide segregations. The problem is little less complicated in case of rootstock breeding as the commonly used disease resistant male parent, Poncirus trifoliata has trifoliate leaves which is dominant over the monofoliate character (other citrus species) and all the hybrids obtained using this male has trifoliate leaves distinct from unifoliate nucellar seedlings. Nucellar polyembryony helps in asexual reproduction by seed as it is vigorous, uniform and free from virus.

7.8.3.1 Techniques

The mature flower buds on the female parents are emasculated early in the morning on the day of opening and bagged. The flowers to be used as male parent are bagged in the previous evening and the next morning as the day warms up. The anthers dehisce releasing the pollen grain when these flowers can be plucked to pollinate the receptive stigma of emasculated flowers. The pollinated flowers are bagged, open after about a week and allowed to mature into ripe fruits. In some cases, especially when the trifoliate orange is used as male parent, difficulties are encountered as its flowering is over before the citrus variety flowers. Therefore, pollen has to be stored at low humidity and temperature. Soost and Cameron (1954) successfully stored P. trifoliata pollen for 36 days at 4°C. Flowering in P. trifoliata and Citrus species can also be synchronized by adjusting the irrigation schedules. Seeds from mating fruits are extracted and sown immediately in sterilized sand and soil mixture when seedlings are about 15 cm height. Hybrid seedlings are identified, particularly those showing some morphological characters of male parents and others are rejected. Electphoretic methods can be used for identification of zygotic seedling. Identification of hybrid seedlings having *P. trifoliata* as male parent is easily done by looking for trifoliate characters. The hybrid seedlings are grown to mature trees in the field and the seedlings raised from the fruits are evaluated for resistance to various diseases, insect pests, nematode and for suitability as scion or root stock.

Generally, much of hybridization in citrus has been inter-specific rather than intra-specific, realizing the importance of hybridization in improvement of citrus varieties. USA has attempted hybridisation on a large scale during early part of the century and reaped a considerable success in obtaining a good number of promising hybrids.

Examples

- Citrange *C. sinensis* x *P. trifoliata*
- Tangor *C. reticulata* x *C. sinensis*
- Tangelo *C. paradisi* x *C. reticulata*

7.8.4 Mutation

It is another important technique to induce mutations for disease resistance, seed lessness etc. by using physical and chemical mutagens. Bud variation is an abnormal variation in stem, leaves and fruit that are capable of perpetuating through bud propagation. Such variations are in the somatic tissues of the plant are termed as bud sports or bud mutation. Many of the fruit variations in the citrus have been the result of individual fruit variation in otherwise normal trees.

Examples

- Naval orange- Washington Navel late, Naveline
- Mandarin- Clausellin
- Grape fruit- Foster, Hudson, Star Ruby, Red Blush, Flame

7.8.5 Polyploid breeding

Most of the species and varieties of Citrus are diploid but occurrence of polyploidy has been reported in many cultivars. The Hongkong wild Kumquat, *Fortunella hindsii* may have been first reported tetraploid (Soost, 1987). Polyploid breeding seems to offer the prospect to obtain large sized fruit with dwarf plant types. Production of triploids by crossing tetraploid with diploids may be useful in obtaining seedless varieties. The seedless lime (*C. latifolia*) is suitable example of advantage of triploidy. Triploids have favourable characteristics and yield well but they are sterile. The development of triploid through breeding is very limited. Production of 3x is achieved by crossing of 4x with 2x is often not feasible for want of sexual parents. The reciprocal cross (2x. x 4x) produces many tetraploid individual (Essen and Soost, 1973). Occurrence of these due to frequent production of 2x megagametophytes as well as failure of most triploid embryos.

7.8.6 Biotechnology in improvement of citrus

Transformation of fruit species by biotechnological tools is potential approach to develop disease free cultivars. Woody plants are known to be difficult to work in vitro than herbaceous plants but citrus is exceptional. Though nucellar embryony in citrus is of great value for producing vigorous, uniform and virus free plants, but it is an obstacle in hybridization. In polyembryonic cultivars the vigorous growth of nucellar embryos inhibit the growth of the zygotic embryo and causes its degeneration prior to seed maturation. Such abortive embryos can be rescued by tissue culture. Tissue culture has effectively been used in obtaining hybrid Poncirus plantlet from polyembryonic citrus cultivars. Poncirus trifoliata not only carries a genetic marker, but also possesses resistance to tristeza, phytophthora nematode and cold stress. Inter-generic hybridization with the aid of cell/tissue culture offers possibility of incorporation of multiple desirable parts found in different genera for improvement of citrus root stocks and scion cultivars.

Cell and tissue culture and specially protoplast manipulations have effectively been explored in citrus improvement by regeneration of citrus trees from protoplast, somatic hybridization (cybridization) and organelle transfer. In an attempt to develop protoplast derived plants in the last one decade, Israel and Florida have shown protoplast system in a dozen genera and interestingly citrus is the only woody plants among them. Efficient protocols have been developed to obtain protoplasts with cell diversion capability from all major citrus cultivars and some of their wild relatives. The availability of protoplast to tree system in citrus and relative genera provides the means to establish novel citrus types through hybridization. It can also be used to produce transgenic citrus tree by integrating appropriate recombinant DNA or by inducing mutations.

7.9. Important species and cultivars

7.9.1 Mandarin group

Citrus reticulata

Loose skinned orange, though mandarin and tangerine are names used more or less interchangeably to designate the whole group, tangarine is applied more strictly to those varieties which produce deep orange or scarlet fruits. Webber (1948) has separated the mandarin oranges into (a) King group (b) Satsuma group (c) Mandarin group (d) Tangerine group (e) Mandarin- Lime group (f) Mitis group.

Calamondin (*C. madurensis*)

Tanaka has recognized it as loose skinned orange group. It is very cold resistant for a true citrus fruit as hardy as Satsuma. Fruit colour is orange to deep orange, smooth and glossy surface, pitted shape oblate to deep orange, size small with flattened base having 7-10 segments.

Clementine (Algerian Tangerine)

It is a tangerine and is probably an accidental hybrid of the mandarin and sour orange which is considered to be originated in Algeria. Fruit colour deep orange, shape globose to elliptical, size medium with depressed apex, rind thick, segments 8-12 adhered slightly. It is an early variety.

Cleoptara (*C. reshni*)

It is originated in China. Plant is thornless with dense top. Fruits are produced singly or in bunch, fruit colour dark orange red, shape oblate, flattened at both ends, size small and segments 12-15.

Coorg orange

It is important variety of South India particularly in Coorg and Wynad tracts. Fruits are medium to large, bright orange colour, oblate to globose in shape, finally papillate and winkled, glossy, segments 9-11.

Dancy tangerine

In USA, the Dancy is the best known and highly prized of all the mandarin oranges. Tree large, nearly thornless and has upright growth. Fruit colour is deep orange red to scarlet, rind thin, loose, easily separable, segments 10-14. It is a late variety.

Deshi mandarin (Pathankot)

This variety is mainly grown in Punjab hills. The tree is large with semi upright growth habit with compact foliage and spineless. Fruit is ovoid to sub globose, colour uniformly cadmium, surface pitted, semi glossy and finally wrinkled, rind medium, adherence slight, segments 7-10.

Khasi mandarin

It is commercial variety of Assam. Fruit is depressed, globose to oblate, orange yellow to bright orange, surface smooth, glossy, base even or obtuse, rind thin, soft, segments 8-10.

CRS-4

It is promising genotype among twelve elite genotypes of Khasi mandarin.

King mandarin (*C. nobilis*)

Swingle believed the king mandarin as a tangor, a hybrid between mandarin and sweet orange. King mandarin was first introduced from Cochin china into California in 1882. King mandarin is cultivated in Assam. The King is a prolific bearer, frost resistant and produces high quality fruit.

Willow leaf mandarin (*C. deliciosa*)

The tree is willowy in growth, almost thornless, fruits usually borne singly at the tip of slender branches. Fruit colour orange, surface smooth, glossy, frequently slightly lobed, necked base, apex depressed, wrinkled, rind thin, with 10-12 segments. It is an early variety.

Kinnow mandarin

It is first generation hybrid between the king and willow leaf mandarin and developed by H.B.Frost at the California Citrus Experiment station in 1915. It was introduced into Punjab from USA. Tree is vigorous, large, top erect, dense symmetrical with few scattered thorn. Fruit colour resembles of King, deep yellowish orange, surface smooth, glossy, very shallow pitted, shape slightly oblate, size medium with flattened base, rind thin, peel tough and leathery, segment 9-10 easily separable, seed 12-24. It is a late variety.

Nagpur santra

This variety occupies premier position in Indian market and is one the finest mandarins grown in the world. It is also known as Ponkan. Tree is large, vigorous, spineless with compact foliage. Fruit size is medium, cadmium colour, smooth surface, glossy, rind thin, soft, slightly adhered with 10-12 segments.

Satsuma orange (*C. unshiu*)

It is a Japanese variety introduced into Florida in 1876. It is frost resistant and useful breeding material. It is also resistant to canker, gummosis, scaly bark, melanose. Plant is thornless having spreading growth habit, orange fruit colour, rough surface, oblate to spherical shape, medium size, thin and easily separable rind, flavour rich and seedless.

Temple mandarin

It is a hybrid between Tangerin and sweet orange. Temple mandarin is most beautiful and highly flavoured fruit of the citrus group. Tree is medium, thorny, spreading with deep orange to reddish fruit colour, rugose glossy surface, medium to large in size, depressed or nearly flat apex, loose rind, solid axis with 10-12 segments, orange pulp. It is late in maturity.

7.9.2 Lemon (*C. lemon*)

Eureka

It is seedling selection of Sicilian lemons. Tree is medium, spreading and thornless. Its fruit color is lemon yellow, surface rugose, pitted, shape obovate, size medium, apex round, rind medium thin axis small, solid, segments 8-10, juice acidic with excellent flavour and quality. Eureka is heavy yielder and begins bearing at early age. It has tendency of tip bearing.

Lisbon

Its appearance and yield is superior to Eureka. It is resistant to frost, heat and high wind velocity. Tree is large and vigorous with spreading shoots. It has upright thorny growth, lemon yellow fruit colour, smooth surface, medium size, pitted rind, small axis, solid, 6-10 segments with 0-8 seeds.

Lucknow Seedless

It is hardy, medium, vigourous, spreading drooping, dense foliage, thorny, fruit colour yellow, smooth, nippled apex, base round, thin rind, hollow axis, segments 10-12 maturity during November- February.

Pant lemon

Fruit size medium, juicy, heavy fruiting, tolerant to pests and diseases.

Hill lemon (Galgal) *C. seudolimon*

It is indigenous to North India grown in sub himalayan region. Plant is tall, vigorous, upright and spreading with an irregular and loose crown, fruit ovate oblong, yellow, apex nipped, base rounded or nippled, rind medium, axis hollow, segments 7-11, seed 25-50. Ripening during October- December.

Vill Franca

It belongs to Eureka group and was introduced in to Florida from Europe in about 1875. Tree is vigorous, spreading, thorny, spreading, erect, fruit oval to oblong, size medium to large, colour bright lemon yellow, apex pointed, base rounded, rind thin, smooth, segments 8-12, flesh fine grained, juice colourless, seed 25-30.

Nepali oblong

Plant medium sized, hardy, spreading drooping with irregular crown, fruit shape oblong colour lemon yellow, segments 10-12, the fruit ripen during December-January.

Meyer lemon

Tree semi dwarf, thornless, spreading, cold resistant, fruit colour light orange, surface smooth, finally pitted, shape obolate or oblong base rounded, rind thin, axis small, segments 8-10, seeds 8-12.

Rasraj

Rasraj is an inter-specific polyembryonic hybrid developed for resistance to bacterial canker disease by breeding. The hybrid derives genes for canker resistance from the Nepali Round lemon parent and other characters from the acid lime parent. The fruit is yellow coloured, weighs on an average about 55 g containing 70% of juice and 12 seeds. The rind is thicker than lime. The acidity is 6% and TSS is around 8°B.

7.9.3 Acid lime (*C. aurantifolia*)

It is native of India and widely cultivated in the tropics. It is tenderest species among all the citrus species. Tree medium sized, hardy, semi vigorous, upright growth, thorny, fruit round to oblong, yellow apex rounded and slightly nippled, base round, rind thin, papery, segments 8-10, seeds 8-10.

Vikram

It was developed at MAU, Parbhani, fruit medium size, heavy fruiting, fruit colour golden.

Pramalini

It was also developed at MAU, Parbhani, high yielder, golden fruit colour, tolerant to Canker.

Sai Sarbati

It was developed at MPKV, Rahuri, high yielder, suitable for summer cropping, tolerant to canker and tristeza.

Jai devi

It was developed at FRS, Periakulum, high yield, juicy, thin peel, and pleasant aroma.

Balaji

This variety is identified as canker tolerant selection.

NRCC Acid Lime - 7

It is released by the Institute CCRI, Nagpur. It is suitable to Maharashtra, Gujarat, Tamil Nadu and Madhya Pradesh. The fruit yield is 540 q/ha and fruit weight is 48 g.

NRCC Acid Lime - 8

It is released by the Institute CCRI, Nagpur. It is suitable to Maharashtra, Gujarat, Tamil Nadu and Madhya Pradesh. Yield is 590 q/ha and 216 kg/tree (4327 fruits). Fruit weight: 50 g Juice: 51.53%. TSS:7-8.

Petlur Selection-1

It is a cluster bearer and high yielder than local varieties. It is tolerance to bacterial canker disease. Recorded high juice percentage (55.8%), high citric acid (7.3 mg/100g) in fruits than other released varieties. Found to be adoptable to climatic conditions in Andhra Pradesh and Telangana. Yields high during summer season (210 to 220 kg fruits/Plant/Year).

Phule Sharbati

It is released by ICAR-AICRP on Fruits Rahuri centre. The yield is 450-500 q/ha. It has precocious bearing on 3rd year. Juice content is 52-53 %.

Pusa Abinav

A promising clonal selection having medium vigorous tree, dense foliage and attractive bright yellow round shape fruit. It has round the year fruiting with peak harvesting during summer. It is moderately susceptible to citrus canker.

Pusa Udit

It is also clonal selection developed from IARI, New Delhi. Peak harvesting time is August to September and February to March. It is moderately susceptible to canker. It is suitable to kitchen garden.

7.9.4 Sweet Orange

Batavian

This variety is popular in the North coastal districts of the state, this is similar to sathgudi. It develops yellow patches on the green rind of the fruit when it is basketed to protect itself from the sucking moth.

Blood Red

Trees are larger in size than Jaffa. Fruits are medium-sized. Deep orange color Fruits at Maturity. Red color Flesh with 1-2% acidity.9-12 seeds is present per fruit. Juice color is Reddish. The average fruit yield is 43 kg per plant.

Hamlin

It is an exotic variety which is performing well in Punjab, Haryana and Rajasthan. It is an early season variety.

Jaffa

Tree has a similar shape to pineapple tree. Bigger sized fruits, round in shape with yellowish orange skin color. 9-12 seeds are present in single fruit. Fruit ripens in the month of December. The average fruit yield is 54 kg per plant.

Mousambi

Trees are medium in size. Fruits are medium in size. Juice is totally acid less with 10% TSS. Fruit ripens in the month of November. 20-25 Fruits are present per fruit. The average fruit yield is 40 kg per plant.

Phule Mosambi

It is released in Rahuri centre with high yield potential of 180-200 q/ha (70-75kg/plant). It has 47-48 % juice content 0.40 % Acidity and TSS: 9-10°B.

Pineapple

Trees are medium in size having spreading habit. Medium-size and Juicy Fruits. 10-20 Seeds are present per fruit. Fruit Ripens in the months of December-January. The average fruit yield is 38 kg per plant.

Pusa Sharad

The plants are moderately vigorous and it has bigger size uniform fruits (227.6g). Yield of the tree is about 2.6 times higher than popular variety Jaffa and 1.8 times than Valencia. It has higher juice content (50.12%) and high TSS content (9.20%). It is resistance to citrus granulation.

Pusa round

It has uniform bigger size fruits (268.7g) and higher juice content (48.26%). It yields more fruits than popular varieties Jaffa and Valencia. It has high TSS content (10.14%).

Satgudi

It is the important cultivar of sweet orange and it is commercially grown in Andhra Pradesh (South India). The Satgudi having hermaphrodite type of flower and it contain more than four stamens per petal in each flower. The fruiting season is from July to end of September. This variety is popular among cultivators in Andhra Pradesh because of higher yield potential, wider adaptability and better consumer acceptance. The fruit is spherical in shape with a smooth surface, the rind is thin, semi-glossy, finely pitted and orange coloured when fully ripe. It has 10 to 12 segments, abundant juice (45%), T.S.S. (Total Soluble Solids) 6.4%, acidity 0.69% and an ascorbic acid content of 44 mg/100g.

Valencia

It is a late season fruit and therefore a popular variety when the navel oranges are out of season. Fruits are more juicy, few seeds and thin skin. Its excellent taste and internal colour make it desirable for the fresh markets. At ripening the color of skin turns to golden yellow. Fruit ripens late up to the month of February. 2-7 Seeds are present per fruit. The average fruit yield is 40 kg per plant.

Tahiti lime (Persian lime) *C. latifolia*

It is large fruited acid lime. The plants are large, spreading, cold resistant, thornless, fruit large in size, seedless, triploid, and produce non-viable pollen. It is considered as hybrid between lime and lemon. Fruit colour orange yellow, smooth surface, segments 8-10. It is late variety.

Rangpur lime (*C. limonia*)

It is indigenous to India and is commonly used as root stock. Rangpur lime is mainly grown for home consumption and ornamental purpose. Fruits are used for making limeada. It is also known as Marmalade orange. It has loose rind, easily separable segments and pulp is light orange yellow.

Sweet lime (*C. limettoides*)

Generally, sweet lime is grown as aroot stock and for its non-acidic fruits.

Pummelo (*C. grandis*)

It is native of Polinasia and Malaysia and commonly grown in South China. Fruit is pyriform, largest fruit size among citrus fruits, rind thick, juice is acid bitter, juice sacs easily separable. Seeds are monoembryonic. Fruits are of two types (a) elongated pear shaped with neck (b) Oblate or globose, flattened and neckless. In India there is no improved cultivar except Nagpur chakotra.

7.9.5 Grape fruit (*C. paradisi*)

Dunkan

It was developed through chance seedling in Florida. It is the hardiest variety, fruit colour yellow, surface smooth, shape oblate to globose, size large, basal area depressed, apex round, rind medium thick, firm, axis medium in size segments 12-14, seeds 25-50.

Foster

It belongs to pink or red pulp group and originated as bud sport of Walters grape fruit by R.B. Foster in 1906-07. Fruit colour is light yellow, surface smooth, oblate or globose shape, size medium large, base rounded, apex round, rind medium thick, segments 12-15 seeds 50 or more. It matures during November- December.

Ruby

It belongs to pink or red pulp group. It originated as bud sport from Thompson. Deep red colour which uniformly distributed throughout pulp.

Marsh Seedless

It comes under pallid pulp group of grape fruit. It is bud sport from cultivar Marsh. Fruit colour light yellow, surface smooth, shape oblate to globose, size medium to large, basal area small, rind thin, segments 12-14, seeds 2-5. It is late cultivar.

Saharanpur Special

Fruit round to oblate, empire yellow, surface small, segments 10-15, seeds 50-100. Fruits ripen in November to February.

Thompson

It originated from bud sport of Marsh. Fruit colour light yellow, surface smooth, segments 10-12, seeds 2-5.

7.10. Citrus Hybrids

7.10.1 Hybrid of Poncirus

Citranges It is hybrid between Poncirus trifoliata and Citrus sinensis and is hardy than sweet orange. It is nearly deciduous, fruit colour varied from yellow to deep orange, surface rough, wrinkled, ribbed or smooth, rind thin, juicy pulp, highly acidic. It is used as dwarfing root stock for grape fruit, Satsuma, sweet orange and lemon.

Varieties Coleman, Etonia, Mortan, Rusk, Cunnigham, Rustic, Sanford, Savage, Troyer.

Citrangequats It is trigeneric hybrid of Citrange (Sweet orange x trifoliate orange) and Kumquat. Citranges are subjected to spring frost injury due to its tenderness while Kumquat is tolerant to the spring frost because of their winter dormancy.

Citrangedin - It is trispecific hybrid of citrange with Calamondin (C.mitis)

Cirangors - A back cross hybrid between Citrage x Sweet orange.

Cicitranges - Citrange x trifoliate orange.

Citrumelo - Trifoliate orange x grape fruit.

Citrandarins - Trifoliate orange x mandarin.

Citremons - Trifoliate orange x lemon.

Citradias - Trifoliate orange x sour orange.

Citrumquats -Trifoliate orange x Kumquat

7.10.2 Hybrids of Fortunella

Poncimequat - *Fortunella japonica* x *C. aurantifolia* x *F. hindsii*

Limequats - *C. aurantifolia* x *F. japonica*

Orangequat - *C. reticulata* cv. *Satsuma* x *F. japonica* x *F. morgarita* cv. *Meiwa*

7.10.3 Interspecific hybrids

Limonage - *C. lemon* x *C. sinensis*

Lemonimes - *C. limon* x *C. aurantifolia*

Lemandarins - *C. limon* x *C. reticulata*

Tangors - *C. sinensis* x *C. reticulata*

Tangelo - *C. reticulata* x *C. paradisi*

7.11. Future thrust

- Development of varieties suitable for export.
- Evolving varieties resistant to biotic and abiotic stresses.
- Exploitation of heterosis / hybrid vigour.
- Exploitation of biotechnological approaches.

7.12. References

Agrawal, P.K. (1988) Cytomorphology of Citrus indica Second All India Confr. Cytol. Genet., Warangal, p 2.

Anonymous (1989) Annual Report AICRP on Tropical Fruits. IIHR, Bangalore, pp 234.

Bhattacharya, S.C. and Datta, (1951) Citrus variety of Assam. *Indian J. Genet.*, **11**: 57-62.

Bhattacharya, S.C. and Dutta, S. (1956) A preliminary observation on root stocks for citrus in Assam. *Indian J. Hort.*, **9**:1-10.

Chadha, K.L. and Singh H.P. (1990) Citriculture Scenario of India. In citriculture in North Western India (K.S. Gill, J.S.Kanwar and R.Singh Ed.), PAU, Ludhiana, pp 21-64.

Chakrawar, V.R., Patil V.K. and Sindha, G.S. (1988) Horticultural development in Maharashtra region. Scientific Seminar, Aurangabad, pp13.

Essen, A. and Soost, R.K. (1973) *American J. Bot.*, **60**: 448-62.

Ghosh, S. P. (1985) Citrus. In: Fruits of India - Tropical and Sub tropical (Ed. T.K. Bose), Naya Prokash, Calcutta, pp 162-218.

Rajput, C.B.S. and Hari Babu, R.S. (1985) Citriculture. Kalyani publishers, Ludhiana

Singh, B. (1981) Establishment of fruit gene sanctuary for citrus in Garo hills. Concept Publ. Co. New Delhi, pp182.

Soost P.K. and Cameron, J.W. (1954) Production of hybrids by use of stored trifoliate orange pollen. *Proc. Amer. Soc. Hort. Sci.*, **63**: 234-238.

Soost, R. K. (1987) Improving vegetatively propagated crops (Ed. Simmonds N.W.) Longman, London, pp. 301-24.

Soost, R.K. and Cameron, J.W. (1975) Citrus, In: Advances in Fruit Breeding (Eds. J. Jainick, and J.N. Moore), Purdue Univ. Press, Indiana, pp 507-40.

Tanaka, T. (1958) The origin and dispersal of citrus fruits having their centre of origin in India. *Indian J. Hort.*, **15**:101-115.

Webber, H.J. (1948) Edited by L.D. Batcheler, and H.J. Webber, University of California Press Berkeley and Los Angles.

8

Breeding of Grape

Botanical name: *Vitis vinifera* L.

Family: Vitaceae

Chromosome number 2n=2x = 38

Scientific classification

Kingdom	:	Plantae- Plants
Division	:	Magnoliophyta–Flowering plants
Class	:	Magnoliopsida – Dicotyledons
Subclass	:	Rosidae
Order	:	Rhamnales
Family	:	Vitaceae – Grape family
Genus	:	*Vitis*
Species	:	*vinifera L.*

Grapes are being grown commercially in the tropics. The region between the tropics of Cancer and Capricorn, which was once considered to be unsuitable for growing quality grapes (Winkler, 1962) is producing good quality grapes. Tropical countries, where grapes are grown are South–East Asia (India, Indonesia, Philippines, Thailand, Taiwan, South China and Australia), Central and South America (Brazil, Columbia, Ecuador, Mexico, Venezuela and West Indies) and Africa (Kenya, Nigeria and Zimbabwe). In India, about 90% of grapes are being produced in the tropical region of the peninsular states i.e. Maharashtra, Andhra Pradesh, Karnataka and Tamil Nadu. However, now-a-days grape is becoming popular in North India due to good income. The vines of tropical region grape are evergreen in nature and do not shed their leaves. In this condition, pruning is done twice in a year during April and September–October. But under North Indian conditions, grapes shed their leaves during winter and pruning is done in the month of February.

8.1. Centre of diversity

European grape (Vitis vinifera L.) is considered to have originated primarily between Caspian and Black seas region (Vavilov, 1951). American grapes, belonging to a large number Euvitis and Muscadinia species have been originated in North America, referred to as 'Vine land' by Zielinski (1955).

8.2. Germplasm resources

Field gene banks of grapes are maintained at Division of Fruits and Horticultural Technology, IARI, New Delhi, Indian Institute of Horticulture Research, Bangalore, Ganesh Khand Botanical Garden, Pune etc. Further, 616 genotypes of grapes are maintained at IIHR, Bangalore (Anon., 1996).

8.3. Objectives

8.3.1 Objectives of breeding for North Indian grapes

- To develop early maturing, seedless and sweet cultivars for table purpose.
- To induce the resistance to anthracnose and chafer beetle.
- To develop varieties with medium vigour and productive basal bud, which can be trained on head system of training (Pandey and Pandey, 1996).

8.3.2 Objectives of breeding for the Tropics

- To develop high yielding and high-quality varieties with increased fruitfulness of basal buds, less degree of apical dominance, suitability for different purpose such as table, raisin, wine and juice and resistance to disease.
- To develop root stocks resistant to salinity, nematodes and drought (Negi, 1992).

8.4. Botany

Grape belongs to the family Vitaecae. It contains about 12 genera (Olmo, 1976) and 600 species. The most important genera are Vitis, Ampelocissus, Cissus, Parthenocissus, Tetrastigma, Cayratia and Leea. Earlier Vitis genus was divided into two subgenera i.e. Euvitis and Muscadinia, but now Muscadinia has been given the generic status (Olmo, 1976). Genus Muscadinia includes three species i.e. *M. rotundifolia* Michaux (Muscadine grape), *M. munsoniana* (Simpson or little muscadine grape) and *M. popenoeifennel* (Mexican muscadine grape). *Vitis vinifera* L. is considered to be hybrid between two American species, *V. vulpina* L. and *V. labrusca* L. (Regel, 1973), which resemble *V. parviflora*

Roxb. and *V. lanata*, found in the Himalayan region (Das and Nath, 1965) indicating that Himalayan zone may be the secondary centre of origin of *V. vinifera* L. Zukovskij (1962) reported that European grapes (wine grape) is supposed to be originated from wild *V. vinifera* which is unknown now a days. The immediate ancestor of *V. vinifera* is supposed to be *V. vinifera* Linn. *spp. sylvestris* Gmel (Negrul, 1940).

8.5. Floral biology and pollination

Genus Vitis has three types of flower i.e. male, female and hermaphrodite. *Vitis rotundifolia* (*M. rotundifolia*) is dioecious in nature. Further, Vitis species had been originally dioecious sub-dioecious but later on transformed to hermaphrodite (Olmo, 1976). At present, most cultivars of *Vitis vinifera* are hermaphrodite in nature (Negi and Olmo, 1971). Inflorescence of cultivated grape is a cyme (Pratt, 1971). Perfect flowers have five partly fused sepals and five petals joint at the top. Thus, it opens from the base through its detachment as a result of outward pressure exerted by the stamens and pistil at the time of blooming termed as calyptra. The flowers have five stamens and pistil with two locule, short style and stigma. (Pandey and Pandey, 1996). Based on the fertility level of pollen, grape can be grouped in to three categories. Self-fertile group have erect stamens with oblong blunt and quite symmetrical pollen, size of pollen grain is very small i.e. 0.0254 mm in diameter (Pandey and Pandey, 1996). However, self sterile and partly self sterile is characterized by the presence of the impotent pollen. The stamens of self sterile are reflexed type e.g. Hur, Katakurghan etc.

Time of anthesis is also influenced by the environmental factors, hence varied reports with respect to anthesis are available. According to Jauhari et. al. (1969) peak time of anthesis under North Indian conditions was between 7.0 to 9.0

AM. However, Chelam and Satyanarayana (1968), Nalwadi et. al. (1972) and Negi *et al.* (1974) observed the peak time between 8.0 to 10.0 AM. Prasad (1971) observed that only 7% of the flowers opened after 10 AM. Randhawa and Negi (1965), Prasad (1969), Kanwar and Nauriyal (1969) observed 8.0 and 9.0 AM was the peak period of anthesis. However, in general, peak time of anthesis is between 7.0 to 8.0 AM in North India and between 8.0 to 9.0 AM in South India.

Pollination, fertilization and seed development results the good berry set. In most of the seedless cultivars of grape pollination is followed by the fertilization but subsequently embryo gets aborted and leads to seedlessness termed as stenospermocarpy (e.g. Perlette, Beauty seedless, Pusa seedless, Delight, Thompson seedless etc.). However, in some cultivars parthenocarpy is reported e.g. Black Corinth which requires stimulus from pollen for berry set. Self pollination is rule in *V. vinifera* grape (Nalwadi et. al., 1972, Veeprappa and Nalwadi, 1969), but cross pollination is required in reflexed anther cultivars e.g. Hur, Banqui–Abyad, Kathakurghan, Hybrid No. 1613, *V. parviflora, V. champinii*, Angoor Kalan (Neema, 1978, Jindal, 1990). For Muscadine grape pollinizer is required. Insects are reported to be helpful for pollination in M. rotundifolia grape (Datijen, 1917).

8.6. Inheritance pattern

Seedlessness in grape is governed by the necessary factor (Pandey and Pandey, 1996). Glabrous young shoot, thicker cane, longer cane and internodes, short stiff hair on leaf surface, smaller size of leaf, main vein length and petiole length, vigorous and drooping habit of the vine, pentagonal leaf shape, shorter maturity period, low bunch weight, round berry and oval shape of seed show dominance character (Yadav and Singh, 1977a, 1977b). However, Singh and Jalikop (1986) reported high phenotypic and genotypic coefficients of variation for number of bunches per vine but both these measures were low for 50 seed weight and seed per 50 berries. Wide differences were observed between phenotypic coefficient of variation and the corresponding genotypic coefficient of variation in respect of bunches per vine, bunch weight, volume of bunch and number of berries per bunch. All these characters are greatly influenced by environmental factors. The number of bunches per vine and berry weight well respond to selection and that these are mostly controlled by additive gene action, since good amount of heritability accompanied by high genetic advance was observed. The high heritability was observed for seeds per berry and seed weight was not accompanied with genetic advance indicating that these characters can be modified through hybridization and selection. For leaf size, main vein length, petiole length, short stiff hair on the lower surface of leaf and petiole, pleiotropic gene action was noticed. Further, reflex stamen character shows monogenic inheritance. Reflex stamen in Hur is supposed to be homozygous and recessive character (Singh et. al., 1972). Presence of loose bunch in the progeny has a direct correlation with the presence of reflex stamen in the seedling.

Yadav and Singh (1977a) noticed that bark colour, cane length, internode length, cane thickness, colour of young leaf, young shoot and tendril, leaf size, main-vein pigmentation, maturity period from sprouting, bunch number per vine, bunch weight, rachis length, berry weight and seed number per berry are governed by number of factors with complementary gene action. Berry colours are not sharply differentiated. The observations on heritability showed that black colour of berry is dominant over red and white, and red is dominant over white (Sandhu and Uppal, 1988, Sikka, 1958). Resistance to Plasmopara viticola Burk and Curt. and Uncinula necator Schw. Burr. are governed monogenically by a dominant gene (Filippenko and Shtin, 1975). Presence of dominant allele at all three independent loci (Pdi, Pdz and Pd3) is essential for pierce's disease resistance (Mortensen, 1968). For the inheritance of flower types so many hypothesis have been proposed from time to time (Olmo, 1970; Mukherjee et. al., 1970; Negi and Olmo, 1970, 1971a). Mukherjee et. al. (1970) reported that flower types is controlled by two alleles, female by (Sf) and Hermaphrodite by (Sh) and Sh > Sf. Further, two types of hermaphrodite are reported i.e. Sh Sh and Sh Sf. The character of reflexed anther type is governed by homozygous pairs of genic complexes. The dominant 'So' or SoSp / SoSp suppresses the normal female organ development while So results normal female organ development and the development of male organs is suppressed by recessive factors complex Sp.

Olmo (1970) reported that male vine can be denoted by Mm and derived types MM or MMh, hermaphrodite vine by Mh or MhMh and female vine by mm. The primitive hermaphrodite is denoted by Su+ Su+ and another type of hermaphrodite is denoted by Su+Sum. Su F suppresses the ovary development and promotes the maleness. Sum, results in reflexed anther type with sterile pollen grains and functional female. In natural population, males (SuFSum) and females (SumSum), occur in equal proportion. The male is determined by dominant gene, female by homozygous recessive gene and hermaphrodite vine is either homozygous or heterozygous. The dominance relation of 3 alleles are as M > mh > m or Su F > Su+ > Sum (Pandey and Pandey, 1996). The derived type rarely occurs. It is observed during sex conversion of male into hermaphrodite on the basis of that Vitis is considered sub dioecious (Pandey and Pandey, 1996).

8.7. Breeding methods and achievements

8.7.1 Introduction

Grapes are reported to introduce in Tropical India about 2600 years ago in 620 BC (Olmo, 1976). Commercial cultivation did not start until the beginning of 20th Century. During 1930, Shree R.S.Pillay, identified Anab-e-Shahi from the collections of Nawab Baquer Ali Khan and subsequently its commercial cultivation picked up in South India, Bhokri and Cheema Sahebi in Maharashtra, Bhokri and Muscat Hamberg in Tamil Nadu and Bangalore Blue in Karnataka.

The commercial varieties of grapes were introduced in India mostly by invaders of Iran and Afghanistan (Thaper, 1960). Muhammad Bin Tughlaq introduced, Bhokri, Fakhri and Sahebi cultivars in Aurangabad (Daulatabad) in 1338 (Pillay,

1968). Large scale introduction in a planned manner were initiated at Lyallpur as early as 1928, when S.B.S. Lal Singh, was Head of Department of Horticulture, introduced as many as 116 grape varieties from different grape growing countries (Singh and Singh, 1940, 1942). The earlier promising introductions include, Thompson Seedless, Perlette, Beauty Seedless, from USA, Kishmish Beli and Kishmish Charni from USSR (Singh and Singh, 1972). The cultivars like Ruby Seedless, Gordo Blano, (Reisling, MS 18–55, MS 19 –77, MS 16-2, Wortly Hall hybrids from Australia, Totlocha from Brazil, Flame Seedling 1281, Dogridge, Pride, Dixie, Wedor and Black Corinth–2 from USA, Surnak Kitabiskij, Pozdrijwir and Shirajx-6 from USSR, Malvasiafina (Douro), Boal De Alicante, Tinta Deira Preta, Jampal, Tinta Roriz, from Portugal and 0912 Horizon (SW), 0913 Leon Millet, Foch and 0912 Swanson Red from Canada for wine, raisin and table purposes have been introduced and are under evaluation. Further, number of *Vitis sp.* have been introduced for resistant to biotic and abiotic stresses e.g. *V. gigas, V. caribea, V. munsoniana, V. smalliana, V. cineraria, V. shuttleworthi, V. arizonica* and *V. monticola* from USA (Singh and Rana, 1993).

8.7.2 Selection

Open pollinated seedling segregates for a large number of characters and hence the population of seedlings from open pollinated seeds is a potential source for selection of desirable type e.g. Cheema Sahebi (Sel – 7), Selection – 49 (Phadnis et. al., 1968). Some promising seedlings from open pollinated population of Pandhari Sahebi and Kabul Monukka were also selected (Gupta, 1971, Phadnis and Gupta, 1972).

Clonal selection is also one of the methods of fruit improvement. Due to natural mutation in existing cultivars, considerable variation occurs between individual that helps in varietal improvement through clonal selection.

Table 1: Promising clonal selection of grapes

S. No.	Cultivars	Clonal Parent	Characteristics
Selection of grape varieties by grower:			
1	Tas–e–Ganesh	Thompson Seedless	Developed by Mr Arue of Borgoan in Sangli district of Maharashtra. Its berries are quite elongated and respond to GA3 treatment.
2	Rao Sahebi	Cheema Sahebi (Sel–7)	Isolated by Rao Saheb Kadlag of Sangamner in Nasik district of Maharashtra. Fruits have longer berries with stronger attachment to rachis which is a major problem in Cheema Sahebi.
3	Sonaka and Manik Champa	Thompson Seedless	Sonaka has much elongated berries as compared to Tas-e-Ganesh. It gives better response to GA3.
4	Dilkhus	Anab-e-Shahi	Selection was made at Hyderabad. It produces golden yellow elongated seeded berries in attractive bunches. The yield potential is almost same as in parent.
Selection made by Institute:			
1	Pusa Seedless	Thompson Seedless	Developed at IARI, New Delhi. It differs from the parent cultivar in respect to having more elongated berries. Its vine is vigorous and heavy yielder on bower and trellis system. The bunch of the fruit is medium with TSS about 22-24%. Acidity is 0.77% and juice 65%. It ripens in middle of June.
2	HS 37-6	Perlette	Developed at HAU, Hisar (Singh et. al., 1974). This cultivar is 15 days earlier in maturity than the parent.

8.8. Hybridization

Grapes are highly heterozygous and are propagated asexually at commercial scale. Inbreeding results in rapid loss of vigour and fertility of vine, even in first generation. The crossing of unrelated parents with good combining ability followed by raising a large number of hybrid seedlings in each combination and rigorous selection may result good ideotype of commercial use.

In India, hybridization work was started in 1958 at IARI, New Delhi. The purpose of hybridization at IARI, New Delhi was to develop early maturity, high yielding, better quality, seedless varieties with resistant to biotic stresses. However, IIHR, Bangalore, started breeding programme in 1968, with objective to develop superior varieties for table, raisins, wine and juice. On the basis of types of parent used, it can be grouped into two (a) Inter specific/Intergeneric hybridization and (b) Intra specific or Inter varietal hybridization.

8.8.1 Interspecific / Intergeneric hybridization

Muscadinia is a rich source of resistance to disease and pests and also possesses a unique and delightful flavour and aroma. The crosses between Vitis and Muscadinia (earlier Euvitis and Muscadinia as subgenera) which differs in chromosome number are made with difficulty, but most of the resulting hybrids remain sterile. The pollen of *M. rotundifolia* will fertilize the egg cell of *V. vinifera* but the reciprocal cross is less successful. Partly, fertile F1 hybrids (2x=39) can cross reciprocally between themselves or with V. vinifera. Some of the selections made from crosses of *V. vinifera* x *M. rotundifolia* have been further improved by back crossing with V. vinifera, resulting in some fertile vines that produce acceptable quality table wines (Olmo, 1971). Crossing within Muscadinia has given outstanding self-fertile cultivars like Tarheel (*M. rotundifolia* x *M. munsoniana*), South land, Magron, Regale (Cold hardy and pierce's disease resistant) Sterling (cold hardy) and Triumph (bronze cloured with big berry 7.9g).

Some of the outstanding contributions of inter specific hybridization are cold and mildew resistant. 'Amur' hybrid (*V. vinifera* x *V. amurensis* Ruprecht), commercial cultivar of Japan, Delaware (*V. labrusca* x *V. aestivalis* Michaux vzr. Bourquiniana x *V. vinifera*), Pierce's disease resistant cultivar. Lake Emerald (*V. simpsoni* Munson cv. Pixiola x *V. vinifera* cv. Golden Muscat) and some important rootstocks of grape i.e. hybrid 1613 (Solonis x Othello) resistant to nematode and Phylloxera, Telki 5A (*V. berlendieri* x *V. riparia*) highly resistant to Phylloxera, tolerant to lime soils and moderately resistant to nematodes, Harmony (1613 x *V. champinii* Planchon cv. Dogridge) have been developed with the results of inter specific hybridization (Pandey and Pandey, 1996).

8.8.2 Intervarietal hybridization

Mukherjee et al. (1965) made some crosses using Gros Colmn, Bhokri, Bharat Early, Pusa Seedless, Black Muscat and Pearl of Casaba cultivars to develop seedless sweet and high yielding varieties with Muscat flavour. Later on, Mukherjee et al. (1970) reported ten hybrids from parental combination of Hur X Pusa Seedless (62-6,62-56,62 –80,62-109) Bharat Early x Bhokri (62-20,62-54), Hur x Bharat Early (62-36,63-76 and 63-65) and Bhokri x Pearl-of-Casaba (63-12).

Singh et al. (1972) made a large number of intervarietal crosses involving 12 varieties at the IARI, New Delhi and seven promising hybrids were identified e.g. Hybrid 62-37 (Hur x Pusa Seedless), H62-65 (Hur X Pusa Seedless), H-62-20 (Hur x Black Hamburg), H-62-27 (Hur x Bharat Early), H-63-10 (Bhokri x Pearl of Casaba), H-63-32 (Bhokri x Pearly-of-Csaba). Jindal et al. (1982) reported 12 potential hybrids for table purpose. Later on Jindal and Singh (1984) identified 84 F1 hybrid which were classified into five categories i.e. Seedless (20), Seedless often with rudimentary seed (41), Empty seeded (12), Sub seedless (8) and Seeded (3).

In 1996, cultivars Pusa Navrang (Madeleine Angevine x Rubired) and in 1997 Pusa Urvashi (Hur X Beauty Seedless) was released from IARI, New Delhi (Pandey and Pandey, 1996).

At the Indian Institute of Horticultural Research, Bangalore, grape breeding programme has been in progress since 1968 (Negi, 1992). The promising hybrids developed at this centre are Arkawati (Black Champa x Thompson Seedless), Arka Kanchan (Anab-e- Shahi x Queen of the Vine Yards), Arka Shyam (Bangalore Blue x Black Champa), Arka Hans (Bangalore Blue x Anab-e-Shahi), Arka Chitra (Angur Kalan x Anab-e-Shahi), Arka Krishna (Black Champa x Thompson Seedless), Arka Majestic (Angur Kalan x Black Champa), Arka Neelmani (Black Champa x Thompson Seedless), Arka Soma (Anab-e-Shahi x Queen of the Vine Yards), Arka Thrishna (Bangalore Blue x Convent Large Black), Arka Shweta Syn. Shweta Seedless (Anab-e- Shahi x Thompson Seedless).

8.8.3 Hybridization technique

It includes the choices of parents, emasculation, pollination, shortening of breeding cycle for early assessment, growing of hybrid seeds and planting in the field for assessment and selection.

(i) *Choice of parents*

In order to incorporate the desirable characters of one cultivar into other through hybridization, the knowledge of inheritance pattern and general and specific combining ability of the cultivars is very essential for making choice of parents in restricting the cross-combination and more seedling population for better selection. The viability and germination ability of the hybrid seeds are also important factors in deciding the parents to be used in hybridization. It has been found that in some cultivars when used as female parents or selfed, the seed germination is poor and sometime do not germinate e.g. Cordinal. If such cultivars are required in hybridization, they should be used as male parents in order to induce seedlessness in the progeny. It would be better to select the variety having high seed index as female parents (Jindal et al., 1984). Cultivar Banqui-abyad, Katha Kurghan and Hur have high seed index (Table 1). Cultivar Angoor Kalan can also be used as female parent for earliness, seedlessness and good quality, but for the same purpose cultivars Beauty Seedless, Perlette and Pusa Seedless should be used as male parents (Jindal, 1990).

Seed index of potential female parents in grape hybridisation (Jindal et al., 1984)

Table 2: Seed Index = (Berry weight/seed weight)

S.No.	Varieties/hybrids	100 berries weight (g)	Number of seeds in 100 berries	Seed weight (g)	Seed Index
1.	Banqui-abyad	610	240	18	33.89
2.	Katha Kurgan	630	260	20	31.50
3.	Hur	456	264	16	28.50
4.	Madeleine Angevine	288	300	14	20.57
5.	Angoor Kalan	300	270	20	15
6.	*Vitis parviflora*	173	220	18.67	9.27
7.	*Vitis parviflora* x Pearl-of-Casaba74-7	65	260	14	4.64
8.	*Vitis parviflora* x Pearl-of-Casaba 74-15	84	176	8	10.50
9.	*Vitis parviflora* x Pearl-of-Casaba 74-23	120	288	15.20	7.89
10.	Hur x Bharat early 63-36	340	210	18	18.89
11.	Hur x Pusa Seedless 62-80	290	140	12	24.17

(ii) *Emasculation and pollination*

Emasculation of small flowers of grapes is a tedious job. Since the grape is self fertile, therefore, its emasculation is most essential for making desired crosses. Use of reflexed stamens and functionally female cultivars like Hur, Angoor Kalan, Banqui-abyad, Katta Kurgan as female parents can helps in eliminating the tedious task of emasculation (Singh et al., 1972, Jindal et al., 1984). Use of MH 500 ppm spray, twice at S III stage, gives complete pollen sterility (Uppal and Mukherjee, 1968). Iyer and Randhawa (1966) reported that aqueous solutions of maleic hydrazide (MH) at 400 to 750 ppm, 2,3,4 Tri Iado-benzoic acid (TIBA) at 400 to 500 ppm and 1,2 dichloro-iso-butyrate (FW-450) at 0.30%, applied twice to 13 to 15 days old inflorescence induced pollen sterility. When emasculation is completed the emasculated bunches are bagged and pollinated with desired male parents very next day.

8.9. Shortening the breeding cycle

Shortening of juvenile period in hybrid progenies is one of the typical problems of grape selection. The efforts have been made in the past to shorten the breeding cycle so as to hasten the improvement. The important strategies are as under:

(a) *Optimum growing condition*

The length of juvenile period appears to be influenced by environmental factors. It has been stated that the bearing of a tree is inseparably connected with the achievement of certain amount of vegetative growth and maturity. As a consequence, a rapid and continuous growth of seedling results in precocious flowering and fruiting. The dormant period of 3-4 month can be avoided by growing the seedlings under glass houses. Seedling so grown come to flowering and fruiting within 18 month of planting (Wagner, 1967). Hoagland solution under glass house condition can be used for growing seedlings of grape. The latent buds were forced after 3-4 months of sowing by removal of lateral bunds. These treatments resulted in fruiting within 9-10 months of sowing (Huglin and Julliard, 1964).

(b) *Budding/grafting (Top working)*

Budding or grafting on old root stock or on bearing tree reduces the juvenile period. Synder and Harmon (1937) reported that budding may reduce the pre-bearing period in grapes. According to him, the basal buds of 3-4 months old seedling were budded on a vigorous rootstock in the month of May and the shoot was pruned to 60cm length at the time of winter pruning. Shoots arising from these canes flowered and gave fruits in the same season and the fruiting was obtained within 18 months of planting. This technique was also used by Negi and Olmo (1971b). Characters like fruit size, colour, quality, seedlessness and flavour could be assessed, while yield trials can be taken up from well grown plants. This technique also makes possible, to bud many seedlings per vine at the same time (like orchard on a tree). This is economical where land is a limiting factor.

(c) *Fruiting cutting*

This technique was developed by Mullins in 1966. In this technique, fruiting test plants were developed from hard wood cuttings. The inflorescence initials are often seen in hard wood cuttings, but usually they shrivel and die soon after bud burst, as the vegetative bud becomes a powerful sink (Mullins, 1968), which is aggravated by the fact that bud burst precedes root development which are the source of cytokinin (Skene, 1975). Keeping these facts in mind, Mullins and Rajasekaran (1981) developed a comprehensive method. In this method dormant cuttings are induced to root by growing in heated containers (260C at the base of cutting) in a cold room (40C). The buds remain dormant due to the prevailing low temperature. After 4 weeks, the rooted cuttings are transferred to glass houses (270/260 C day and night temperature). At the time of bud burst, leaves basal and adjacent to inflorescence are removed and the shoot tips are excised. A lateral shoot is allowed to grow from an axillary bud of the main axis which feeds the growing inflorescences. When four to five leaves are produced, the lateral shoot is tipped and disbudded. Ripe berries are smaller than the size of normal berries but the seed is highly germinable (Ottenwaelter et al., 1974). This technique is especially useful in preliminary evaluation of new introductions which saves money and labour. Most of the Vitis sp. and commercial cultivars have responded to this technique.

(d) *Conversion of tendril into inflorescence*

Now it is concept that tendril is modified form of inflorescence (Mullins, 1968, Negi and Olmo, 1970, Srinivasan and Mullins, 1976). The tendrils are morphologically homologous to inflorescence and both of them arise form uncommitted meristems called anlagen. Critical evaluation of the hormonal balance revealed that abundance of inhibitors and gibberellins promotes tendril formation while cytokinins promotes inflorescence productions. This fact had been used by Srinivasan and Mullins (1976, 1978 and 1979) to force inflorescence production in grapes. When the rooted cultivars produced shoots with 10 leaves they are decapitated to 6 leaves and the laterals were allowed to grow. The tips of these laterals were treated with 10-20μl of 6-(Benzyl amino)-9-(2-tetrahydropuranyl)-9H-purine [PBA, 50-200 μm] daily once for 20 days. The tendril which were to be produced, developed into inflorescence and produced fruits. Seeds from these fruits showed about 95% germination. Seedlings of three months age were also forced to produce inflorescence and subsequently fruits. A wide range of Vitis species were found to respond to cytokinin treatment, among which Vitis vinifera and V. rupestris and their hybrids were found more responsive (Srinivasan and Mullins, 1980). Excised tendril could also be transformed to inflorescence in vitro (Srinivasan and Mullins,1978).

(e) *Overcoming seed dormancy*

Generally, for the normal germination, seeds of grapes require stratification of 3-4 months. Therefore, for the sowing of seeds, 6-7 months are required after harvesting mean time if temperature of out side is not fit for proper growth that also affect the germination of the seeds. However, Pandey and Singh (1988) reported that the dormancy of grape can be overcome in 70 days by soaking the seeds in a solution containing 1500 ppm GA3 and 2500 ppm kinetin for one day. The seedlings grew about 30-40 cm before entering into dormancy and hence these seedling come to bearing one year in advances as compared to the normal method.

(f) *Pre selection criteria*

Jindal and Rao (1992) reported that in coloured cultivars the senesced leaves before dormancy, show pink or purple colour. The intensity of the colour, also corresponds to the intensity of the berry colour. During and Scienza (1980) studied the properties associated with drought tolerance as a guide to early selection. A high degree of succulence, greater reduction in stomatal frequency under humid conditions, a smaller change in stomatal resistance, per unit change in capacity to rehydrate after stress, high water use efficiency and relatively small response in leaf to excision are associated with drought tolerance and can be utilized for selecting drought tolerance genotypes.

(g) *In vitro techniques*

It can also be used for study of salt resistance (Jindal and Rao, 1988). To screen downy mildew resistance plants, dual culture method developed by Barlass et al. (1986) can be used. Further, embryo rescue technique can be utilized to obtained the hybrid seeds before abortion of embryo through in vitro culture for further evaluation (Cain, 1980).

8.10 Mutation

Mutation breeding may be attempted as a complementary tool in grape breeding for one or more important characters, without altering the whole genetical setup. The important mutagens used in grape breeding are physical mutagens (x ray and ã rays) and chemical mutagens (Ethyl Methane Sulphonate (EES), N-Nitroso-N-Methyl Urethane (NMUT) and N-Nitrose-N-Methyl-Urea (NMU) (Das and Mukherjee, 1968, Sharma, 1970, Upadhyaya, 1973).

Further, induced mutations have resulted in a few improved varieties, New Perlette (Loose Perlette) with comparatively loose bunch has been evolved with X-rays (2.5 KR) treatment on Perlette. Self-thinning property of New Perlette is a result of meiotic irregularities caused by chromosomal translocation. Red Niagara having red fruit from Niagara and Robin Cordinal an early maturing variety form Cardinal are other important induced mutants of grapes.

8.11 Polyploidy

Polyploidy breeding has immense importance in the improvement of table grapes. The chief benefit from polyploidy is the increase in berry size. However, autotetraploids are found to be considerably sterile and are less productive than the parents. The crossing of diploid with induced tetraploids may help in evolving new triploid seedless grapes. The triploids are highly sterile. Allotetraploids even between infertile species, have been more desirable as commercial varieties (Olmo, 1952). Colchicine is generally used as an aqueous solution of 0.25-0.5.0% with 5-10% glycerine to induce polyploidy (Das and Mukherjee, 1967). Perle, Marvel Seedless from Delight, Early Niabel (Campbell x Niagra) Lonetto, Early Giant from campbell, Muscat Common Hall from Muscat Alexandria, Black King from Campbell, Wallis Giant from Concord,

Case from Sultana etc are important examples of polyploidy (Pandey and Pandey, 1996).

8.12 Biotechnological tools

8.12.1 Embryo rescue technique

Seedlessness is a desirable character for table and raisin grapes. Inheritance of seed lessness is postulated to depend on two complementary recessive genes and only about 7.5% of the total progeny from crosses between Seeded x Seedless grapes produced fruits without noticeable seed traces. The embryo rescues theoretically increases the proportion of seedless progeny as it makes possible to cross two seedless varieties. Ovules are excised before abortion and are cultured on either filter paper in liquid medium (Emershad and Ramming, 1984) or solid medium (Spiegel Roy et al., 1985).

8.12.2 Genetic engineering/ plant transformation

Some encouraging preliminary results have been obtained on Agrobacterium-mediated transformation of grape vines (Baribault et al., 1989) but the production of genetically transformed grape vines which express a marker gene is yet to be reported.

8.12.3 Protoplast culture

Protoplasts are of great importance as tool for genetic amelioration and somatic hybridization. But regeneration of grape vines from protoplasts has not yet been successful.

8.12.4 Anther culture

Anther culture can result into haploid grape vines which can then be developed into homozygous diploids by doubling chromosomes. These homozygous diploids will be very useful for producing F1 hybrids and for making genetical studies. But there is low success rate of regeneration of grape vines from anther and only one case of haploid has been reported in grape (Cersosimo, 1989).

8.12.5 Somaclonal variation

Somaclonal variations induced by tissue culture are suggested to be utilized as tools to improve the existing grape cultivars. (Bessis, 1986).

8.13. Important species and cultivars of grape

Table 3: Important species of grapes with their vernacular / common name are as under (Jindal, 1990, Olmo, 1970)

S.No.	Name of the species	Characteristics
1	Middle Asian and Mediterranean grape, wine grape *Vitis vinifera* (European grape)	Susceptible to most of the pest and disease but moderately tolerant to salinity, produce best quality grape, most of best cultivars of grape belong to this species. Pulp of the fruit is adherent to skin.
2	North America *Vitis labrusca* (Fox grape, New world grape)	It is cold hardy, blooms earlier than vinifera. It is resistant to many pests and diseases, Concord, Isabella and Royers etc. are the important cultivars. The skin of fruits is thick and slips from the pulp at maturity.
3	*Vitis riparia* (*V. vulpina*) Frost grape or River bank grape	It is cold hardy and first to flower among the Vitis species.
4	*Vitis aestivalis* (Summer grape or Pigeon grape)	It is resistant to many fungal diseases but it is susceptible to Phylloxera. Cultivar Delaware, Herbemont and Denoir have been developed by crossing V. aestivalis and V. vinifera. It flowers after V. labrusca, fruits are small in size.
5	*Vitis berlandieri* (Spanish grape or Winter grape)	It is very difficult to propagate vegetatively. It is resistant to Phylloxera and high lime content in soil.

8.13.1 Other species of American grapes

Vitis argentifolia (Silver leaf grape), *V. arizonica* (Canyon grape), *V. baileyana* (Possum grape), *V. californica* (Pacific grape), *V. candicans* (Mustang grape), *V. champini* (Calcarie grape), *V. cinerea* (Gray back grape), *V. cordifolia* (Winter grape), *V. doaniana* (Panhandle grape), *V. gigas* (Florida blue grape), *V. helleri* (Round leaf grape), *V. illex* (Manatee grape), *V. lincecumii* (Post– oak grape), *V. longii* (Bush grape), *V. monticola* (Sweet mountain grape), *V. novae– anglia* (Pilgrim grape), *V. palmata rubra* (Cat grape), *Vitis rufotomentosa* (Red shank grape), *V. rupestris* (Sand grape), *V. shuttleworthii* (Calloosa grape), *V. smalliana* (Fig-leaf grape), *V. simpsoni* (Current grape), *V. sola* (Curtiss grape), *V. treleasii* (Gulch grape).

8.13.2 Asiatic species

V. amurensis (Resistant to cold and frost, it produces edible fruits, native to China), *V. coignetiae*, *V. flexusa*, *V. pentagona*, *V. embergeri*, *V. betuliflora*, *V. reticulata*, *V. armata*, *V. davidii*, *V. lanata*, *V. pedicellata*.

8.13.3 Carribean grape

V. indica (It is also known as trifoliata or Carribean.

8.13.4 Genus Muscadinia

This genus has three species they are *M. rotundifolia* (Muscadine grape), berries of this species drop down at the time of ripening. It is resistant to Phylloxera, downey mildew, anthracnose and nematodes. Other species of this genus are *M. munsoniana* (Little muscadine grape) and *M. popenoei* (Mexican muscadine grape).

8.13.5 Important native species of grape

V. lanata Roxb, *V.barbata* Wall, *V. parviflora* Roxb., *V. tomentasa* and *V.himalayana* Br.

Table 4: Distinguishing characteristics of Euvitis and Muscadinia (Source: Pandey and Pandey, 1996)

S.No.	Characteristics	*Muscadinia*	*Euvitis*
1.	Chromosome number(2n)	40	38
2.	Nature of tendril	Unifid	Forked
3.	Bark	Tight	Loose
4.	Presence of lenticels	Present	Absent
5.	Seed shape	Beak absent	Beaked and pyriform
6.	Presence of diaphragm at the node of shoot, interrupting pith tissue	Absent	Present

8.13.6 New cultivars of grape (*Source*: Pandey *et al.*, 2002)

Arkawati (Black Champa x Thompson Seedless)

Bunch is medium in size, yellowish green berry, sweet, TSS 22-25%, seedless berry, suitable for raisin making, fresh table use and making good quality dry and white table and dessert wine. Released in 1980.

Arka Chitra (Angur Kalan x Anab-e-Shahi)

Released in 1994, this is tolerant to powdery mildew, moderately vigorous vine, good yield potential (34Kg/vine), Bunch is well filled, medium to large (310g), berry very attractive, golden yellow with pink blush, slightly elongated, large (3.18g), sweet, TSS 20-21o B, acidity 0.4-0.6%, suitable for table purpose, all the buds are fruitful, suitable for head system of training and gives two crops in a year.

Arka Hans (Bangalore Blue x Anab-e-Shasi)

Released in 1980, bunch is medium in size, berry yellowish green, sweet, TSS 18-21 °B, having foxy flavour, seeded cultivar, suitable for making quality wine, resistant to anthracnose.

Arka Kanchan (Anab-e-Shahi x Queen of the Vineyards)

Bunch is large, golden yellow colour berry, ellipsoidal to ovoid, sweet, TSS 17-20 °B, having muscat flavour, seeded cultivar, suitable for fresh table use and dry white table and dessert wines released in 1980.

Arka Krishna (Black Champa x Thompson Seedless)

Vigrous vine, good yield potential (30Kg/vine), Bunch is well filled, berry dark coloured, seedless, sweet TSS 20-21 °B, acidity 0.6-0.7%, suitable for head system of training and gives two crops in a year, suitable for juice making released in 1980.

Arka Majestic (Angur Kalan x Black Champa)

Released in 1994, vine is vigorous, high yield potential (34Kg/vine), Bunch is well filled, medium to large (370g), berry deep tan colour, sweet, TSS 18-20 °B, acidity 0.4-0.6%, all the buds are fruitful, suitable for head system of training, it gives two crops in a year. Tolerant to anthracnose.

Arka Neelamani (Black Champa x Thompson Seedless)

Vigrous vine with high yield potential (25Kg/vine), bunch is well filled, sweet, TSS 20-22 °B, suitable for head system of training, it is good for table purpose and making red dessert wine, tolerant to anthracnose, released in 1980.

Arka Shyam (Bangalore Blue x Black Champa)

It was released in 1980, bunch is medium in size, berry bluish black, spherical to obovoid, sweet, TSS 20-25 °B, having mild foxy flavour and seeded, it is good for fresh table use and making dry table, dessert wines and juice, resistant to anthracnose disease.

Arka Soma (Anab-e-Shahi x Queen of the Vineyards)

This variety was released in 1994, vine is vigorous with heavy yield potential (36Kg/vine), bunch is well filled, large (410g), berry greenish yellow, round to ovoid, large (3.8g), sweet, TSS 20-21°B, acidity 0.5%, having muscat flavour, gives two crops in a year, suitable for good white dessert wine.

Arka Trishna (Bangalore Blue x Convent Large Black)

It was released in 1994, it is improvement over variety Bangalore Blue, vine is vigorous, having high yield potential, medium bunch, well filled, very sweet, TSS 22-23 °B, acidity 0.3-0.4%, it is male sterile hybrid, good for wine making, suitable for head system of training, resistant to anthracnose and tolerant to downy mildew.

Arka Sheweta or Shweta Seedless (Anab-e-Shahi x Thompson Seedless):

It was released in 1994, moderately vigorous vine, yield potential is about 28Kg/vine, bunch is medium, it responds to GA3 application for berry thinning and enlargement, berry is seedless, sweet, TSS 18-19 °B, acidity 0.5-0.6%, berry greenish yellow.

Degrasset

This variety was collected at ARI (MACS), Pune from the grape germplasm collection maintained at Ganesh Khind Botanical Garden, Pune in 1976. It is a clone of Vitis champini. Vine shows vigorous, spreading and prostrate growth having deep root system. It remains dormant during winter season after October pruning and again grows in February-March under climatic conditions of Maharashtra. This is potential root stock for growing grape under saline and drought conditions.

Pusa Navrang (Madeleine Angevine x Rubired)

It was released in 1996, it is basal bearing, tenturier (peel and pulp both coloured), seeded cultivar, it is early maturing, suitable for making coloured juice and wine bunch is loose and medium, TSS 19°B, resistant to anthracnose.

Pusa Urvashi (Hur x Beauty Seedless)

It was released in 1997, it comes in early group, basal buds are productive, bunch is medium, berry is greenish yellow, TSS 20-2°B.

Clonal Selection A- 17/3

It is released by NRC for Grapes in the year 2006

Manjari Naveen and Red Globe

It is a clonal selection released by National Research Center for Grapes, Pune.

Arka Neelkiran

This hybrid is a cross between Black Champa and Thompson seedless. The plant is vigorous with a yield potential of 28 tons/ha and 25 kg/vine. Time required for harvesting from pruning is 150-155 days. The bunches are well filled to slightly compact, weighing an average of 360 g, black, seedless and average berry weight is 3.2 g with crispy pulp. T.S.S. varies from 20-22 °B. It is possible to take two crops in a year and it is also suitable for head system of training. It is tolerant to anthracnose disease. It is very good for table and also for red dessert wine making.

Table 5: Suitable varieties for different growing regions

S. No.	Region	Varieties
1	Maharashtra, Northern Karnataka and Hyderabad	Thompson Seedless and its mutants like Tas – A – Ganesh, Sonaka and Manik Chaman and A 17/3 found promising, however, yet to be released; colored seedless varieties like Fantasy Seedless, Sharad Seedless and Crimson Seedless; seeded varieties like Red Globe (found promising but yet to be recommended).
2	South Interior Karnataka	Thompson Seedless, Sonaka, Flame Seedless, Sharad Seedless, Crimson Seedless and Red Globe
3	Tamil Nadu	Thompson Seedless (for Table Purpose), Gulabi, Bangalore Blue (Juice purpose)
4	North India	Flame Seedless, Perlette and Beauty Seedless

Table 6: Rootstocks suggested based on global data for different situation

S.No.	Situation/problem	Rootstock
1.	Water shortage	1103 P, 140 RU, 110 R, 420 A, SO 4, 99 R, St. George, Dogridge
2.	Soil EC more than 2 m mohs/cm and water EC more than 1 m mohs/cm	Ramsey, Dogridge, 140 RU, 99 R, 110 R.
3.	Soil ESP more than 15 per cent and/or water SAR more than 8.	140 RU, 1613, Ramsey, Dogridge.
4.	Free calcium content of soil is more than 12%	140 RU, SO 4, 420 A.
5.	Chloride content of water is more than 4 meq/litre	Ramsey, Dogridge B, 140 RU. Teleki 5-C
6.	Poor vigor of the variety without any soil/water problem	Dogridge, St. George, SO 4, 140 RU.
7.	For increased nitrogen, potassium uptake.	Dogridge, St. George, 34 EM, Ramsey.

8.14. References

Anonymous, (1996) Annual Report. IIHR Bangalore.

Baribault, T.J., Skene, K.G.M. and Scott, N.S. (1989). Genetic transformation of grape vine cells. *Plant Cell Rep.*, **8**:137-140.

Barlass, M., Miller, R.M. and Antclift, A.N. (1986) Development of methods for screening grape vines for resistance to infection by downy mildew. *Amer. J. Enol. Vitic.*, **37**: 61-66.

Bessis, R. (1986) Grape vine physiology: The contribution of culture *in vitro Experientia*, **42**: 927-933.

Cain, D.W. (1980) In vitro culture of seeded and abortive grape ovules. *Hort. Sci.,* **15** : 415.

Cersosimo, A. (1989) Aims and problems of grape vine anther culture. *Quad. Vitic. Enol. Univ. Torino*, **13**: 75-105.

Chelam, G.V. and Satyanarayana, G. (1968) Some studies on floral biology of grapes (Vitis sp.). *Andhra Agric.J.*, **15** (6):154-162.

Das, P.K. and Mukherjee, S.K. (1967) Induction of autotetraploidy in grapes. *Indian J. Genet. Plant Breed.*, **27**(1):107-116.

Das, P.K. and Mukherjee, S.K. (1968). Effect of gamma radiation and ethyl methane sulphonate on seeds, cuttings and pollen in grapes. *Indian J. genet. Plant Breed.*, **28 (3)**: 347-351.

Das, P.K. and Nath, N. (1965) Origin and cytogenetical studies in grape. *Punjab Hort. J.*, **5:** 96-105.

Datigen, L.R. (1917) Pollination of rotundifolia grape. *J. El. Mitchell Sci. Soc.*, **33:** 121-127.

During, H. and Scienza, A. (1980) Studies on drought resistance of Vitis sp. and cultivars. Proc. Int. symp. Grape. B reed. paper 22.

Emershad, R.L. and Ramming, D.W. (1984) Invitro embryo culture of Vitis vinifera L. Cv. Thompson seedless. *American J. Bot.*, **71**:873-877.

Filippenko, I.M. and Shtin, L.T. (1975) Breeding disease resistant varieties of grapes on genetic principles. *Tr. Tsen. Genet. Lab. Im. I.V. Michurina,* **16**:56-62 (Russian).

Gupta, O.P. (1971) Seedling selection in grape vine (*Vitis vinifera* L.). M.Sc. Thesis, Mahatma Phule Krishi Vidyapeeth, Rahuri.

Huglin, P. and Julliard, B. (1964) The question of vigorous vine seedlings that come into breeding rapidly and their importance in the genetical improvement of vine. *Ann. Amel. Plantes.*, **14**: 229-244.

Iyer C.P.A. and Randhawa, G.S. (1966) Inducing pollen sterility in grape (*Vitis vinifera*). *Vitis*, **5**:433-445.

Jauhari, S. Nand, D and Srivastava, G.P. (1969) Studies on flowering, pollination and fruit set in grapes. *South Indian Hort.*, **17**:21-24.

Jindal, P.C. (1990) Grape. In: Fruits Tropical and Subtropical (Eds. T.K. Bose and S.K. Mitra), Naya Prokash Publication Calcutta, pp 186-251.

Jindal, P.C. and Rao, M.S. (1988) Present status and future prospects in improving salt tolerance in grapes. Draksha Vritta Souvenir, pp.12-17.

Jindal, P.C. and Rao, M.S. (1992) Strategies for shortening breeding cycle in grapes an overview in: Proc. Int. Symp. on Viticulture and Oenology Hyderabad. pp. 92-99.

Jindal, P.C. and Singh, K. (1984) Research Reports Fruit Research Workshop, Lucknow, August 29 AICFIP, Lucknow, pp. 10-14.

Jindal, P.C., Pandey, S.N. and Singh, K. (1982) Improvements of grapes through breeding. Research Reports, Fruit Research Work Shop, Nagpur Feb. 9-14, AICFIP, Lucknow. pp. 87-92.

Jindal, P.C., Singh, K and Pandey, S.N. (1984) A note on choice of parents in grape hybridisation for seedlessness. *Haryana. J. Hort. Sci.*, **13** (1&2):33-34.

Kanwar, J.S. and Nauriyal, J.P. (1969) Studies on floral biology of some varieties of grapes (*Vitis vinife ra* L.). *J. Res. Ludhiana,* **6**:829-837.

Mortensen J.A. (1968) The inheritance of resistance to pierce's disease in Vitis. *Proc. Am. Soc. Hort. Sci.,* **92**: 331-337.

Mukherjee, S.K., Nath, N. and Yadav, I.S. (1965) Selection and improvement of grapes at IARI New Delhi. *Punjab Hort. J.*, **5:** 172-176.

Mukherjee, S.K., Nath, N., Yadav, I.S. and Singh, A.N. (1970) Grape breeding, some promising hybrids. *Indian J. Genet. Plant, Breed.*, **30(1)**: 36-46.

Mullins, M.G. (1966) Test plant for investigation of the physiology in *Vitis vinifera* L. *Nature*, **239**: 419-420.

Mullins, M.G. (1968) Regulation of inflorescences growth in cutting of the grape vine (Vitis vinifera L.). *Aust. Exp. Bot.*, **19**: 532-543.

Mullins, M.G. and Rajasekaran, K. (1981) Fruiting cuttings. Revised method for producing test plants of grape vine cultivars. *Amer. J. Enol. Vitic.*, **32**:35-40.

Nalwadi, U.G., Nalini, A.S. and Farooqui, A.A. (1972) Studies on the floral biology in some varieties of grapes (Vitis vinifera L.) under Dharwar conditions. *South Indian Hort.*, **20 (1/4):** 29-36.

Neema, M.K. (1978) Taxonomic study in grapes. Ph.D. Thesis, P.G. School, IARI, New Delhi.

Negi, S.S. (1992) Strategies for grape breeding in tropics. In: Proc. Inter. Symp. on Viticulture, Hyderabad India, pp 83-90.

Negi, S.S. and Olmo, H.P. (1971a) Conservation and determination of sex in Vitis. *Vitis*, **9:** 265-279.

Negi, S.S., Randhawa, G.S., Mahaswarappa, R.G. and Suresh, E.R. (1974) Some studies on grape hybridization in South India. *Indian J. Hort.*, **31(1):** 1-8.

Negi,S.S. and Olmo, H.P. (1971b) Induction of sex conversion in male Vitis. Vitis, **10:** 1-14.

Negi. S.S. and Olmo, H.P. (1970) Studies on sex conversion in male *Vitis vinifera* L. *Vitis*, **9:** 89 96.

Negrul, A.M. (1940) The evolution of cultivated grapes. Priroda (*Nature*) **4:**37-46.

Olmo, H.P. (1952) Breeding tetraploid grapes. *Proc. Am. Soc. Hort. Sci.*, **59:** 285-290.

Olmo, H.P. (1970) Grape culture. UNDP/FAO Bull. No TA2825, pp 99.

Olmo, H.P. (1971) V. rotundifolia hybrid and wine grapes. *Am. J. Enol Vitic.*, **22 (2)**: 87-91.

Olmo, H.P. (1976) Grapes. Vitis, Muscadinia (Vitaceae). Evolution of crop plants, (Ed. N.W. Simmonds), Wah Cheong Printing Press Ltd. Hong Kong, pp 294-298.

Ottenwaelter, M.M., Boussion, C., Doazan, J.P. and Rives (1974) A technique for improving the germinability of grape seeds for breeding purposes. *Vitis*, **13**: 1-3.

Pandey, R.M. and Pandey, S.N. (1996) The grape in India. ICAR Publication.

Pandey, S.N. and Singh. R., (1988) Germination of seeds extracted form immature berries of grapes. *Indian J. Hort.*, **45 (1/2)**: 56-60.

Pandey, S.N., Pandey, D., Chandra, R. and Singh, R. (2002) Fruit varieties released by centres of AICRP on subtropical fruits. AICRP (STF) Technical bulletin No. 2, CISH, Lucknow.

Phadnis, N.A. and Gupta O.P. (1972) Seedling selection in the grape vine (*Vitis vinifera* L.). Third International Symposium on Subtropical Horticulture, HIS, Bangalore, pp.14.

Phadnis, N.A., Kunte, Y.N., Chougare, I.B. and Bagde, T.R. (1968) History of development of Selection-7, a promising variety of grapes for the Deccan. *Punjab Hort. J.*, **8 (2)** :65-69.

Pillay, R.S. (1968) History of grape growing in the Deccan and South India Grape. Andhra Pradesh Grape Growers, Association, Hyderabad, pp 1-3.

Prasad, A. (1969) Investigations on floral biology, pollination and fruit quality of gapes. *Hort. Sci.*, **1**:5-21.

Prasad, A. (1971) Investigations on blossom biology and fruiting behaviour in grapes. i. flowering habit, panicle and bud development and anthesis. *Prog. Hort.*, **3(2)**: 70-78.

Pratt, C. (1971) Reproductive anatomy in cultivated grapes a review. *Am. J. Enol. Vitic.*, **22(2)**: 92-102.

Randhawa, G.S. and Negi, S.S. (1965) Further studies on flowering and pollination in grapes. *Indian J. Hort.,* **22**(3/4):287-308.

Regel, E. (1873) Comspectus specierum generis Vitis regions amaricane boreales chinae borealis et japonica habitautium. *Acta Hort. Petrop.,* **2:** 289-319.

Sandhu, A.S. and Uppal, D.K. (1988) Inheritance of berry colour in grapes (*Vitis vinifera*). *J. Res. Ludhiana,* **25**(1) :37-39.

Sharma,R.L. (1970) Studies on mutation in grapes. Ph.D. Thesis IARI, New Delhi.

Sikka, S.M. (1958) Breeding of fruit plants in India. *Indian J. Hort.,* **15** (1) :10-18.

Singh, B.P. and Rana, R.S. (1993) Promising fruit introductions. In: Advance in Horticulture Vol I (Eds. K.L. Chadha and O.P. Pareek) Malhotra Publishing House New Delhi, pp. 43-66.

Singh, J.P. and Singh I.S. (1972) Seedless grapes that excel in Northern India. *Indian Hort.,* **16**:4-6.

Singh, J.P., Daulta, B.S. and Godara, N.R. (1974) An early clone of Perlette grapes. *Haryana J. Hort. Sci.,* **31**(1/2): 92-93.

Singh, L. and Singh S. (1942) Grape varieties at Lyallpur. *Indian Farming,* **3:**372-375.

Singh, L. and Singh, S. (1940) Distinguishing characters and behaviour of some grapes vine varieties introduced at Lyallpur in the Punjab. Indian *J. Agric. Sci.,* **10**:552-560.

Singh, R. and Jalikop, S.H. (1986) Studies on variability in grapes. *Indian J. Hort.,* **43**(3/4): 207-209.

Singh, R.N., Yadav, I.S. and Pandey, S.N. (1972) Grape improvement work at the IARI. In: Viticulture in Tropics, Proc. Working group on Viticulture in SE Asia, Bangalore, February 8-14.

Skene, K.G.M. (1975) Cytokinin production by roots as factor in the control of plant growth. In: the development and functions of roots. (Eds. J.G. Torrey and D.T. Clarkson). Academic Press London, pp. 395-396.

Spiegel- Roy, P., Sahar, N., Baron, J. and Lavi. V. (1985) In vitro cutlrue and plant formation from grape cultivars with abortive ovules and seeds. *J. Amer. Soc. Hort. Sci.,* **110**:109-112.

Srinivasan, C. and Mullins, M.G. (1976) Reproductive anatomy of the grapevine (*Vitis vinifera* L.) origin and development of the anlagen and its derivatives. Ann. Bot., 40:1079-1084.

Srinivasan, C. and Mullins, M.G. (1978) Control of flowering in grape vine (*Vitis vinifera* L.) formation of inflorescences in vitro by isolated tendrils. *Plant Phy.,* **61** :127-130.

Srinivasan, C. and Mullins, M.G. (1979) Flowering in Vitis conversion of tendrils in to inflorescence and bunches of grapes. *Planta,* **145**: 187-192.

Srinivasan, C. and Mullins, M.G. (1980) Effect of temperature and growth regulators on formation of anlagen, tendrils and inflorescences in *Vitis vinifera* L. *Ann. Bot.,* **45**: 439-446.

Synder, E. and Horman, F.N. (1937) Hastening the production of fruits in grape hybridisation work. *Proc. Amer. Soc. Hort. Sci.,* **34**: 426-427.

Thaper, A.R. (1960) Horticulture in the Hill regions of North India. Directorate of Extension, Ministry of Food and Agriculture, New Delhi, pp 150-154.

Upadhyay, N.P. (1973) Mutation studies in grapes. Ph.D. Thesis IARI, New Delhi.

Uppal, D.K. and Mukherjee, S.K. (1968) Induction of male sterility in grape. Proc. Growth Regulator Symp, Calcutta University, pp 381-388.

Vavilov. N.I. (1951) Origin, variation, immunity and breedings of cultivated plants. *Chron. Bot.,* **1(6):** 364.

Veeprappa, K.B. and Nalwadi, U.G. (1969) Study of flowering and anthesis in two grape varieties Gulabi and Bhokri. *Punjab Hort. J.,* **9**:28-34.

Wagner, R. (1967) Preliminary selection of vine seedling under glass. *Ann. Amer. Plantes*, **17**:155-173.
Winkler, A.J. (1962) General viticulture, University of California Press, Berkeley and Los Angeles.
Yadav, I.S. and Singh R.N. (1977a) Mode of inheritance of certain vegetative characters in grapes (V. vinifera L.). *Indian J. Hort.*, **34(4)**: 342-349.
Yadav, I.S. and Singh R.N. (1977b) Mode of inheritance of certain fruit characters in grapes (Vitis vinifera) In: Fruit breeding in India (Ed. G.S. Nijjar), Oxford and IBH Publishing Co. New Delhi, pp153-166.
Zielinski, Q.B. (1955) Modern systematic pomology, Wm.C. Brown Co. Dubuque. Lowa.
Zukovskij, P.M. (1962) Cultivated plants and their wild relatives. Common Wealth Agriculture Bureau, Farnham Royal, London.

9

Breeding of Guava

Botanical name: *Psidium guajava* L.

Family: Myrtaceae

Chromosome number: 2n=2x=22

Scientific classification

Kingdom	:	Plantae- Plants
Division	:	Magnoliophyta–Flowering plants
Class	:	Magnoliopsida – Dicotyledons
Subclass	:	Rosidae
Order	:	Myrtales
Family	:	Myrtaceae – Myrtle family
Genus	:	*Psidium*
Species	:	*guajava L.*

Guava is also known as the "Apple of the Tropics". It is very rich and cheap source of vitamin C and also contains a fair amount of calcium. Important guava growing states in the country are Uttar Pradesh, Bihar, Madhya Pradesh and Maharashtra. Allahabad district of Uttar Pradesh has the reputation of growing the best quality of guava fruits in the world (Mitra and Bose, 1990). The importance of guava is due to the fact that it is the hardy fruit which can be grown in alkaline and poorly drained soil.

9.1. Centre of diversity

Tropical America is supposed to be centre of origin of guava where it is found in wild as well as cultivated forms. Guava came to India at a very early time before 17th century (Hayes, 1970).

9.2. Germplasm resources

Guava is mainly a self-pollinated crop but occurrence of cross pollination results in great variation in the seedling population. About 103 genotypes are available in the Indian collections (Iyer and Subramanian, 1987) while Yadav (1990) has listed 153 genotypes including Psidium species, cultivars and hybrids mainly at HETC, Basti, CISH, Lucknow, IIHR, Bangalore, NDUAT, Faizabad, and HAU, Hisar. Guava germplasm is being maintained at several centres in the country in field gene banks which are often not systematically maintained (Pathak and Ojha, 1993).

9.3. Objectives

- Dwarf plant stature.
- Tenete branching habit for more fruiting area.
- Better fruit quality in terms of shape, size, colour, firmness, thick pulp and pleasing aroma.
- To develop varieties with less number of soft seeds.
- Resistant to biotic and abiotic stresses.

9.4. Botany

Most of the cultivars of Indian guava belong to the genus Psidium and species guajava. Botanically guava belongs to the family Myrtaceae to which Eucalyptus, cloves, nutmeg, allspice, cinnamon and javaplum belong. Based on the shape of common guava fruits, they are classified into two groups (De Candolle, 1904) i.e. *Psidium pyriferum*, *Psidium pomiferum*. Genus Psidium contains about 150 species (Hayes, 1970). All cultivated varieties of guava are either diploid 2n=2x=22 or triploid 2n=3x=33 (Atchinson, 1947).

9.5. Floral biology and pollination

Guava bears flower solitary or in cyme of two to three flowers, on the current season growth in the axil of the leaves. About one month is required from flower bud differentiation to complete development upto calyx cracking stage (Prakash, 1976). However, Braganza (1990) observed this period from 45-51 days. Sehgal and Singh (1967) reported that the calyx splits any time between 13- 26 hours before flower opening.

The flower structure of guava comprises of five lobed calyx, 6-10 petals (corolla arranged in two whorls), androecium 160 to 400, bilobed anthers. Gynoecium consists of inferior ovary, syncarpus, with axile placentation. In India there are three flowering seasons in guava i.e Ambe bahar, Mrig bahar, and Hastha bahar.

Peak time of anthesis is between 5.00 - 6.30 AM in most of the varieties of guava. (Balasubramanyam, 1959, Arvindakshan, 1960). The dehiscence of anthers starts 15-30 minutes after anthesis and continue for two hours. The pollen fertility is high in almost all the cultivars. The pollen fertility is 78% and 91% in Allahabad Round and Lucknow Safeda, respectively (Balasubramanyam, 1959). Although, the guava is self-pollinated crop but to some extent cross pollination takes place by insect.

9.6. Inheritance pattern

- Bold seed is found to be dominant over soft seed and governed monogenically.
- Red flesh colour is dominant to white pulp colour and also governed monogenically.
- Red fleshed cultivars are supposed to be heterozygous.
- There is linkage between red flesh colour and bold seed size.
- Triploidy and some other genetic factors are responsible for female sterility (Subramanyam and Iyer, 1982).

9.7. Breeding methods and achievements

9.7.1 Introduction

There are number of varieties and species introduced from different countries from time to time for specific purposes.

Examples

S. No	Name of the cultivars/species	Name of the country
1.	Beaumont and Indonesian seedless	Australia
2.	Acerapera	Brazil
3.	Verdie USA	USA
4.	*Psidium sartorianum*	USA
5.	*Psidium littorale var. longipes*	USA
6.	*Psidium longipes*	USA

(*Source*: Singh and Rana, 1993)

9.7.2 Selection

Ganesh Khand Fruit Research Station, Pune started the improvement work in guava for the first time in the country in 1907. The seeds of guava were collected from Lucknow, Sindh, Nasik, Kothrud and Dharwad. From these seeds, about 600 seedlings were raised and evaluated for fruit yield and quality. (Cheema and Deshmukh, 1927). One strain from open pollinated seedlings of Allahabad Safeda collected from Lucknow was selected and released as Lucknow-49, which is also known as Sardar guava.

At Horticulture Research Station Saharanpur, the work was initiated by Singh (1953) resulted in a superior selection S-1 having good fruit shape, less number of seeds, sweet in taste and high yield (Singh, 1959). Twelve strains were collected from Aurangabad and Bhir districts of Marathwada, out of which ABD-3, BHR-3 and BHR-5 were observed to be superior (Thonte and Chakrawar,1981). At IIHR, Bangalore, out of 200 open pollinated seedlings of Allahabad Safeda collected from Uttar Pradesh, seedling Selection-8 was found to be promising. The plant was dwarf in growth, prolific bearer with medium size fruit having white pulp and few soft seeds and excellent fruit quality (Iyer and Subramanyam, 1988).

Achievements

S. No.	Varieties	Important characters
1.	L-49	Developed at GFES, Pune, seedling selection of Allahabad Safeda, semi dwarf tree, high yielder and white flesh.
2.	Banarsi Surkha	It is selection from local red fleshed type, heavy bearer, large fruits, flesh soft and pink.
3.	CISHG-1	Developed at CISH Lucknow. Fruit skin colour is deep red, TSS 15° Brix, soft seeds.
4.	Bangalore Local	It is local selection, with white flesh and soft seeds, fruit is large.
5.	Arka Mridula (Sel-8)	Developed at IIHR, Bangalore, it is selection from Allahabad seedling, soft seeds and white flesh.
6.	Lalit (CISHG-3)	Developed at CISH, Lucknow,it is selection from Apple colour seedling, skin and flesh colour is pink with good acid sugar blend.
7.	Pant Prabhat	Seedling selection from GBPUAT, Pantnagar, prolific bearer, soft seed with good quality.

9.7.3 Hybridization

(i) *Intervarietal hybridization*

IARI, New Delhi

The work was started with the objective to evolve a cultivar with less seed content and more productivity. Since most of the cultivars of guava are diploid in nature except seedless guava i.e. triploid and shy bearer, the crosses were made between the triploid guava and Allahabad Safeda. From the hybrid seeds, 73 F1 were raised out of that 26 diploid, 09 trisomic, 05 double trisomic and

14 tetrasomic progenies were obtained which were distinct with respect to plant growth, leaf and fruit characters. From this aneuploidy breeding, 3 trisomic plants had dwarf stature and normal size and shape of fruits with less number of seeds. (Majumder and Mukherjee 1972a, 1972b, and Mukherjee, 1977).

HETC, Basti

The objective of breeding was to obtain a variety with higher vitamin C, attractive skin and flesh colour. For this purpose, crosses were made between Seedless x Allahabad Safeda, Seedless x L-49, Allahabad Safeda x Patillo, L-49 x Patillo, Apple colour x Kothkhurd and Apple colour x Red Flesh (Anon., 1982).

IIHR, Bangalore

Hybrid Arka Amulya (Seedless x Allahabad Safeda) has been released with soft seeds and good keeping quality. Hybrid16-1 (Apple Colour x Allahabad Safeda) has been developed. The plants are semi vigorous, moderate yielding, fruit skin bright red, with few seeds, high TSS and good keeping quality (Subramanyam and Iyer, 1993).

FRS, Sangareddy

Two hybrids Safed Jam (Allahabad Safeda x Kohir) and Kohir Safeda (Kohir x Allahabad Safeda) have been released for commercial cultivation in Telengana and Rayalseema region (Subramanyam and Iyer, 1993).

Bihar Agricultural University, Sabour

Many cross combinations viz. Chittidar x Apple Colour, Chittidar x Sardar, Chittidar x Red Flesh, Chittidar x Allahabad Safeda, Sardar x Allahabad Safeda, Sardar x Chittidar, Apple Colour x Sardar, Allahabad Safeda x Apple Colour and Allahabad Safeda x Sardar were used in hybridization programme and 210 F1 seedlings were raised for evaluation purpose (Subramanyam and Iyer, 1993).

HAU, Hisar

Hisar Safeda (Allahabad Safeda x Seedless) – It is having upright tree habit, compact crown, TSS 13.4%, acidity 0.38%, average fruit weight 92g and white flesh

Hisar Surkha - (Apple Colour x Bararsi Surkha) the tree crown is broad to compact, TSS,13.6%, acidity 0.46%, pink flesh and average fruit weight of 86g are released for commercial cultivation.

CISH, Lucknow

H-21 and H-136 have been isolated for red pulp and soft seeds with high TSS.

FRS, Ananjharajupet (Andhra Pradesh)

Six hybrids are under evaluation for yield and yield attributes i.e. H-1 (Red Fleshed x Saharanpur Seedless), H-2 (Smooth Green x Nagpur Seedless), H-3 (Allahabad Safeda x Red Fleshed), H-4 (Smooth Green x Saharanpur Seedless), H-5 (Red Fleshed x Nagpur Seedless) and H-6 (Banarsi x Allahabad Safeda).

9.7.4 Breeding for wilt resistance

Resistant species of guava can be utilized for imparting the wilt resistant character. It was observed that *Psidium guajava* and *Psidium chinensis* are compatible. However, cross between *Psidium guajava* and *Psidium molle* was incompatible but reciprocal cross was a compatible combination (Subramanyam and Iyer, 1982). Work at CISH, Lucknow has shown that Chittidar, Portugal, Seedless and Spear Acid are tolerant to wilt. The hybrids of *Psidium molle* x *Psidium guajava* are in progress.

9.7.5 Mutation

It can be a good tool for altering some specific characters. Ram Kumar (1975), induced tetraploidy in cultivar Allahabad Safeda by treating the shoot tip with 0.1% aqueous solution of colchicine.

9.7.6 Biotechnological tools

Micropropagation, haploid production, somatic embryogenesis, protoplast fusion, genetic engineering etc can be utilized in improvement of guava.

9.8 Classification of guava cultivars

Although the guava is mainly self-pollinated but cross pollination is also common which results in the great variation in seedling population. The nomenclature of guava cultivars grown in India has not yet been well established. Most of the cultivars of guava are named on the basis of shape, colour, and place etc (Pathak and Ojha, 1993). The general classification of guava cultivars is as follows:

a. According to presence of seeds
   - (i) Seeded cultivars - L-49, Allahabad Safeda, Lalit, Arka Mridula .
   - (ii) Seedless cultivars - Triploid guava, Nagpur Seedless, Saharanpur Seedless.

b. On the basis of shape - Pear shape, Karela.

c. On the basis of place - Allahabad Safeda, Nasik, Dharwar, Behat Coconut.

d. On the basis of colour - Chittidar, Red flesh, Apple colour.

e. On the basis of surface of fruit - Dharidar.

f. According to breeding method

   (i) Selection - L-49, Lalit, Arka Mridula .
   (ii) Hybridization - Safed Jam, Kohir Safed, Arka Amulya .

g. On the basis of state growing area

   (i) Andhra Pradesh - Smooth White, Smooth Green, Seedless, L-49, Hafsi, Chittidar, Banarsi, Anakapalli, Allahabad Safeda.
   (ii) Assam - Am-Sophri, Madhuri-Am, Safrior Payene.
   (iii) Bihar - Allahabad Safeda, Chittidar, Hafsi (Red Fleshed), Harijha, Seedless.
   (iv) Maharashtra - Dharwar, Dholka, Kothkhurd, L-24, L-49, Nasik, Seedless, Sindh.
   (v) Gujarat - Sindh, L-49, Kothkhurd, Dholka.
   (vi) Tamil Nadu - Anakapalli, Banarsi, Bangalore, Chittidar, Hafsi, Nagpur Seedless, Smooth Green.
   (vii) Uttar Pradesh - Allahabad Safeda, Apple Colour, Chittidar, Red Fleshed, Banarsi Surkha, Lalit, L-49, Mirzapuri, Seedless.
   (viii) West Bengal - Barimpur, Allahabad Safeda, Apple Colour, L-49, Chittidar.

9.8.1 Characteristics of important species and cultivars

Description of some important species are as under

Psidium guineense

This is also known as the Guinea guava or Brazilian guava. The plants are like shrub or small tree. The leaves are green in colour, broad, oblong-oval, acute or obtuse, 8-12 cm long with lower surface pubescent. Red hairs are found on the mid veins. Generally, 8-10 pairs of primary veins are found in one leaf, corolla is of creamy colour with 5-7 in number, stamens are 150-185 in number, stigma is medium in size with 3-8 chamber in ovary. Flowering comes throughout the year. Average fruit weight is about to 6 g and vitamin C content is about 24.7 mg/100g of pulp with poor quality fruits.

Psidium montanum

Plants are just like shrub attain a height of about 1.5 m, flat round branches. It is found in mountains of Jamaica. Fruits are round with very poor quality (Bailey,1919).

Psidium fredrichsthalianum

It is known as Chinese guava. Plants are tall (7-11 m height), fruits are small and globose in shape with high acid content. It can be used for jelly making. Plants are tolerant to guava wilt.

Psidium cattleianum

It is known as the Cattley guava or Strawberry guava. It is a shrub or small tree (3-6 m in height), fruits are small, deep scarlet in colour, and globose in shape. This species is more tolerant to low temperature than Psidium guajava.

Psidium cattleianum var.lucidum

Tree height is more than Cattley guava in Hawai Island. Height of plant is noticed up to 12m. Generally, it is propagated through seeds. Fruits are yellow in colour and used for jelly making.

Psidium molle

Tree is medium in height, leaves are green and oval in shape. Apex of leaf is pointed; lower part of leaves is velvety in appearance. Red hairs are found on the central veins. In one leaf 6-8 pairs of primary veins are found. Petals are 5-11, stamens are 196-239, stigma is long with big ovary of 3-5 chambers. Fruits are small in size, average fruit weight is 13g. It contains vitamin C about 70 mg / 100g of pulp (Pandey and Mishra, 1984).

Psidium pumilum

It is also known as Chinese guava. Tree is like pyramidal in shape, leaves are light in colour, small in size, non-pubescent, having 13-17 pairs of primary veins. Petals 7, smooth and creamy colour which drop immediately after anthesis, Stamens are 252-327 in number, small stigma with medium size of ovary having 4-5 chambers. It flowers twice in a year. It takes about 130 days for attaining the maturity of fruits. Average fruit weight is about 19g and average vitamin C content is 171mg/100g of pulp.

Psidium cujavilis

Growth characters and flowering habit of the plants is just like *Psidium guineense*. The size of fruit is small to medium, average weight is 30-50g, and sour in taste.

Psidium policarpum

The growth characters are similar to *Psidium guajava* except the shape of the fruits is pyriform. Average fruit weight is about 200-250g. Flavonoid patterns show close affinity between *P. guajava* and *P. molle* (Dass and Prakash, 1981). However, inspite of the morphological similarities in *P. molle* and *P. guineense*, they showed minute differences in flavonoid pattern.

9.8.2 Important varieties

Sardar guava (Syn. Sardar, L - 49 or Lucknow-49)

This cultivar was developed by Cheema and Deshmukh 1927, through the selection from Allahabad Safeda. It is a semi dwarf tree 2-3.5m tall, profuse branching type, somewhat flattened top, dense foliage, large leaf, elliptic-ovate to oblong in shape, apex obtuse, base round, fruit medium to large, average fruit weight is about 175g, skin colour primrose with occasional red spot on the skin, taste sweet and keeping quality excellent.

Allahabad Safeda

This is the famous cultivar of Uttar Pradesh. Plant is dwarf with compact, globose, round crown. It has profuse branching with dense foliage. Fruits are medium to big in size, skin smooth, flesh white soft and firm, roundish fruit shape, the average fruit weight is about 185g, contains approximately 300 to 315 viable seeds. Taste is sweet with excellent fruit quality. It takes about 60 to 75 days for fruit maturity, keeping quality of fruit is good.

Chittidar

This cultivar is similar to Allahabad Safeda but the fruits of this cultivar are highly red dotted on the surface. Plant is tall with round crown and spreading branches. The cultivar was developed in Allahabad region. Its leaves are large, elliptic-ovate, to oblong-elliptic in shape, apex acutely pointed with roundish base. Fruits are almost round, white flesh with smooth skin. Keeping quality of fruit is good.

Sebia (Syn. Apple colour)

This cultivar is famous for its skin colour, but not commercially cultivated because of poor yield. Tree is medium in height, with broad crown and spreading growth habit. This cultivar has also originated in Allahabad district of U.P. Fruits are pink in colour with small dots on the surface. Shape is spherical, sweet in taste with good keeping quality. This can be used as a donor parent for fruit colour in breeding programme.

Red Fleshed (Syn. Red Fleshed Local, Hafsi)

Tree is medium tall with spreading branches. The flesh of the fruit is red with good taste and flavour. The number of seeds are less, fruit size is medium, and fruit shape is spherical to roundish ovate. The keeping quality of fruit is poor.

Behat Coconut

The cultivar is developed from the Behat place of Distt. Saharanpur Uttar Pradesh. Tree is medium in height with spreading growth habit. It has profuse branching habit. The fruits are round, aureolin in colour, taste is sweet and keeping quality is good.

Supreme Mild Fleshed

This cultivar is best suitable for jelly making because of good amount of pectin content. The fruits are acidic in taste, medium in size and oval in shape.

Seedless

In these group two cultivars are famous i.e. Nagpur Seedless and Saharanpur Seedless, there may be possibility that both the cultivars are same. In this cultivar two types of fruits are found i.e. complete seedless and partial seedless. The trees are tall with upright branches. Its fruits are oblong to globose in shape, straw yellow in colour with thick and creamy white flesh.

Anakapalli

Fruits are medium sized with red pulp having 393 mg / 100g of pulp vitamin C. The number of seeds in fruit are more.

Pear Shaped

Tree is tall, erect with upright growth habit. It has medium branching habit. Fruits are pyriform in shape and straw yellow in colour. Keeping quality of the fruit is good.

Baruipur

Tree is medium to tall, with spreading growth habit. Its crown is broad and compact. Fruit is roundish shape, yellow in colour with white flesh. Keeping quality of the fruit is medium. This is the most important commercial cultivar of West Bengal.

Harijha

Plants are prolific bearer; tree is medium in vigour and sparse branching habit. Fruits are round in shape, yellow in colour, sweet in taste and excellent in keeping quality.

Dhawal

It is released by CISH, Lucknow. It is heavy bearer (about 20% higher than Allahabad safeda), fruits develop light yellow colour on ripening, soft seeded and white pulp, taste sweet with muskiness. Yield is about 384q/ha. Cropping duration is 100 to 120 days.

Lalima

It is released by CISH, Lucknow. The yield is about 340q/ha. Attractive crimson coloured fruits, proportion of coloured fruits are higher, good yield with higher shelf life.

Lalit-1999

Selection from half sib population of apple guava suitable both for fresh as well as processing purposes. It was released by Central Institute of Tropical Horticulture, Lucknow. It yields about 100kg/plant by 6 years of planting under improved cultural practices. Saffron yellow coloured fruits with red blush, pulp firm pink, good blend of sugar and acid. It has 250mg/ 100g vitamin C content.

Shweta-2005

Selection from half sib population of apple guava variety with globose fruits, medium size, weight 225g, creamy white exocarp, with snow white flesh. High TSS (12.5 – 13.2%) and vitamin C (300mg /100g pulp) with good keeping quality. High yield, good quality and attractive fruit appearance, medium size, creamy white exocarp with red spots or blush. It yields 90kg/ plant at the age of 6 years.

Pusa Srijan

It was released by IARI, New Delhi on 2004. Extremely dwarf rootstock, hence, recommended for high density planting.

Arka Amulya

It is a progeny from the cross Allahabad Safeda x Triploid. Plants are medium in vigor and spreading type. Fruits are round in shape. Skin is smooth and yellow in colour. Fruits on an average weigh about 180 - 200 g, flesh is white in colour and firm. TSS is around 12 Brix, soft seeded, weight of 100 seeds is 1.80 g. Keeping quality is good.

Arka Kiran

It is a pink pulp variety with average fruit weight of 200-220g with medium soft seeds (9 kg/cm^2). Fruits have TSS of 11-12^0 brix and lycopene content of 7.14 mg/100g. Crop will come to harvest after two years of planting. Economic yield starts after five years with average fruit yield of 20 t/acre in a spacing of 4m x 3m.

Arka Mridula

It is a selection from open pollinated seedlings of Allahabad Safeda. Plants are semi-vigorous in nature and spreading. Fruits are round in shape and weigh about 180 g. Skin is yellow in colour and smooth. Flesh is white in colour. TSS is around 12 & Brix. Fruits are soft seeded, weight of 100 seeds is 1.60 g. Keeping quality is good. The pectin content is 1.041%. It is good for jelly making.

Arka Poorna

This variety is developed from the progeny selection of the cross Purple local X Allahabad Safeda. The plants are semi- vigorous in growth habit with prolific bearing, hence suitable for medium to high density planting. The fruits are round, medium to big in size (200-230g) with smooth, shiny pericarp. The pulp is firm, white with thick outer rind, good flavour, TSS (10-12 ^{0}B) , ascorbic acid (190-198 mg/100g) , medium soft seeds (10 .0 to 12.0 kg/cm^2) and keeping quality.

Arka Rashmi

This variety is hybrid between Kamsari and Purple Local. It is heavy bearer and precocious than its parents. Fuits are round in shape, weighs about 180-200 g, dark pink flesh, seeds are soft, TSS 12° Brix. Fruits are very tasty and rich in Lycopene and ascorbic acid is about 235 mg/100g pulp. The yield is 30-35 tonnes/ ha.

CISHG-1

Fruits are deep red coloured with soft seeds, attractive shape, high TSS (15 °Brix) and long shelf life. Fruits of this selection recorded higher anthocyanin content. The spatial distribution of the pigments in various portions of the same fruits has shown less variation in this selection.

CISHG-5

A selection from open pollinated seedling guava population raised at Lucknow has been found promising (fruit weight 190 g & TSS 13.7 °Brix). It has attractive crimson colour fruits, higher proportion of coloured fruit, good yield and responsiveness to pruning.

Safed Jam and Kohir Safeda are some other promising varieties of guava.

9.9 References

Anonymous, (1982) Report of Fruit Research Workshop on Sub-tropical and Temperate fruits, Nagpur, p.119.

Aravindakshan, M. (1960) Studies on certain aspects of growth and flowering in some varieties of guava. M.Sc. (Ag.) Thesis, Agriculture College and Training Institute, Coimbatore.

Atchinson, E. (1947) Chromosome number in the Myrtaceae. *Amer. J. Bot.*, **34**: 159-164.

Bailey, L.H. (1919) Standard Encyclopedia of Horticulture. Mac. Millan, New York, P-2, pp. 2847-2849.

Balasubramanyam, V.R. (1959) Studies on blossom biology of guava (*P. guajava* L.). *Indian J. Hort.*, **16**: 69-75.

Braganza, M.A. (1990) Floral biology studies and varietal evaluation in the genus *Psidium*. M.Sc. (Ag.) Thesis, University of Agricultural Sciences, Bangalore.

Cheema, G.S. and Deshmukh, G.B. (1927) Culture of guava and its improvement by selection in western India. *Bombay Dept. Agric. Bull.*, **148**: 20-30.

Dass, H.C. and Prakash, D. (1981) Phylogenetic affinities in *Psidium* species as studied by flavonoid patterns. National Symposium on Tropical and Subtropical Fruit Crops, Bangalore. pp-15.

De Candolle, A.P. (1904) Origin of cultivated plants. Kegal Paul, London.

Hayes, W.B. (1970) Fruit growing in India. Kitabistan Allahabad.

Iyer, C.P.A. and Subramanian, T.R. (1987) Genetic resources activities concerning tropical fruit plants. In: Plant genetic resources, Indian perspective (Eds. R.S. Paroda, R.K. Arora and K.P.S. Chandel), NBPGR, New Delhi, pp.310- 319.

Iyer, C.P.A. and Subramanyam, M.D. (1988) IIHR Selection – 8, an important guava cultivar. *South Indian Hort.*, **36**(5): 258-259.

Majumder, P.K. and Mukherjee, S.K. (1972b). Aneuploidy in guava II. Occurrence of trisomics, tetrasomics and higher aneuploid in the progeny of triploid. *Nucleus*, **13**:42-47.

Majumder, P.K. and Mukherjee, S.K. (1972a) Aneuploidy in guava. Mechanism of variation in number of chromosomes. *Cytologia*, **37**: 541-548.

Mitra, S.K. and Bose, T.K. (1990) Guava. In: Fruits Tropical and Subtropical (Eds. T.K. Bose and S.K. Mitra), Naya Prokash Calcutta, pp.278-303.

Mukherjee, S.K (1977) Improvement of mango, grape and guava. In: Fruit breeding in India (Ed.G.S. Nijjar), Oxford and IHB Publication, New Delhi, pp 15-20.

Pandey, R.M. and Mishra, K.K. (1984) Guava. ICAR Publication.

Pathak, R.K. and Ojha, C.M. (1993) Genetic resources of guava. In: Advances in Horticulture Vol. I (Eds K.L. Chadha and O.P. Pareek.), Malhotra Publishing House, New Delhi, pp 143-147.

Prakash, N.A. (1976) Studies on growth and fruiting in Sardar guava (*P. guajava* L.). M.Sc. (Ag.) Thesis, Univ. of Agric. Sciences, Dharwad.

Ram Kumar (1975) Inducing polyploidy and cytological studies in guava. *Indian J. Hort.*, **32** (3-4): 128-130.

Sehgal, O.P. and Singh, R. (1967) Studies on the blossom biology of guava (*Psidium guajava* L.) Flowering season, flowering habit, floral bud development anthesis and dehiscence. *Indian J. Hort.*, **24**: 118-126.

Singh, B. P. and Rana, R. S. (1993). Promising fruit introductions. In: Advances in Horticulture Vol. I (Eds. K. L. Chadha and O. P. Pareek), Malhotra Publishing House, New Delhi, pp 43-66.

Singh, L.B. (1959). S – 1 A new promising selection of guava (*P. guajava*). Annual Report, Fruit Research Station, Saharanpur. pp. 58-60.

Singh, R.L. (1953). Annual Report. Fruit Research Station, Saharanpur.

Subramanyam, M.D. and Iyer, C.P.A. (1982) Report. Fruit Research Workshop, Nagpur, pp. 117-118.

Subramanyam, M.D. and Iyer, C.P.A. (1993) Improvement of guava. In: Advance in Horticulture Vol. I (Eds. K.L. Chadha and O.P. Pareek), Malhotra Publishing House, New Delhi, pp.337-347.

Thonte, G.T. and Chakrawar, V.R. (1981) The variability and correlation studies of guava strains. National Symposium on Subtropical Fruit Crops, Bangalore, pp. 17.

Yadav, I.S. (1990) Germplasm collection of mango, grape, guava and litchi in India. AICRP on Subtropical fruits. CIHNP, Lucknow.

10

Breeding of Ber

Botanical name: *Ziziphus mauritiana* L.

Family: Rhamnaceae

Chromosome number: 2n=4x=48

Scientific classification

Kingdom	:	Plantae- Plants
Division	:	Magnoliophyta–Flowering plants
Class	:	Magnoliopsida – Dicotyledons
Subclass	:	Rosidae
Order	:	Rhamnales
Family	:	Rhamnaceae – Buckthorn family
Genus	:	*Ziziphus*
Species	:	*mauritiana Lam.*

Ber (chinese date, chinese fig or bore) is an important minor fruit of India. It is hardy fruit tree cultivated all over India and also known as poor man's fruit. It is most widely cultivated in Punjab, Haryana, Rajasthan, Uttar Pardesh, Madhya Pradesh, Bihar, Gujarat, Maharashtra etc. Ber can do well even under marginal growing conditions and provides quality yields at low cost (Shukla, 1996). Fruits are mostly eaten fresh but the other forms, such as dried, candied, squash etc can also be prepared from ber. Stem bark, root and leaves have some medicinal values. Leaves of ber are used as fodder in dry regions. Ber tree (*Ziziphus xylopyrus*) is a host plant for rearing the *Tachardia laccad* a lac insect (Sharma and Kore, 1990).

10.1. Centre of diversity

Ziziphus mauritiana L. is supposed to be native of India, while Ziziphus jujuba Mill. is native to China. Liu and Cheng (1995) reported that Indo-Malaysia region is the centre of both evolution and distribution of the genus Ziziphus. However, De Candolle (1886) stated that Myanmar (Burma) and India are the home of the ber.

Distribution

In India a number of ber cultivars have been developed largely by the growers through selection in different regions. Maximum variability of ber is observed in Rajasthan, Gujarat, Haryana, Punjab and Madhya Pradesh. In Gujarat, Mehsana, Anand, Panchmahal, Patan and Sabarkatha districts are having rich diversity of ber (Shukla et al., 2003). It is also reported that the smaller fruits of Umran is known as Chameli in Gujarat. Further, there are many unexplored areas of ber variability from where promising types could be collected for further characterization and evaluation. Historically central part of the country is supposed to be the richest area in the wild ber germplasm.

10.2. Germplasm resources

Since ber is a cross pollinated and heterozygous fruit crop, therefore, growing of the plants through seeds may results the maximum variability. The collection and evaluation of the ber germplasm has been in progress at various places. More than 320 genotypes of ber have been conserved in field gene bank of CIAH, Bikaner (Anon., 2001). However, at Hisar and Bawal in Haryana, Bangalore in Karnataka, Udaipur, Jobner and Jodhpur in Rajasthan, Rahuri in Maharashtra, Bahadurgarh and Ludhiana in Punjab, Varanasi and Faizabad in Uttar Pradesh, S.K. Nagar and Godhra in Gujarat and Arupukottai in Tamil Nadu are also maintaining the field gene banks of ber.

10.3. Objectives

- To breed early maturing varieties.
- Evolved drought resistant varieties.
- To develop superior quality, high yielding varieties.
- To develop varieties resistant to fruit fly and powdery mildew.
- Development of dwarf statured less thorny crop suitable for high density planting.

10.4. Botany

Genus Ziziphus is having more than 600 species (Bailey, 1947). Maheshwari and Singh (1965) recognized six economically important species in India namely *Z. mauritiana* L., *Z nummularia* (Burm. f.) wight. and Arn, *Z. oenoplia* mill, *Z.*

rugosa Lam and *Z. xylocarpus* wild. It is supposed that cultivated Jujube is derived from the Z. sativa Gaertn. (Pathak and Pathak, 1993). Ber cultivars, in general are polyploid in nature. Cytological studies have revealed that Z. lotus and *Z. jujuba* (Chinese jujube) are diploids, *Z. mauritiana* and *Z. oenoplia* are tetaploids and *Z. rotundifolia* is both tetraploid and hexaploid. Nehra et al. (1983 and 1984) reported that cultivar Illaichi is an octaploid (2n=96) and Umran is a tetraploid (2n=48). Khoshoo and Singh (1963) reported that 33 cultivars of ber namely Banarsi, Banarsi Pewandi, Chhuhara, Dandan, Desi Alwar, Gola, Gorwa, Kaithli, Kala Gola, Katha Bombay, Kathaphal, Laddu, Mirchia, Nalagarhi, Narikeli, Nazuk, Noki, Pathani, Sandhura Narnaul, Sanaur, Sanaur -2, Sanaur -3, Sanaur- 4, Sanaur -6, Seo, Safeda Selected, 28/1, Umran, Wilayti, ZG 2, ZG-3 and Hoshiarpur were tetraploid (2n=4x=48), cultivars Illaichi and Mithianwali were octaploids (2n=8x=96), and one cultivar Kalianwali was pentaploid (2n=5x=60). Further, Pareek (1998) reported that Umran and Chonchal are tetraploid while Gola rootstock and Boradi is octaploid. According to Qu et al. (1986) out of 111 genotypes *Ziziphus sativa*, *Ziziphus spinosus*, *Ziziphus jujuba* Var. Tortuosa, *Ziziphus jujuba Var. logeniformis* were diploid (2n=2x=24) while Ziziphus jujuba Var. Zanhu-angla was triploid (2n=3x=36).

Chinese jujube is a small upright tree with glabrous bright green leaves, which sheds leaves during winter and starts vegetative growth late in the spring to avoid danger of frost. Flowering period is April to June and fruit matures in Autumn (August to October). While Indian jujube is evergreen, spreading tree having drooping branches. It sheds leaves in summer, August to October is flowering time and fruits start ripening during January to March.

Table 1: The important species of Ziziphus

Sl.No.	Name of the species	Distribution	Remarks
1	*Z. rugosa*	In India it is found in U.P. Bihar, M.P. and Western Peninsula, Burma and Ceylon	It produces edible fruits
2	*Z. nummularia*	Found in North West India and A.P.	Bark can be used for tanning, produces edible small size fruits, leaves used for fodder.
3	*Z. oenoplia*	North and peninsular India	Bark can be used for tanning and it produces edible fruits.
4	*Z. sativa*	Punjab and Bengal	Diploid in nature, produces edible fruits.
5	*Z. xylopyrus*	M.P.	Host for Lac Insect, can be used for tanning

Contd.

6	*Z. heloia*	South Africa	-
7	*Z. mucronata*	South Africa	-
8	*Z. joazeiro*	Brazil	-
9	*Z. mistol*	Argentina	-
10	*Z. spina-christi*	Egypt, Syria and Palestine etc	-

10.5. Floral biology

In ber one to ten branchlets develop from the nodes of primary woody branches which bear leaves and flowers. Teaotia and Chauhan (1963) and Desai and Patil (1978) reported that flower buds appear both on mature and current season's growth. The flowers are born in the axil of leaves on the main as well as lateral shoots and the inflorescence is an axillary cyme (Vashishtha and Pareek, 1978). Ber flower buds in axillary cymes are borne mainly on the current season's growth.

The number of flowers per cluster has been reported to be 12 to 14 Vashishta and Pareek (1978), 16 by Teaotia and Chauhan (1963). There are various reports regarding the full development of flower. Vashishtha and Pareek (1978) observed that 17-20 days were required for the smallest bud (0.13 cm diameter) to attain the largest size (0.35 cm) a day before anthesis. Teaotia and Chauhan (1963) and Josan et al. (1980) reported that 21 days are required for development of flower passing through eight stage, the growth of first four stages are slower, whereas Desai et al. (1986) observed that flower bud development occurred in seven stages and took, on an average of 19.2, 19.3 and 19.85 days in Chhuhara, Sanaur 1 and Shamber cultivars, respectively. Godara (1980) reported that the number of days required for bud development varied between 12.7 and 30.3 days. Singh and Jindal (1982) from studies on Gola Gurgaon, Muria, Kaithali, Chhuhara and Umran reported that Umran had the highest number of hermaphrodite flowers (22.2%).

Teaotia and Chauhan, (1963) observed that the time of anthesis varied in different cultivars. It occurred between 7.30 and 8.30 hr. in cultivars Seb, Jogia,, Ponda, Aliganj and Illaichi and between 12.00 and 13.00 hr. in Gola and Mundia. In Banarsi Pewandi and Banarsi Karaka flowers opened at 12.00 hr. Desai and Patil (1978) found that in Umran and Illaichi, anthesis was completed between 5.30 and 6.45 hr. Singh et al. (1970) observed that in Banarsi Karaka anthesis started after 12.00 hr. and stopped at 15.00 hr. being maximum between 13.00 and 14.00 hr. The anthesis was followed by dehiscence with in one or two hours and continued for 4 to 6 hours.

The stigmatic receptivity has also been studied in different varieties. Vashishtha and Pareek (1978) observed that stigma receptivity was maximum on the day of anthesis in Gola, Mundia, Seb, Jogia, Ponda, Aliganj and Illaichi cultivars.

The flowers, pollinated 24 hours before anthesis completely failed to set fruits whereas, some success was obtained when pollination was done 24 hours after anthesis. Prasad (1964) observed that stigma becomes receptive 4 hrs. prior to anthesis and it remained at peak at the time of anthesis and continued up to 12 hours after wards.

10.6. Pollination

Generally, cross pollination is rule in ber. As the pollens of ber are sticky, they are not transferred by wind. During flowering time, a lot of insect activities are found in ber orchards. Prominent pollinating insect agents are honey bees, house flies and yellow wasp. The life of individual flower is very short and in an inflorescence many flowers remained unpollinated during their receptive period, further the unfertilised flowers drop down. It was also observed that branches of eastern side gave maximum set of 8.47 % compared with minimum set of 2.04% on northern side (Yamdagni et al., 1968). Teaotia and Chauhan (1963) reported that pollen size in Banarsi Pewandi and Thornless were 42μ x 24μ and 35μ x 20.3 μ, respectively. Desai et al. (1986) reported that yellow and sticky pollen grains remained in anther unless disturbed, and were 29.34, 26.99 and 30.35 μ in size in Chhuhara, Sanaur -1 and Shamber cultivars, respectively with viability of 79.95, 82.40 and 91.46%. Khoshoo and Singh (1963) observed pollen sterility in ber owing to polyploidy and it was up to 90.5 % in the octaploid cultivar Illaichi. Teaotia and Chauhan (1964) observed that Banarsi Pewandi, Thornless and Banarsi Karaka were reciprocally cross incompatible. Godara (1980) found that cultivars Banarsi Karaka, Illaichi, Kakrola Gola, Kaithali, Kathaphal, Mundia Murhara, Reshmi, Sandhura Narnaul, Safeda Selected and Umran were self incompatible. Illaichi x Kakrol Gola gave the highest fruit set, while Umran showed the best combining ability when used as a male or female parent.

10.7. Breeding methods and achievements

10.7.1 Selection

Most of the common cultivars are the result of selection made by local people in different regions. The promising cultivars under commercial cultivation are Gola, Seb, Banarsi Karaka, Banarsi Peondi, ZG.1, Sanour-1, 2, 3, 4, Katha, Umran, Mundia etc.

Gomakirti is clonal selection from the Umran which is 25 days early than Umran. Recently CIAH Sel-1 of ber is developed by the CIAH through selection from local materials collected from Bhusawar area of Rajasthan. This selection is more juicy, sweet and early bearer. It is resistant to fruit rot (Anon., 2002, Shukla et al., 2003).

10.7.2 Hybridization

At HAU Hisar 72 hybrids were raised which are under evaluation, work was also initiated at Jodhpur by making cross of Katha with Seb and Seb x Tikadi (Pareek, 2001). At the CIAH, Bikaner, CIAH hybrid 1 (Seb x Katha) was found promising, precocious prolific bearer, early maturity with good fruit set in at high temperature under arid environment (Shukla et al., 2003). For fruit fly resistance BCF1 (Seb x Tikadi F1 x Seb) is under evaluation (Anon., 2002).

10.7.3 Biotechnological tools

Use of DNA finger printing to identify gene for resistance to fruit fly, powdery mildew, salinity/ alkalinity, drought and high ascorbic acid content and their transfer in the recipient parent through somatic hybridization may be another approach. Standardization of callus culture and organogenesis from mature as well as juvenile plant parts of Ziziphus mauritiana (Mathur et al., 1993) and Ziziphus jujuba (Cheong and Kim, 1984) have been attempted and have been successful.

Goma Kirti: Goma Kirti is an early-maturing ber. It matures 3 weeks earlier than Umran. Its fruits have good keeping quality. It is high –yielding with 25.8-38.2% increase in Yield over the control. Its fruits have good keeping quality. The physiological loss in weight is just 12%, whereas it is 27% in early-maturing Gola over 3 days of storage. The shelf-life is also higher (6 days) as compared to Gola (3 days). Thus, it can fetch good price in distant market.

Thar Sevika: The variety is released by CIAH, Bikaner on 2006. Developed by the hybridization from a cross Seb x Katha. Thar Sevika is an early maturing variety. Average fruit yield is 30-32 Kg/tree.

Thar Bhubharaj: This variety was released by CIAH, Bikaner on 2007. A selection from local material of Bhusavar area of Bharatpur district of Rajasthan, having an average yield potential of 30-36 kg/tree. The fruits are very juicy, sweet with a TSS content of 22-23%.

Other important cultivars: Maharwali, Mehrun, Narikeli, Narma, Noki, Nimaj, Ponda, Pathan, Safed Rohatak, Sanaur-2, Sanaur-3, Tikadi, Thornless, Triloki, Vikas-2, ZG-3, Madhuri, Mirchia etc.

10.8. Characteristics of important cultivars of ber

Cultivar	Growth habit	Flowering period		Fruit set period		Time of anthesis		Fruit characters	
		Initiation	Peak	End	Initiation	Peak	End		
1	2	3	4	5	6	7	8	9	10
Ajmeri	Semi-erect	Aug.	Sept.	Oct.	Early Sept.	Late Sept.	Late Oct.	1 to 2.30 PM	Late in maturity, TSS 18^0 Brix, Fruit weight 32g Fruit size 4.7x3.5 cm, fruit shape oval, Styler end round.
Akhorata	Sprea-ding	End of Aug.	Mid Sept.	Late Oct.	Early Sept.	End Sept.	End of Oct.	1 to 2.30 PM	Mid season cultivar, fruit weight 20g, fruit size 3.4 x 3.4 cm, TSS 20^0 Brix, fruit shape obovate, styler end flat.
Aliganj	Sprea-ding	End of Aug.	First week of Sept.	End of Oct.	First week of Sept.	Mid of Sept.	End of Oct.	1 to 2.30 PM	Mid season crop, fruit weight 21 g, fruit size 3.8x2.9 cm, TSS 17^0 Brix , Fruit obovate, styler end round with depression.
Bhagwadi	Sprea-ding	End of Aug.	Second week of Sept.	Mid of Oct.	First week of Sept.	Mid of Sept.	Mid of Oct.	1 to 2.30 PM	Mid season cultivar, fruit size 3.68x3.15 cm, fruit weight about 20g, TSS 15^0 Brix, fruit oval styler end round.

Contd.

Bahad-urgarh	Semi-erect	End of Aug.	Mid of Sept.	End of Oct.	Second week of Sept.	End of Sept.	End of Oct.	1 to 2.30 PM	Mid season cultivar, fruit size 3.93 x 2.70 cm, fruit weight 17g, TSS 23^0 Brix.
Banarsi Karaka	Erect	End of Aug.	First week of Sept.	Mid Oct.	First week of Sept.	End of Sept.	Mid of Oct.	1 to 2.30 PM	Mid season cultivar, fruit size 5.0 x 3.0 cm, TSS 17^0 Brix, fruit weight 28 g, fruit oval, styler end obliquely pointed.
Banarsi Pebandi	Semi- spreading habit	Mid Aug.	Early Sept.	First week of Oct.	End of Aug.	Mid Sept.	First week of Oct.	1 to 2.30 PM	Mid season cultivar, oblong fruit, styler end obliquely pointed necked, fruit weight 35 to 40 g, fruit size 5.5 x 3.8 cm to 6.0 x 4.0 cm, TSS 16^0 Brix.
Chhuhara	Erect	End of Aug.	Mid Sept.	End of Oct.	First week of Sept.	End of Sept.	End of Oct.	7 to 8 AM	Mid season cultivar, oval fruit, styler end round, stem end broad, fruit weight 22 g, fruit size 4.5 x 2.9 cm, TSS 23^0 Brix.
Chonchal	Semi- erect	Early Sept.	Mid Sept.	End of Oct.	End of Sept.	Mid of Oct.	End of Oct.	7.0 to 8.0 AM	Mid season cultivar, fruit oblong, fruit weight 25 g, fruit size 5.4 x 3.3 cm, TSS 20^0 Brix.

Contd.

Dandan	Semi- erect	End of Aug.	Mid Sept.	End Oct.	Mid Sept.	Early Oct.	End of Oct.	7.0 to 8.0 AM	Mid season cultivar, fruit oblong, styler end bluntly tapering with ridges, fruit size 5.65 x 3.45 cm fruit weight 27 g, TSS 17^0Brix.
Gobind-garh Selection	Erect	End of Aug.	Mid Sept.	Mid Oct.	First week of Sept.	Mid Sept.	Early Oct.	1 to 2.30 PM	Mid season cultivar, fruit oblong, styler end obliquely beaked, fruit size 4.2 x 2.5 cm, fruit weight 18 g, TSS 19^0 Brix.
Gola	Sprea-ding	End of Aug.	Early Sept.	End of Oct.	Early Sept.	Mid Sept. to first week of Oct.	End of Oct.	12 to 2.30 PM	Early cultivar, fruit round, styler end flat, stem end round, fruit size 3.8 x 3.6 cm, fruit weight 28 g, TSS 22^0 Brix.
Goma Kirti	Semi- erect	End of Aug.	Mid Sept.	Mid Oct.	End of Aug.	First week of Sept.	Mid Oct.	1 to 2.30 PM	Maturity is early than Umran, Fruit oval TSS, 18^0 Brix, fruit weight 35 g.
Illaichi	Erect	End of Aug.	Mid Sept.	First week of Nov.	Mid of Sept.	Early Oct.	First week of Nov.	7.0 to 8.0 AM	Octaploid ber, mid season cultivar, fruit oblate, styler end flat ridged, stem end round, fruit size small (2.1 x 2.2 cm), fruit weight 8.5 g, TSS 20^0 Brix.

Contd.

Jhajjar selection	Sprea-ding	Early Aug.	Early Sept.	End of Oct.	End of Aug.	Mid Sept.	End of Oct.	7.0 to 8.0 AM	Mid season cultivar, fruit obovate, styler and stem end round, fruit weight 21 g, fruit size 3.9 x 3.1 cm, TSS 17^0 Brix.
Kaithali	Erect	First week of Sept.	End of Sept.	First week of Nov.	Mid Sept.	Mid Oct.	First week of Nov.	1 to 2.30 PM	Mid season cultivar, fruit is oblong, styler end round, stem end shallow, mid season cultivar, fruit size 4.8 x 3.5 cm, fruit weight 30 to 35 g, TSS 18^0 Brix.
Katha	Semi- erect	Mid Aug.	Mid Sept.	End of Oct.	First week of Sept.	End of Sept.	End of Oct.	1 to 2.30 PM	Late cultivar, oval fruit, styler end round, stem end round and riged, fruit weight 39 g, size 5.2 x 4.3 cm, TSS19^0 Brix.
Kathaphal	Sprea-ding	End of Aug.	Mid Sept.	End of Oct.	First week of Sept.	End of Sept.	End of Oct.	1 to 2.30 PM	Fruit obovate, styler end round, stem end shallow, fruit size 3.2 x 2.8cm, fruit weight 15 g, TSS17^0 Brix, mid season cultivar.

Contd.

Mundia	Semi-erect	End of Aug.	End of Sept.	First week of Nov.	First week of Sept.	End of Sept.	First week of Nov.	1 to 2.30 PM	Fruit oblong, styler end round, stem end narrow with shallow notch, early in maturity, fruit weight 40 g, fruit size 5.8 x 3.6 cm, TSS 22^0 Brix.
Seb	Erect	End of July	Mid Aug.	End of Oct.	Mid Aug.	Mid of Sept.	End of Oct.	7.0 to 8.0 AM	Mid season cultivar, fruit is round, fruit size 3.8 x 3.4 cm, fruit weight 29 g TSS 17^0 Brix.
Umran	Semi-erect	End of Aug.	Mid Sept.	End of Oct.	First week of Sept.	End of Sept.	End of Oct.	1 to 2.30 PM	Late season cultivar, fruit shape oval, styler round, stem end round and ridged, fruit size 4.8 x 3.5 cm, fruit weight 45 g, TSS 19^0 Brix.

10.9. References

Anonymous, (2001) Annual Report. CIAH, Bikaner.

Anonymous, (2002) Annual Report. CIAH, Bikaner.

Bailey, L.H. (1947) The Standard Cyclopaedia of Horticulture. Macmillan and Co., New York, pp. 3547-3548.

Cheong, S.T. and Kim, J.S. (1984) Characteristics of the fruit as affected by the developmental stages, embryo germination and in vitro propagation of seedling *Ziziphus jujuba* Mill. *J.Kotean Soc. Hort. Sci.,* **25** (3) :241-249.

De Candolle, A. (1886) Origin of cultivated plants. Bibliotheque Scientifique Internationale, 43, Paris.

Desai, U.T. and Patil, A.V. (1978) Studies on floral biology of ber (*Ziziphus mauritiana* Lamk.) cultivars Umran and Illaichi. *Udyanika* **2**:26-32.

Desai, U.T., Randhawa, D.B. and Warhal, K.N. (1986) Floral biology of ber. *J. Maharashtra, Agric. Univ*., **11**:76.

Godara, N.R. (1980) Studies on floral biology and compatibility behaviour in ber (*Ziziphus mauritiana* Lamk.) Ph.D. Thesis. Haryana Agricultural University, Hisar.

Josan, J.S., Jawanda, J. S., Bal, J.S. and Singh, R. (1980) Studies on the floral biology of ber. Flowering habit, flower bud development, time and duration of flowering, floral morphology, time of anthesis and dehiscence. *Punjab Hort J*., **20**:156-161.

Khoshoo, T.N. and Singh, N. (1963) Cytology for North Western Indian trees. *Ziziphus jujuba* and *Z. rotundifolia. Silvae Genetica*, 12 Heft :141-180.

Liu, M. J. and Cheng, C.Y. (1995) A taxonomic study of the genus *Ziziphus*. *Acta Hort.*, **390**

Maheshwari, P. and Singh, U. (1965) Dictionary of economic plants in India. ICAR, New Delhi.

Mathur, N., Ramawat, K.G. and Sonie, K.C. (1993) Plantlet regeneration from seedling explant of *Ziziphus* and silver nitrate and nutrient requirment for callus and morphogenesis. *Gartenbavwissenschaft,* **58**(6):255-260.

Nehra, N.S., Chitkara, S.D. and Singh, K. (1984) Studies on morphological characters of some wild forms and cultivated varieties of ber. *Punjab Hort J.*, **24** (1/4): 49-59.

Nehra, N.S., Sareen, P.K. and Chitkara, S.D. (1983) Cytological studies in genus *Ziziphus*. *Cytologia,* **48**: 103-107.

Pareek, O.P. (1998) AICRP (arid fruits) In: 50 years of horticulture research in India (Eds. S.P. Ghosh, P.S. Bhatnagar and N.P. Sukumaran), Division of Horticulture,ICAR, New Delhi, pp 150-158.

Pareek, O.P. (2001) ber. International Centre for Underutilised Crops, Southampton, UK.

Pathak, R.K. and Pathak, R.A. (1993) Improvement of minor fruits. In: Advances in Horticulture Vol. I (Eds. K.L. Chadha and O.P. Pareek) Malhotra Publishing House, New Delhi, pp 412-416.

Prasad, A. (1964) Studies on floral biology of jujube (*Ziziphus mauritiana* Lamk) *Agra Univ. J. Res*., (Science) **13**:41-48.

Qu, Z.Z., Wang, Y.H., Lu, Z.R. and Yan, G.J. (1986) Studies on the chromosome number of chinese jujube. *Acta. Horti. Sini.* **13** (4):232-236.

Sharma, V.P. and Kore, V.N. (1990) Ber. In: Fruits Tropical and Subtropical (Eds.T.K. Bose and S.K. Mitra), Naya Prokash Publishing House Calcutta, pp 592-615.

Shukla, A.K. (1996) Efficacy of drip and mulching on young plantation of aonla + ber cropping model in sodic soil. M. Sc (Ag) Thesis NDUAT, Faizabad.

Shukla, A.K., Dhandar, D.G., Vashishtha, B.B. and Umamaheswari, C. (2003) New promising ber varieties from CIAH. *Indian Horticulture.*

Shukla, Anil Kumar, Shukla, Arun Kumar, Dhandar, D.G. and Patel J.I. (2003) Genetic variability of ber and aonla in Gujarat. National symposium on Agroforestry and Sustainable Production Jhansi 7-9 November2003.

Singh D., Bakhshi, J.C. and Singh, K. (1970) Flowering and fruiting behaviour of ber, variety Banarsi Karaka (*Z. mauritiana* Lamk.). *Punjab Hort. J.*, **10**:21-28.

Singh, D. and Jindal, P.C. (1982) Studies on flowering and sex ratio in some ber (*Z. mauritiana* Lamk) cultivars. *Hariyana Agric. Univ. J. Res.*, **12**:292-294.

Teaotia, S.S. and Chauhan, R.S. (1963) Flowering, pollination, fruit set and fruit drop studies in ber (*Ziziphus mauritiana* Lamk) 1. floral biology. *Punjab Hort. J.* **3** : 60-70.

Teaotia, S.S. and Chauhan, R.S. (1964) Flowering. pollination, fruit set and fruit drop studies in ber (*Z. mauritiana* Lamk) II pollination, fruit set , fruit development and fruit drop. *Indian J. Hort.*, **2**:40-45.

Vashishtha, B.B. and Pareek, O.P. (1978) Flower morphology, fruit set and fruit drop in some ber (*Ziziphus mauritiana* Lamk) cultivars. *Annals Arid Zone*, **18**:165-169.

Yamadgani, R., Bajpai, P.N. and Mishra, R.S. (1968) Pollination, fruit set and fruit development studies in ber (*Ziziphus mauritiana* Lamk) *Labdev. Sci. Tech.*, **6**:101-103.

11

Breeding of Aonla

Botanical name: *Emblica officinalis* G.

Family: Euphorbiaceae

Chromosome number: 2n=2x=18

Scientific classification

Kingdom	:	Plantae- Plants
Division	:	Magnoliophyta–Flowering plants
Class	:	Magnoliopsida – Dicotyledons
Subclass	:	Rosidae
Order	:	Euphorbiales
Family	:	Euphorbiaceae – Spurge family
Genus	:	*Emblica*
Species	:	*officinalis* G.

Aonla has been regarded as a sacred tree in India. The tree is worshipped as mother earth and is believed to nurture humankind because the fruits are very nourishing. It is also known as Amrit phal. Aonla is mainly recognized due to its high nutritive, medicinal and therapeutic properties. It contains high amount of vitamin C (500-1500 mg/100 g), it is also rich in pectin, iron, calcium and phosphorous. Dried fruits of aonla can be is used in curing hemorrhage, diarrhea, chronic dysentery, diabetes, jaundice, dyspepsia, cough etc. Recently, aonla has gained more popularity in arid and semi-arid region of the country because of its hardy nature, prolific bearing capacity and ability to grow under various adversities as a remunerative crop (Shukla et al.2002). It may be an important fruit in future due to its high productivity, suitability in wasteland particularly salt affected soil and sandy soil of arid ecosystem. Aonla is the main ingredient of Chavanprash and Triphala powder which is used for curing different abnormalities. The fruit contains a chemical substance gallic acid and leucoanthocyanin which have antioxidant property.

11.1. Centre of diversity

Aonla is a hardy fruit plant which can be successfully grown under various agroedaphic situations. It is said to be indigenous to tropical South-Eastern Asia particularly Central and Southern India (Firminger, 1947). It is also reported to be the native of India, Sri Lanka, Malaysia and China. Aonla thrives well throughout tropical India and met with wild or cultivated species in the region extending from the base of Himalaya to Sri Lanka and from Malaysia to South China.

11.2 Distribution

Aonla is naturally growing from India, Sri Lanka, Cuba, Puerto Rico, Hawaii, Florida, Iran, Iraq, Java, West Indies, Trinidad, Singapore, southern Thailand, Pakistan, Malaya and China and Panama Canal regions. However, In India its cultivation is more common and widely grown in Uttar Pradesh, Gujarat, Rajasthan, Madhya Pradesh, Karnataka and Tamil Nadu etc. The seedling aonla plants are commonly observed in Vindhyan hills, North-Eastern hills region (Garo, Khasi and Jayantia hills) and also in North-Western Himalayan region up to the elevation of 1350 m ASL.

11.3. Germplasm resources

In aonla, variability exists in seedling population in respect of fruit size, maturity period and yield per plant. In fact, seed germination is not a problem therefore, great genetic variability is being found for desirable economic traits in aonla growing areas of the country. Despite of cultivated species of Emblica (Phyllanthus), some other available species can be utilized in aonla breeding programme to infuse hardiness into the offspring for adverse habitat. In India, major center where germplasm resources are being maintained are CIAH, Bikaner, NDUAT, Faizabad, GAU, Anand and RBS College, Agra. At present 43 genotypes of aonla are conserved and maintained at national field repository at CIAH, Bikaner for further evaluation with respect to frost resistant, yield and quality parameters (Dhandar and Shukla, 2003). Recently, an exploration was conducted in parts of hill ecosystem of Himachal Pradesh, where it was found that there is ample scope for selection of frost tolerant types, although the fruit size was smaller but plants are more hardy than existing commercial varieties (Shukla and Dhandar, 2003). Further, Chomu, Pushkar valley and Badighati region of Rajasthan is hunting ground for aonla variability.

11.4 Breeding problems

- Since, aonla is highly heterozygous plant, therefore, large size of population is required for selection.

- It has long generation cycle i.e. 2-8 years, depending upon species and varieties.
- Lack of recombination
- Long juvenile phase prohibiting early assessment of strain.
- Presence of self-incompatibility.
- Frost Susceptibility.
- Lack of knowledge on inheritance pattern.

11.5. Objectives

- To breed varieties having wider geographic adaptability.
- To develop varieties suitable for export.
- To evolve coloured varieties based on market demand.
- To breed varieties resistant to frost.
- To develop varieties resistant against biotic (e.g. rust disease) and abiotic stresses (e.g. drought hardiness).
- Exploitation of available hybrid vigour (heterosis) for yield and quality attributes.
- To breed varieties having high yield with good quality fruits.
- Varieties with less fibre content.
- Good pollinating variety.
- Varieties with high sex ratio with more number of female flowers.

11.6. Botany

Aonla (*Emblica officinalis* Gaertn), which is also known as Indian gooseberry, myrobalan emblic, belongs to the family Euphorbiaceae. Aonla (Phyllanthus) comprises about 350 (Hooker, 1973) to 500 species (Bailey,1917) mostly shrub, few herb or tree. It bears two types of shoot. On the basis of growth characteristics, these have been categorized as long or indeterminate and short or determinate (Bajpai,1965). The indeterminate shoots are longer and continue to put forth new growth in the season. The determinate shoots appear on the nodes of indeterminate shoots and their number at each node may vary from 3-5 in different cultivars. These determinate shoots bear small sized leaves (9-14 x 2-4 mm) which are of simple type. Flowering in aonla takes place during March-April under hot arid conditions and in the month of February in Uttar

Pradesh on newly emerged determinate shoots. However, there are two prominent cropping seasons in South India i.e. July-August and April-May (Naik, 1963). The fertilized ovary of aonla is unlike those of other fruit plants which remains dormant for 3.5 month and resume growth following division in endosperm and zygote nuclei in the month of August. The fruits mature sometimes in December under hot arid ecosystem of Rajasthan. Somatic chromosomes number of aonla is 2n = 2x = 18 (Bajpai and Shukla, 1985). Other related species Phyllanthus acidus commonly known as star gooseberry or Otaheite gooseberry is grown for ornamental purposes.

11.7. Floral biology

In North India, senescence / leaf shedding of aonla plants begin in the month of February and complete shedding of determinate shoots takes place in the end of February to mid of March. The new shoots emerge out during first week of April (Bajpai, 1965, Ram, 1971 and Anon., 2001). The flowering period varied in different varieties from 17-26 days. In South India, flowering occurs twice a year, February-March and June-July (Naik, 1963). In aonla male flower appears first in the form of cluster at the basal part of the determinate shoot followed by female flower in the axil of leaves at the distal end of same shoot. The percentage of female flower varies from 1.12 to 8.33 depending upon varieties. According to Dhar (1979) the ratio of female to male flowers may vary from 1:109 to 1:501 depending upon the cultivars. However, the sex ratio in different cultivars varies from 1:12 to 1: 89 under hot arid ecosystem of Rajasthan (Anon., 2001). The time of anthesis in aonla is between 5.30 PM-7.30 PM with the peak anthesis timing 6.10 PM to 7.30 PM, after 8.00 PM flowers do not open. The anther dehiscence takes place 10-15 minutes after anthesis. The female flowers open in different stages taking about 70 hours. Stigma becomes receptive on 3rd day of anthesis and remain receptive for 48 hrs. Aonla is highly cross-pollinated crop due to its monoecious nature and pollination takes place through wind, honey bees and gravity (Reddy and Janki Bai,1979). The common visiting hours of honey bees is late evening and morning (Bajpai, 1957).

It is reported that since there is no self-incompatibility in aonla, therefore, the causes of poor fruit set seem to be high percentage of staminate flowers (Bajpai, 1968). Fruit setting can be increased up to to 18-28% by hand pollination. However, Pathak and Pathak (1993) did not notice any fruit set for a number of years in some isolated Banarasi aonla plant even after profuse flowering. It indicates the presence of self-incompatibility in this cultivar. A gene controlled natural incompatibility system is held responsible for unfruitfulness in some aonla cultivars. Information related to narrow and broad-based heritability of desirable traits is still lacking. Preliminary observations indicated that yield and yield attributing traits are polygenically inherited as crosses between large and small fruited types failed to produce any set segregation pattern in the progenies with respect to fruit size and related attributes.

11.8. Breeding methods and achievements

In India, breeding in aonla has so far been limited to selection only through which several high yielding varieties were identified. There are some important breeding methods which may be useful for making the aonla improvement programme a success.

11.8.1 Introduction

It is one of the oldest methods for improvement of fruit crop. It is bringing or exchange of germplasm / genetic material from one place where it was not known previously. Presently, germplasm exchange is being done in different crops through NBPGR, New Delhi. This method may be an important tool to bring exotic materials from foreign country for further evaluation and incorporation of specific gene lacking in indigenous aonla.

11.8.2 Selection

While selecting new ideotypes, plant height, vigour, growth habit, precocity, fruiting intensity, fruit size etc. are kept in mind. There are sufficient variations in fruit size and number of fruit/ determinate shoot, which directly affect the fruit yield and provide ample scope for selecting superior types. Most of the Indian varieties of aonla are developed through selection. Major research work in this direction is being done at NDUAT, Faizabad (NA-4, NA-5, NA-6, NA-7, NA10), GAU, Gujarat (Anand-1, Anand-2, Anand-3), RBS College (Balwant), Agra etc. Recently, some coloured and cluster bearing genotypes have been identified through exploration in Rajasthan, which will be further evaluated at national repository of aonla at CIAH, Bikaner (Dhandar and Shukla, 2003).

11.8.3 Hybridization

Hybridization is crossing of two parents which are genetically dissimilar. Not a single variety has been bred so far through this method. In order to get combination of highly desirable traits in one cultivar, a systematic crossing programme is essential. There is an apparent need to create some genetic combination by fusion of segregating gametes of the seedling population. While sexual progenies may have comparatively low mean performance related to the parental clone, the range in the performance of the progeny can be greater with some individual exceeding the performance of the best parent. Such transgressive segregants having both good genes and better combination of genes can be selected and propagated asexually. These superior individuals can become variety or may be used as superior parental clones in future breeding programme.

Occurrence of xenia effect between Chakaiya x Krishna, Banarasi x NA-9, Francis x NA-7, Kanchan x NA-6 and NA-6 x NA-9 for fruit size and weight were reported from the crosses (Srivastava and Pathak, 1993).

11.8.4 Mutation

Mutation is sudden heritable change in a character of a plant. In India, research work related to application of mutation in aonla is almost negligible but there is greater prospect to develop coloured varieties through induced mutation and selection from bud sport.

11.8.5 Polyploidy

Exact ploidy level is not known in aonla but it is realized by the scientists that aonla is characterized by polyploidy behaviour in composition of chromosome. The structural and numerical changes in chromosome can be made through application of Colchicine, which is found to be useful for getting small seeded fruit or seedlessness. Keeping in view the usefulness of polyploidy breeding, these principles may be applied in aonla to obtain desirable economic attributes.

11.8.6 Biotechnological tools

Incorporation of desirable genes in aonla plants is possible only with the application biotechnological approach. In fact, there is absolute dearth of information on biotechnological aspect of aonla. Tissue culture, cell culture and genetic manipulation through molecular technique may be useful to get early result in varietal improvement programme. This technique can also be helpful to modify particular traits and in turn provides new avenue for improving both the range and quality of the emblic fruits available for industrial and domestic uses.

11.9 Important cultivars/ varieties of Aonla

In India most of the varieties are developed through selection particularly from NDUAT, Faizabad, GAU, Gujarat and RBS College, Agra.

Banarasi

Developed through seedling selection from Varanasi district of Uttar Pradesh. It is shy bearing in nature due to less number of female flowers and presence of self-incompatibility. It is rich in vitamin C. It is an early maturing variety and mild susceptible to necrosis.

Francis

It is seedling selection popularly known as Hathijhool and originated from Pratapgarh (U.P.). The branches are drooping habit, moderate bearer, greenish yellow fruit, nearly fibreless but highly susceptible to necrosis.

Chakaiya

It is also a seedling selection having spreading growth habit. It contains fibrous fruits with six segments. It is late maturing variety and free from necrosis.

Krishna

This is chance seedling of Banarasi from Pratapgarh district of Uttar Pradesh. It is early maturing variety with shy bearing habit, pinkish green flesh and less fibrous.

Kanchan

Believed to be a chance seedling of Chakaiya. It is having spreading growth habit and bears profusely. Its fruits are yellowish in colour and suitable for pickle preparation.

NA-6

It is chance seedling of Chakaiya. It is having semi-spreading growth habit, light green colour of fruit, good keeping quality, late maturing, moderate bearer and free from necrosis.

NA-7

It is seedling selection from open pollinated strain of Francis. It is precocious and prolific bearer, mid-season, free from necrosis and suitable for preserve making.

NA-8

A chance seedling of Chanakya. It is upright growth habit, late maturing, moderate bearer, susceptible to necrosis and good keeping quality.

NA-9

Seedling selection from open pollinated strain of Banarasi. It is semi-spreading growth habit, moderate keeper, early, shy bearer and susceptible to necrosis.

NA-10

It is chance seedling developed from Banarasi locally known as Agra Bold. It has semi spreading growth habit, fruits yellowish green with pink tinge, highly astringent, good keeping quality, early, moderate bearer and moderately susceptible to necrosis.

Balwant

Developed through selection from Banarasi at RBS College, Agra, Uttar Pradesh. It is prolific bearer, semi-spreading growth habit, moderate bearer and good keeping quality.

Anand–1

It was developed at GAU, Anand. It is moderate bearer.

Anand-2

It was also developed at GAU, Anand. It is prolific bearer, fair keeping quality.

Anand-3

It is moderate bearer, yellowish green fruit, 6-7 segments and developed at GAU, Anand.

Laxmi 52

Identified by the Central Institute for Subtropical Horticulture, Lucknow. This is being cultivated in the private orchard as superior chance seedling of Francis i.e. Laxmi-52. Its tree has semi-erect growth and branches do not droop like its parent Francis. Fruit size large, 40-60 g, diameter 4.0-4.5 cm with 6 ridges. In early part of fruit growth, fruit colour is light pink which disappears on full development. It is mid-season maturing (mid-November to December) and is free from necrosis with yielding potential of 2-2.5 q/tree (after 10 years). Owing to larger fruit size and attractive colour, it fetches higher price in the market. The fruit has been found suitable for preparation of segments in syrup, candy and preserve.

Goma Aishwariya

The variety is released by CIAH, Bikaner on 2016. A selection from plus tree. It is an early and drought tolerant. The average yield potential is 105 kg/tree.

11.10. Future line of work

- Development of varieties with high percentage of female flowers.
- It is desirable to breed coloured variety as its demand in the trade market would be more in comparison to traditional varieties.
- Exploitation of heterosis (hybrid vigour).
- To evolve variety with very small stone size.
- Development of variety suitable for export purpose.
- Development of transgenic aonla through biotechnological tools.
- Breeding for resistant to frost, biotic and abiotic stresses.

11.11. References

Anonymous (2001) Floral biology of aonla, Annual Report, CIAH, Bikaner.

Bailey, L.H. (1917) The Standard Cyclopedia of Horticulture. The Mac Millan Co. London.

Bajpai P.N. (1965) Studies on vegetative growth and development of male and female gametophyte in aonla (*Emblica officinalis* G.). Agra Univ. *J.Res.(Sci.)*, **14**:167-186.

Bajpai P.N. (1968) Studies on flowering and fruit development in aonla. *Hort Adv.*,**7**: 38-67.

Bajpai P.N.(1957) Blossom biology and fruit set in Phyllanthus emblica. *Indian J. Hort.*, **14**: 95-102.

Bajpai, P.N. and Shukla, H.S. (1985) Aonla. In: Fruits of India Tropical and Subtropical (Ed. T.K. Bose), Naya Prokash, Calcutta, pp. 591-600.

Dhandar D.G. and Shukla, A.K. (2003) Varietal improvement in aonla. National seminar on Production and processing of aonla. 7-10 August,2003 at Salem, Tamil Nadu.

Dhar, L. (1979) A note on the sex expression and bearing behaviour of Banarasi aonla. *Prog Hort.*, **11**: 31-33.

Ferminger, T. A. (1947) Ferminger manual of gardening in India. 8th Edn. Thaker Spink & Co. Ltd, Calcutta.

Hooker, J.D. (1973) The Flora of British India Vol 5 Ist Indian Reprint, Periodical Expert Delhi.

Naik, K.C. (1963) South Indian Fruits and their Culture. P. Varadachary and Co., Madras.

Pathak, R.K. and Pathak, R.A. (1993) Improvement of minor fruits. In: Advances in Horticulture Vol. I (Eds. K.L Chadha and O.P. Pareek), Malhotra Publishing House, New Delhi, pp 407-421.

Ram, S. (1971) Studies on physiology of fruit growth in aonla. Ph.D thesis submitted to Kanpur University, Kanpur.

Reddy, L.S. and Janki Bai (1979) Studies on the pollination in *Emblica officinalis. New Bot.,* **4**:155-160.

Shukla A.K. and Dhandar D.G. (2003) Genetic diversity of aonla (*Emblica officinalis* G) in hill ecosystem of Himachal Pradesh. National Symposium on Agroforestry and Sustainable production, November 7-9, IGFRI, Jhansi, pp127-128.

Shukla, Arun Kumar, Shukla, Anil Kumar, Awasthi, O.P. and Vashishtha, B.B. (2002) Shushk Kshetra Mein Aonla Utpadan. *Krishi Chayanika*, **3:** 20-22.

Srivastava, A.K. and Pathak, R.K. (1993) Studies on sex ratio, pollen viability and suitable pollinizer in aonla. Abst. of paper presented in Golden Jubilee Symposium on "Changing Scenario in Horticulture" at IIHR, Bangalore.

12

Breeding of Pomegranate

Botanical name: *Punica granatum* L.

Family: Punicaceae

Chromosome number: 2n = 2x = 18

Scientific classification

Kingdom	:	Plantae- Plants
Division	:	Magnoliophyta–Flowering plants
Class	:	Magnoliopsida – Dicotyledons
Subclass	:	Rosidae
Order	:	Myrtales
Family	:	Punicaceae – Pomegranate family
Genus	:	*Punica*
Species	:	*granatum L.*

Pomegranate is an ancient and important fruit. It is grown for its delicious and juicy pink arils (outer growth of seed) which are eaten fresh (Keskar *et al.*, 1993). Fresh fruit is of exquisite quality while its processed products like bottled juice, syrups and jelly are highly appreciated. The juice is considered useful for patients suffering from leprosy. Fruit juice easily fermented and may be used for the production of wine. Juice of wild pomegranate is used in the manufacture of citric acid and sodium citrate for medicinal purpose. Pomegranate is known best for curing the chronic stomach ailment. Seeds of pomegranate contain oil which has a potential for industrial use. Sour aril of the wild types can be utilized for the preparation of *anardana*.

12.1. Centre of diversity

Pomegranate is native of Iran and cultivated extensively in the Mediterranean countries like Spain, Morocco, Egypt, Iran, Afghanistan and Baluchistan etc. It is also grown to some extent in Burma, China, Japan, USA, USSR and India.

12.2. Germplasm resources

Being a cross pollinated crop, a lot of variability exists in seedling population, which can be utilized in further improvement programme. At present 150 genotypes of pomegranate have been maintained at CIAH, Bikaner (Anon., 2002). Out of these genotypes 55 are deciduous and rest 95 are of evergreen in nature. Field gene banks of pomegranate are maintained at Abohar, Rahuri, Bikaner, Bangalore, Allahabad, Jodhpur and Ludhiana etc.

12.3. Objectives

- The main objective in pomegranate breeding is to obtain suitable types which produce small soft seeds with attractive red (pink) aril.
- To develop easily manageable upright growth habit of the tree.
- To develop thornlessness in the twigs, a desirable character as it helps in cultural management of the tree.
- To develop varieties resistant to fruit borer (*Virachola isocrates*) and fruit rot (*Phomopsis* species).
- To develop varieties free from fruit cracking.

12.4. Botany

Pomegranate belongs to genera *Punica* and family Punicaceae (Chatterjee and Randhawa, 1952, Joshi, 1956). Genus *Punica* has two species i.e. *Punica protopunica* (wild form in Socotro Island) and *Punica granatum* (cultivated type). Further, *Punica granatum* has been classified in to two sub species i.e. *Cholorocarpa* and *Porphyrocarpa*. *Cholorocarpa* is found in Transcaucacus region and *Porphyrocarpa* in Central Asia. *Punica granatum* is deciduous shrub (some times evergreen) or tree. The tree takes a height of 7 M and relatively long lived (up to 40 to 75 years). Leaves petiolated, oblong or oval, lanceolate, 2.5 - 5 cm long lengthwise, obtuse, glabrous and shining. Another species of pomegranate *Punica nana* L. (dwarf pomegranate) is the double flowered habit and on the pacific coast it is grown as a hedge plant. According to Smith (1979) *Punica granatum* has 2n =2x =16, 18 chomosomes. The number of chromosomes in the somatic complements of Dholka, Ganesh, Kandhari, Muscat White and Patiala were found to be 2n=16, while the variety Double flower had 2n=18 (Nath and Randhawa, 1959 a). The Chromosomes number in Vellodu and Kashmiri varieties was found to be 2n=18 (Raman *et al.*, 1963).

12.5. Floral biology

Inflorescence of the pomegranate has been reported to be cyme. Nath and Randhawa (1959b) observed the inflorescence to be dichasial cyme and due to heavy drop of secondary and tertiary buds they appear to be solitary in clusters. In the evergreen pomegranate cultivar, flower buds of the spring flush are borne on mature wood of one-year old shoot whereas, the flower which appear during July-August are borne on the current year's growth. In deciduous cultivars, flowers are borne on the current seasons growth between July and August.

There are two flowering seasons in North India whereas Nalawadi *et al.* (1973) reported three flowering seasons in Western India. There are three types of flowers present on same plant of pomegranate i.e. male, hermaphrodite and intermediate. Ovary of male flower is rudimentary whereas that of intermediate flowers are degenerating type. If fruit set takes place in such flowers they may drop before reaching to maturity, even if some fruits reach maturity that become misshaped. Heterostyly is also common in pomegranate, in this case hermaphrodite flowers are pin eyed and male flowers are thrumb type. The flowers are handsome brilliant orange red, calyx is tubular, persistent, five to ten-lobed, petals five to seven in numbers, are lanceolate, inserted between the calyx lobes. The ovary is embedded in the calyx tube and contains several locules in two series one above the other. The time of dehiscence of anther varies in different cultivars and no general sequence was found at the time of anthesis (Nath and Randhwa, 1959c, Josan *et al.*, 1979). In cultivar Patiala, dehiscence starts prior to anthesis whereas in Muskat White it starts afterward. Dehiscence was found to be affected by temperature and atmospheric humidity. The fertility of pollen grains in hermaphrodite flower was found to be higher than in the male flower (Nath and Randhawa, 1959c). Stigma attained maturity one day before anthesis and receptivity remained at peak on the same and second day after anthesis and it gradually went down till the third day after which it abruptly became non receptive. Josan *et al.* (1979) noted that stigma remained receptive from one day before anthesis to five days after the completion of anthesis and that pollen germination was over 90%. Both self and cross-pollination are reported in pomegranate. The greater percentage of fruit set was observed by hand pollination and pollination under natural condition. i.e. open pollination (Nath and Randhawa, 1959d). According to Singh (1977) pomegranate is often cross-pollinated crop where as Nalawadi *et al.* (1973) reported it as cross pollinated crop.

Table 1: Flowering period of pomegranate in different parts of India are as follows.

S.No.	State/ Zone/Region	Flowering period
1.	Central and Western India	Ambe bahar (January – February)Mrig bahar (June – July)Hast bahar (September – October)
2.	Karnataka	June – August (For 80 to 87 Days)March (For 22 – 30 Days)September (For 20 – 30 Days)
3.	Punjab	April to June
4.	Delhi	Once or twice in a year
5.	Bihar	February – MarchJuly – August
6.	Rajasthan	Mrig bahar (June – July), Ambe and Hast bahar also comes but fruit quality is poor.

12.6. Inheritance pattern

Scanty information is available on the inheritance pattern of the pomegranate. Manohar *et al*. (1981) reported that economic characters like rind weight, acidity percentage, fruit weight, aril per fruit, yield per tree and number of fruits per tree exhibit high heritability and high genetic advance. Therefore, satisfactory benefits can be derived if these characters are used for selecting individual with better yield potential. However, according to Purohit (1987) use of a soft seeded cultivar as a male parent slightly decreased seed hardiness in hard seeded cultivars. Correlation studies at IIHR, Bangalore indicated that smaller fruits possess bright aril colour. A positive and significant relationship between brighter fruit colour and TSS and dark aril colour and T.S.S. was observed. Further, mellowness was associated with less sweetness.

12.7. Breeding methods and achievements

12.7.1 Introduction

Some important cultivars including soft seeded, dark red grained types, viz. Wonderful from the USA, A. Males, A. Be Hastah, A Alah, A Agha, Mohammad Ali, A Post Sephid Sirin from Iran and Ranninj G-1-8-23, Rannyij G-1-3-34, Chereny, Gulsha Red, J G-1-8-7 from USSR and few cultivars from Tunisia have been introduced. At Hisar, cultivar Shirin Anar and Russian Seedling were found resistant to bacterial leaf spot. A pomegranate line of Iranian origin has been identified at Rahuri which has dark pink arils, soft seed and high TSS (Singh and Rana, 1993).

12.7.2 Selection

Many pomegranate types cultivated in India are of seedling origin. They offer a wide range of variability with respect to shape and size of fruits, mellowness of seeds, aril colour, rind colour, sweetness and acidity. On the basis of yield and physico–chemical characters of fruits, number of cultivars have been recommended for commercial cultivation in different states of India, viz. Ganesh, G-137, P-23, P-26, and Muscat in Maharashtra, Bassein Seedless, Jyothi and Madhugiri in Karnataka, Dholka in Gujarat, Jalore Seedless, Jodhpur Red, and Jodhpuri White, in Rajasthan. Kabul Red, Vellodu, Yercaud 1, and Co-1 in Tamil Nadu. Two ornamental types (Japanese Dwarf and Double flower giving red, yellow and white flowers) are planted in the ornamental gardens (Nath and Randhawa, 1959e). Due to considerable variability and their adaptability to existing agro climatic conditions, selection of superior genotypes will be the best approach to get desirable ideotypes. The cultivar GBG-1 is selection from open pollinated population of Alandi in 1932. The name Ganesh was given in 1970. Choudhari and Shirsath (1976) selected elite plants based on fruit size, shape, rind and aril colour, from the seedling plantations of Muskat variety around Kolhar village in Ahmednagar district. Open pollinated fruits from these selected plants were used to raise a seedling progeny at Mahatma Phule Agricultural Univeristy, Rahuri. The promosing individual plant selection are $L_2 P_{21}$, $L_5 P_{11}$, L_4P_{24}, L_4P_{25}, $L_{24}P_{42}$, $L_{20}P_{50}$, No.1056 and No. 778 (Patil, 1976, Bhapkar, 1976, Karale, 1977, Kolhe,1980).

Five Muskat types, namely P-13, P-16, P-23, P-26 and SK-1 were identified by Naik (1975). Further, P-23 and P-26 were released in 1986 for commercial cultivation by MPKV, Rahuri (Anon., 1986, Keskar *et al.*, 1993). At the University of Agricultural Sciences, Bangalore as a result of evaluation of seedling populations raised from Bassein Seedless and Dholka Varieties, GKVK-1 now named as Jyothi was released (Sulladmath, 1985). At Coimbatore self seeded selection Co-1was identified (Khader *et al.*, 1982).

12.7.3 Clonal Selection

Sawant (1973) reported that G-137 is a superior clonal selection over Ganesh, other clones are also superior i.e. G-107, G-132, G-133. Sayed *et al.* (1985) reported a clone Acc. No. 455 which has been renamed as Yercaud-1 and released for commercial cultivation in Tamil Nadu.

12.7.4 Hybridization

In order to incorporate blood red colour of Russian types into Ganesh, several crosses were made at Rahuri in 1976. Out of 122 F_1 hybrids, seven had deep red aril colour but hard seeds and inferior in taste than Ganesh (Kale, 1986). A promosing line from the F_2 population (No.61) combining all desirable quality attributes has been released by the name Mridula (Ganesh x Gulsha Rose Pink). At IIHR Bangalore, F_2 progeny of Muskat and Jyothi were short statured with higher yield potential. At CIAH, Bikaner 16 F_1 progenies of desirable parental combination are under evaluation (Anon., 2002).

12.7.5 Mutation

Use of physical (ã rays) and chemical (N, N-dimethyl N- nitrosourea) mutagen may help in the development of the superior cultivar of soft seeded (Levin, 1990).

12.7.6 Biotechnological tools

Attempt have been made to regenerate the plant by using leaf (Omura *et al.*,1987) and shoot tip explant (Mahishni *et al.*,1991). Enzyme based marker was also used to identify the genetic variability among the existing genotypes. Somatic embryogenesis was also practiced by using petal as explant (Nataraja and Neelambika, 1996).

Table 2: Important characteristics of Ganesh and G-137 (*Source:* Keskar *et al.*, 1993)

S. No.	Characters	Ganesh	G-137
1.	Fruit weight (g)	231	269
2.	Weight of 100 grains (g)	25.0	28.6
3.	Juice (%)	64.4	65.2
4.	TSS (%)	13.6	15.5
5.	Acidity (%)	0.39	.036
6.	Reducing sugar (%)	11.6	13.7
7.	Number of fruit / trees	60	57
8.	Yield / tree (kg)	12.21	11.98

Table 3: Important characteristics of some promising selection raised from open pollinated fruit of F_1 hybrids of pomegranate (Keskar *et al.*,1993)

S.No.	Characteristics	Sel-5 (Ganesh x Shirin Anar)	Sel-130 (Ganesh x Gulsha Rose Pink)	Sel-303 (Ganesh x Gulsha Red)
1.	Fruit colour	Apple red	Greenish brown	Yellowish brown
2.	Fruit weight (g)	130	107	140.0
3.	Fruit size LxB (cm)	6.9x6.5	6.1x6.5	5.5x6.8
4.	Aril colour	Blood red	Dark red	Blood red
5.	Mellowness of seeds	Soft	Soft	Soft
6.	Taste	Sweet	Sweet	Sweet
7.	No. of grains/100g	58	94	45
8.	Grain peel ratio	1.17	1.17	1.80
9.	Juice colour	Dark red	Blood red	Blood red
10.	Juice (%)	80	80	80
11.	TSS (%)	18.4	15.8	19.0
12.	Acidity (%)	0.64	0.80	0.64

Table 4: Commercial varieties from different countries (Jyotsana et al., 2014)

S.No.	Country	Varieties
1.	India	Bhagwa, Ganesh, Ruby,Phule, Arakta, Mridula
2.	Iran	Schahvar, Robab
3.	Israel	Wonderful
4.	Spain	Mollar, Tendral
5.	Turkey	Hicaznar, Beynar
6.	USA	Wonderful, Granada

Table 5: Cultivars of Pomegranate grown in different states

S.No.	Name of the states	Cultivars
1	Rajasthan	Jalore Seedless, Jodhpur Red, Jodhpuri White.
2	Haryana	Ganesh, Muskat Red, Paper Shell.
3	Gujarat	Dholka, Muskat Red, Kandhari, Ganesh.
4	Maharashtra	Ganesh, G137, P23, P26, Muskat, Mridula.
5	Karnataka	Bassein Seedless, Jyothi, Paper Shell, Madhugiri.
6	Tamil Nadu	Co-1, Yercaud, Vellodu, Kabul Red.

12.8 Important varieties and cultivars of Pomegranate

Alandi

Also known as Vidaki, medium fruit size, fleshy testa, blood red or deep pink with sweet, slightly acidic juice with hard seeds.

Dholka

Large fruit size, greenish white rind, fleshy testa, pinkish white or whitish with sweet juice, soft seeds and acidic juice.

Kabul

Large fruit size, rind deep red mixed with pale yellow, thick, fleshy testa dark red, slightly bitter juice.

Kandhari

Fruit large in size, rind deep red, fleshy testa, blood red or deep pink with sweet , slightly acidic juice, hard seeds.

Muskat red

Fruit small to medium in size, rind some what thick, fleshy testa with moderately sweet juice, seeds are semi hard.

Table 6: Important characteristics of some popular varieties in India (Jyotsana *et al*., 2014)

S.No.	Variety	Fruit	Aril/Seed	Juice	Remarks
1	Arakta	Medium sized dark red rind	colour Arils are also dark red soft seeded	Sweet	-
2	Bhagwa	Big size, glossy redrind, thick rind	Bold red arilsand softseeded	Sweet	Matures in 180 to190 days afterblooming
3	Dholka	Large in size, rindgreenish white, fleshy testa	Pinkish whiteor whitishsoft seeds	Sweet acidic taste	-
4	G-137	Surface smooth, yellow with red tinge. Fruits are large sized	Deep pink and bold aril, soft seeds	Sweet Sweet	Clonal selection from ganesh, prolific bearer.
5	Ganesh (GBG- 1)	Medium sized, with yellow, smooth surface and red tinge	Soft with pinkish aril	Sweet	Revolutionized the cultivation of pomegranate in Maharashtra
6	JaloreSeedless	Round fruit, yellowwith red tinge incolour	Aril colourlight pink topink, juicy	Tastesweet, tss15-16Úbrix	
7	Jyoti (GKVK-1)	Medium to large sized, having attractive, yellowish red, testa fleshy	Pink aril. Soft seeded	Very sweet, good in taste	Yields moderately
8	Kandhari	Large size, rind deepred, fleshy testa	Blood red ordeep pink,hard seed	Sweet, slightlyacidic	Temperate variety
9	Mridula	Medium sized, rindsmooth, dark red incolour	Blood-redarils withvery softseeds	Sweet	Plants are dwarf
10	Phule Bhagwa Super	Red rind and redarils			Recently releasedvariety by MPKV, Rahuri. Maturesabout 3 weeks earlier than Bhagwa.
11	Ruby	Rose pink	Soft and redarils	Goodflavour	Plants are dwarf, prolific bearer, providinguniformly redfruits.

Paper Shell

Fruit medium in size, rind thick, fleshy testa, reddish pink with sweet juice and soft seed.

Spanish Ruby

Fruit small to medium in size, rind thin, fleshy testa rose coloured, soft seed.

Ganesh

Prolific bearer, medium fruit size, soft seeds, sweet in taste.

Jyothi

Also known as GKVK-1, attractive yellowish red fruit colour, medium fruit size, red aril colour and soft seeds.

Vellodu

Fruit medium to large in size, rind moderately thick, fleshy testa, juicy, seed moderately hard.

Poona

Fruit large in size, fleshy testa, deep scarlet or pink and red.

Bedana

Fruit medium to large in size, rind brownish or whitish, fleshy testa, pinkish white with sweet juice and soft seeds.

Bhagwa

It is developed by MPKV, Rahuri. It is tolerant to thrips and mites, it is free from blackening of arils and there is no incidence of fruit cracking. Fruits have cherry red bold aril.

Phule Arakta

It is also developed by MPKV, Rahuri. Plant is heavy yielder with bigger fruits and sweet soft seed. It is less susceptible to fruit spots and thrips.

Ruby

It is a multiple hybrid of pomegranate developed by IIHR, Bangalore for aril colour and seed mellowness. A fully-grown plant of Ruby is about 2.72 m tall and less spreading (4.7 m). The mature fruits resemble in size and shape the fruits of cv. Ganesh; however, the skin colour of Ruby is reddish brown with green streaks. The rind is thin (0.24 cm), fruit contains red bold arils (37.2 g/ 100 arils) which yield (72%) thick (1.83 cps) dark red (4.71 at 540 nm) sweet (16°B) juice having tanin content of 173 mg per 100 ml juice. Under Bangalore conditions its performance is comparable with cv. Ganesh in respect of seed softness (2.19 kg/cm2), fruit weight (270 g) and yield (16-18 tonnes/ha). Better adoptability to milder climate.

Goma Khatta

This variety is released in 2010. The variety is developed for Anardana purpose. Yield potential is 6.59 kg/plant and anardana yield is 1.18 kg/plant. Seeds hardness is medium. Fruit having 46.7% of Juice and TSS is 14.5^{0}Brix. Acidity is 7.3%.

12.9. References

Anonymous, (1986) Report of Research Work done on Horticultural Crops. Research Review Sub-Committee. Report Department of Horticulture, MPAU, Rahuri.'

Anonymous, (2002) Annual Report 2001-2002. CIAH, Bikaner.

Bhapkar, S.K. (1976) Seedling selection in Pomegranate (*Punica granatum* L.) cv. Muskat. MSc(Ag) Thesis, MPAU, Rahuri.

Chatterjee, D. and Randhawa, G.S. (1952) Standardized name of cultivated plants in India. I. fruits. *Indian J. Hort*., **9**:24-36.

Choudhari, K.G. and Shirsath, N.S. (1976) Improvement of pomegranate (*Punica granatum* L.) by selection. *South Indian Hort*., **24**:56-59.

Josan, J.S., Jawanda, J.S. and Uppal, D.K. (1979) Studies on floral biology of pomegranate, anthesis, dehiscence, pollen studies and receptivity of stigma. *Punjab Hort. J*., **19** : 66-70.

Joshi, B.C. (1956) A contribution to the morphology of *Punica granatum* L. Thesis, Agra University, Agra.

Jyotsana Sharma, Deodas Tarachand Meshram, Ashis Maity and N. V. Singh (2014) Technical Report of Pomegranate: Cultivation, Marketing and Utilization Technical Bulletin No. NRCP/2014. ICAR-National Research Centre on Pomegranate, Solapur- 413 255 (Maharashtra). Pp.17-20.

Kale, R.H.(1986) Studies on F_1 hybrids of Pomegranate. M.Sc.(Ag) Thesis MPAU Rahuri.

Karale, A.R. (1977) Seedling selection in Pomegranate (*Punica granatum* L.) cv. Muskat M.Sc(Ag) Thsis MPAU, Rahuri.

Keskar, B.G., Karale, A.R. and Kale, P.N. (1993) Improvement of pomegranate. In: Advance in Horticulture Vol. I (Eds. K.L. Chadha and O.P. Pareek), Malhotra Publishing House, New Delhi, pp 399-405.

Khader, Abdul, J.B.M. Md., Kulusekaran, M. and Muthuswami, S. (1982) C0-1 A soft seeded selection. *South Indian Hort*., **30**(4):298.

Kolhe, V.H. (1980) Improvement of Muskat Pomegranate by seedling selection. M.Sc. (Ag.) Thesis MPAU, Rahuri.

Levin, G.M. (1990) Induced mutagenesis in pomegranate. *Refrativnyl Zhurnal*, **10**:3399.

Mahishni, D.M., Muralikrishna, A., Shivashankar, G. and Kulkarni, R.S. (1991) Shoot tip culture method for rapid clonal propagation of pomegranate (*Punica granatum* L.). In: Horticulture- New technologies and applications (Eds. J.Prakash and R.L.M. Pierik), Kluwar Academic Publisher, Dordrecht, Netherlands, pp. 215-217.

Manohar, M.S., Tikka S.B.S. and Nathulal (1981) Phenotypic variation and its heritable components in some biometric characters in Pomegranate (*Punica granatum* L.). *Indian J. Hort.*, **38**:187-190.

Naik, V.B. (1975) Seedling selection in pomegranate (*Punica granatum* L.) cv. Muskat. MSc(Ag) Thesis, MPAU, Rahuri.

Nalawadi, U.G., Farooqui, A.A., Reddy, D., Gubbaiah, M.A.N., Sulkeri, G.S. and Nalini, A.S. (1973) Studies on floral biology of pomegranate. *Mysore J. Agric. Sci.*, **7**(2): 213-225.

Nataraja, K. and Neelambika, G.K. (1996) Somatic embryogenesis and plantlet from petal cultures of pomegranate (*Punica granatum* L.). *Indian J. Exptl. Biol.*, **34** (7): 719-721.

Nath, N. and Randhawa, G.S. (1959a) Studies on cytology of Pomegranate (*Punica granatum* L.). *Indian J. Hort.*, **16**:210-215.

Nath, N. and Randhawa, G.S. (1959b) Studies on floral biology of Pomegranate. I Flowering habit, flowering season, bud development and sex ratio in flowers. *Indian J. Hort.*, **16**:61-68.

Nath, N. and Randhawa, G.S. (1959c) Studies on floral biology of Pomegranate II Anthesis, dehiscence, pollen studies and receptivity of stigma. *Indian J. Hort.*, **16**:121-135.

Nath, N. and Randhawa, G.S. (1959d) Studies on floral biology of pomegranate (III) Pollination, fruit set and seed formation. *Indian J. Hort.*, **16**:136-140.

Nath, N and Randhawa, G.S. (1959 e) Classification and description of some varieties of *Punica granatum* L. *Indian J. Hort.*, **16**:191-201.

Omura, M., Matsuta, N., Moriguchi, T. and Kozaki, I. (1987) Adventitious shoots and plantlet formation from cultured Pomegranate leaf explants. *Hort. Sci.*, **22** (1):133-134.

Patil, B.N. (1976) Seedling selection in Pomegranate cv. Muskat. M.Sc.(Ag) Thesis MPAU, Rahuri.

Purohit, A.G. (1987) Effect of pollen parent on seed hardiness in Pomegranate. *Indian J. Agric. Sci.*, **57**:753-755.

Raman, V.S., Kesavan, P.C., Manimekalai, G., Alikhan, W.M. and Rangaswami, S.R. (1963) Cytological studies in some tropical fruits plants–Banana, Annona, Guava and Pomegranate. *South Indian Hort.*, **11**(3&4): 27-33.

Sayed, S., Ramdas, S., Nanjan, K. and Muthuswami, S. (1985) YCD-1Pomegranate. *South Indian Hort.*, **33**(1): 67.

Singh, B.P. and Rana, R.S. (1993) Promising fruit introductions. In: Advance in Horticulture Vol. I (Eds. K.L. Chadha and O.P. Pareek), Malhotra Publishing House, New Delhi, pp.43-66.

Singh, J. (1977) Studies on floral biology of Pomegranate (*Punica granatum* L.). M.Sc. (Ag) Thesis, PAU, Ludhiana.

Smith , P.M. (1979) Minor crops. In: Evolution of crop plants (Ed. N.W. Simmonds), Longman, p. 320.

Sulladamath, U.V. (1985) Progress Report Third National Workshop Arid Zone Fruit Research, MPAU, Rahuri 5-8 July.

Sawant, R.B. (1973) Clonal selection in Ganesh Pomegranate (*Punica granatum* L.). MSc(Ag.) Thesis MPAU, Rahuri.

13

Breeding of Date Palm

Botanical name: *Phoenix dactylifera* L.

Family: Palmaceae

Chromosome number 2n=2x=36

Scientific classification

Kingdom	:	Plantae- Plants
Division	:	Magnoliophyta–Flowering plants
Class	:	Liliopsida – Monocotyledons
Subclass	:	Arecidae
Order	:	Arecales
Family	:	Arecaceae – Palm family
Genus	:	*Phoenix*
Species	:	*dactylifera L.*

Date palm is one of the oldest fruits and was known to man many thousand years before Christ. It is staple food of the Arabs. Large quantities of date are imported by India from the Arabian countries. Wild date palm are seen growing even up to an altitude of 1500 M (Pareek, 1990), of course they do not fruit at this altitude every year. Raw sugar and a fermented drink 'toddy' are prepared from the sap of this palm. In the Indians valley, date palm is believed to have been introduced by the soldiers of Alexander the great in the fourth century B.C. and some wild groves of seedling date palm are found in the coastal belt in the Kachchh district of Gujarat. Investigations to develop indigenous production of dates in India were initiated during the 1950 by the ICAR at Regional Fruit Research Station, Abohar (Punjab). Some commercial varieties of date palm were introduced from USA, Pakistan and Middle East countries during 1955-1962. The extremely dry areas comprising Jaisalmer, Barmer, Bikaner and Jodhpur in Rajasthan, Abohar in Punjab and Kachchh in Gujarat are the potential regions for the date growing in India.

13.1. Centre of diversity

The exact origin of the date palm (*Phoenix dactylifera* L.) is considered to be lost in Antiquity. However, it is certain that the date palm was cultivated as early as 4000 B.C. Since it was used for the construction of the temple of the moon god near Ur in Southern Iraq-Mesopotamia (Popenoe, 1913,1973) date palm is probably the most ancient cultivated tree in the world. It could be safely assumed that the reason for mentioning dates and date palms in the Jewish, Christian and Islam religions was due mainly to the influence of the Prophet Ibrahim, who was born and raised in the old city of Ur where date palms were grown. (Zaid and Wetide, 1999).

13.2. Distribution

The major date palm growing areas in India are Kutch (Gujarat), Rajasthan, and certain parts of the Punjab, as well as Tamil Nadu State to some extent. In Kutch there are more than 2 million date palms, the majority of them grown from seeds and offshoots, providing a huge biodiversity for experimentation and improvement of products.

13.3. Germplasm resources

There are about 1.4 million seedling palms in the coastal belt of Kachchh (Bhuj), Anjar, Khedoi, Mundra and Mandevi etc. Field genebanks of date palm is maintained at Abohar (Punjab), Bikaner (RAU and CIAH, Rajasthan), and Mundra (Gujarat). At present about 52 date palm accessions are maintained at CIAH, Bikaner (Anonymous, 2002).

13.4. Objectives

The main objective of breeding under Indian agroclimatic condition should be:

- To develop an early maturing and dwarf cultivars resistant to damage by rain and high humidity,
- To develop varieties resistant to Graphiola disease.

13.5. Botany

The botanical name of the date palm, *Phoenix dactylifera* L. derived from a Phoenician name "Phoenix", which means date palm, and dactylifera derived from a Greek Word "daktulos" meaning a finger, illustrating the fruits form. (Linne, 1734).

Another source refers this botanical name to the legendary Egyptian bird "Phoenix" which lived to be 500 Years old and Cost itself into a fire from which it rose with renewed growth (Piliny, 1489, Vanzyl, 1983). This resemblance to the date palm, which can also regrow after fire damage, make the bird and the date palm share this name while "dactylifera" originates from the Hebrew word "dachel" which describes the fruits shape (Popenae, 1938).

Date palm belonging to the Palmaceae family. In this family there are about 200 genera and 1500 species (Dowson,1982). Beside date palm, *P. atlantica* chev., *P. reclinata* Jacg., *P. farinifera* Roxb. *P. humilis* Royle and *P. acaulis* Roxb. bear edible fruits.

According to Chevalier (1952) geographical distribution of 12 species of date palm are as under-

S.No.	Name of the species	Common Name	Distribution
1.	*P. dactylifera* L.	Date palm	Mediterranean countries, Africa and part of Asia, North America and Australia
2.	*P. atlantica* A. chev.	-	Occidental Africa and Canary Islands.
3.	*P. canariensis* Chabeaud	Canarly palm	Canary Islands and Cape Verde
4.	*P. reclinata* Jacq.	Dwarf palm	Tropical Africa (Senegal and Uganda) and Yemen (Asia)
5.	*P. sylvestris* Roxb.	Wild date palm or Sugar palm	India and Pakistan
6.	*P. humilis* Royle	-	India, Burma and China
7.	*P. hanceana* Naudin	-	Meridional China and Thailand
8.	*P. robelinic* O' Brein	-	Ceylon, Toukin, Annam, Laos and Thailand
9.	*P. farinifera* Roxb	Pigmy palm	India, Ceylon and Annam
10.	*P. rupicola* T. Anders	Rocky date palm	India
11.	*P. acaulis* Roxb.	Dwarf palm	Bengal
12.	*P. paludosa* Roxb	Hental or Juliana palm	Bengal, Tenqsherim, Andaman, Niko bar and Thailand.

Close relationship among 12 species is illustrated by the ease of hybridization and pollination (Moore, 1963, Munier, 1973). Several natural hybrids were hence obtained by natural crossing of *P. dactylifera* x *P. sylvestris* (India), *P. dactylifera* x *P. canariensis* (Morocco, Algeria and Israel), *P. dactylifera* x *P. reclinata* (Senegal). Further, *P. dactylifera* L. has 36 chromosomes (n=18,2n=36) (Beal, 1937) but polyploidy cases were reported by A.L. Salih and Al Jarrah (1987) with Iraqui date varieties (2n=64). He also reported for Sayer 2n=32, and Khasab 2n=36.

13.6 Floral biology

Date palm is a dioecious species with male and female flowers being produced in clusters on separate palms. These flowering clusters are produced in axil of the previous year's growth. In rare cases both pistillate and staminate flowers are produced on the same spike while the presence of hermaphrodite flowers in the inflorescence has also been reported (Mason, 1915, Milne, 1918). Palms which carry both unisexual and hermaphrodite flowers are known as polygamous.

Inflorescence spadix also called flower cluster, in its early stages is enclosed in a hard-covering envelope known as spathe which splits open as the flowers mature exposing the entire inflorescence for pollination. The spathe protects the delicate flowers from being shrivelled up by the intense heat until these are mature and ready to perform their functions. The spathe at the beginning is greenish in colour, becoming brown when near splitting. Splitting of spathe is longitudinal. The male spathes are shorter and wider than the female ones. Each spikelet carries a large number of tiny flowers as many as 8000 to 10000 in female and more in male inflorescence (Chandler, 1958). The annual number of spathes born by a palm varies from none to about 25 in female and to even more in male, but the average is a dozen for female and more for male.

Male inflorescence is crowded at the end of the rachis, while branches of the inflorescence of the female cluster are less densely crowded at the end of the rachis. These characteristics allow the identification of the sexes before its opening. Male flower is sweet scented and normally has six stamens, surrounded by a waxy scale- like petals and sepals (3 each). Each stamen is composed of two little yellowish pollen sacs.

Female flower has a diameter of about 3 to 4 mm and has rudimentary stamens and three carpels closely pressed together, ovary is superior (hypogynous). Three sepals and three petals are united together so that only tips diverge. On opening the female flowers show more yellow colour while the male ones show white colour dust, on shaking. The pollen sacs usually open within an hour or two after the bursting of the spathe. Only one ovule per flower is fertilized, leading to the development of one carpel which in turn gives a fruit called date. The other ovules are aborted. The aborted carpels persist as two brown spots in the calyx of ripe fruits.

13.7. Pollination

Being dioecious in nature, date palm sexes are borne by separate individuals. The unisexual flowers are pistillate (female) and staminate (male) type in character. Male palm produces the pollen and the female palm produces the fruit. Natural pollination, by wind, bees and insects is found to yield a fair fruit set in various areas of the date growing countries. Commercial date production necessitates artificial pollination which ensure good fertilization and over comes disadvantages of dichogamy and also reduces the number of male palms.

13.8. Pollination Technique

Depending on the types of pollen available, one of the following three techniques can be effectively used.

(a) Fresh male strands

The most common technique of pollination is to cut the strands of male flowers from a freshly opened male spathe and place two to three of this strand length wise and in an inverted position, between the strands of the female inflorescence. This should be done after some pollen has been shaken over the female inflorescences (Dowson, 1982). In order to keep the male strands in place and also to avoid the entanglement of the female cluster's strands during their rapid growth, it is recommended to use a twine (a strip torn from a palm leaflet or a string) to tie the pollinated female cluster 5 to 7 cm from the outer end.

(b) Pollen suspension

Laboratory and field experiments on three varieties from Saudi Arabia (Khalas, Ruzaiz and Shishi) have shown that the pollen grain suspension, containing 10% sucrose and 20 ppm GA3 could be used for pollination (Ahmed and Jahjab, 1985). Pollination sprays were found to be as good as hand pollination in relation to fruit setting.

(c) Dried pollen

This pollination technique is more economical and allows proper use of the pollen as well as adequate control of the timing of pollination. Important technique of applying dry pollens are as under-

(i) Cotton pieces

The most common technique of using dry pollen is to dust it on cotton pieces, about the size of a walnut and place one or two pieces between the strands of female inflorescence.

(ii) Use of puffer

A small manual insecticide duster, known as "puffer" is also used to apply dry pollen (Nixon, 1966).

(iii) Mechanical pollination

Mechanical pollination was developed mostly in the USA and Israel where laboures are expensive and not always available. It consists of pollinating freshly opened female spathes from the ground with the use of a special apparatus. Mechanical pollination has been one of the most important alternatives when the labour has been reduced by 50-70% (Nixon and Carpenter, 1978, Galeb et al., 1987).

(iv) Aircraft pollination

Experiments for pollinating dates with an aircraft were conducted in the Coachella valley of California on Deglet Noor variety by Brown and Perkins (1972). Results showed that even though temperatures and weather conditions were favorable, both the helicopter and fixed using method of application yielded less fruits set than the hand pollination methods.

13.9. Metaxenia

It is well known that the pollen not only affects the size of the fruit, seed (affected more by fruit thinning) and also the time of ripening (Swingle, 1928). Metaxenia is not to be confused with xenia, which is the effect of the pollen on the endosperm (embryo and albumen).

Metaxenia effect was verified by several investigations in the USA (Nixon and Carpenter, 1978), Israel (Comelly, 1960), Pakistan (Ahmad and Ali, 1960) and Morocco (Pereau-Leroy, 1958). The effect of pollen on the time of fruit ripening was proven to be beneficial and is actually considered as the most important practical application of metaxenia. Producing and selling date fruits at high prices early in the season, along with the aim of having more uniform and short ripening period (avoiding a prolonged harvest) are the two main objective of using a selected pollen of high metaxenia effect. A third useful application of metaxenia is where the development period of the plant is characterized by an insufficient sum total of heat for the fruit ripening of late varieties. The use of Fard 4 male has advanced the maturation stages of various varieties all around the world by two weeks.

13.10. Breeding methods and achievements

All the commercial date cultivars have arisen by selection of chance seedlings based on local needs. The systematic breeding work has been started recently.

13.10.1 Introduction

Most of the cultivar of date palm are introduced from different countries time to time. e.g. Halawy, Barhee, Medjool, Khalas, Sayar, Zahidi (USA), Khadraway

(Iraq), Barshi , Khuneizi, Nagal, Khashab (Oman), Hatemi, Tayar, Ruziz (Saudi Arabia), Amri, Sakloti and Agaloni (Egypt).

13.10.2 Selection

From the rich genetic pool of nearly 1.4 million palms developed from seeds in the coastal belt of Kachchh district of Gujarat, 20 promosing palms have been selected, most of which yield non-astringent Khalal fruit. One of them bears coconut shape fruit. These selections flower twice in a year. An early ripening date seedling has been identified at Abohar.

No breeding work has so far been done on date palm in India except evaluation of cultivars against rain damage and selection of some promising female seedlings from the Kachchh. At Abohar Zahidi cultivar has been found to be resistant to rain damage, Barhee is more tolerant than Shamran. It was also found that Medjool is resistant to rain damage, as it missed rains during fruit ripening (Pareek, and Vishal Nath, 1996).

Reaction of some date palm cultivars to Graphiola phoenicis leaf spot and black scorch disease at varous locations are as under (Pareek and Vishal Nath, 1996)

S.No.	Variety	Graphiola disease			Black scorchdisease
		Bikaner (1983-85)	Jodhpur (1989-91)	Hisar (1981-88)	Jodhpur (1989-91)
1.	Abdul Rehman	R	MS	—	MS
2.	Barhee	—	R	HS	MS
3.	Halawy	MS	S	HS	HS
4.	Hayani	R	R	S	R
5.	Khadrawy	R	R	R	R
6.	Khalas	MS	—	—	—
7.	Medjool	R	S	S	S
8.	Muscat	R	MS	—	MS
9.	Sayer	—	HS	—	MS
10.	Shamran	S	S	S	HS
11.	Zaglool	MS	HS	HS	MS

R= Resistant, S= Susceptible, HS = Highly Susceptible, MS= Moderately Susceptible

13.10.3 Hybridization

In USA, date palm breeding was started in 1948 with the objective of producing female clones adapted to mechanical harvesting and processing, equal or better in fruit quality than the Deglet Noor which is subjected to blacknose disorder and slower in vertical growth than in this cultivar (Carpenter, 1979).

Interspecific crosses of *P. dactylifera* with *P. sylvestris, P. reclinata, P.canariensis, P. roebelenii, P. rupicola* and *P. humilis* are successful and have metaxenic effects (Carpenter and Ream,1976). This offers the scope for improvement through hybridization and selection of promising males. Further, there is scope of interspecific and intergeneric hybridization for improving moderate to slow vertical growth in date palm. Breeding for resistance to fusariose or bayoud disease caused by Fusarium oxysporum f.sp.albeclinis is a major objective in Morocco. After 14 years of observations on 32 cultivars Saaidi et al. (1981) found Bou Sthammi Noire, Iklane, Tadment, Sair-Layalet and Bou Feggous Ou Moussa to be completely resistant. Crosses between selected female parents having either high quality dates or resistant to bayoud and the back crossed varietal male palm have been developed in the USA. This is followed by screening of the resistant progenies against the disease by artificial inoculation.

13.10.4 Biotechnological tool

A new potential tool can be utilized in the improvement of date palm. Attempts have been made to develop protocol for rapid multiplication through micropropagation. There are tremendous scope of molecular markers for the identification and characterization of the genetic variability. Recombinant DNA technology and somatic embryogenesis has also much scope.

13.11. Important cultivars of date palm grown in different countries (Pareek, 1990)

S.No.	Name of the country	Varieties
1.	Algeria	Ghars, Deglet Noor,
2.	Baharain	Murzaban, Khanezi
3.	UAE	Angal
4.	Egypt	Haiyani, Saidy, Zagloul, Samani
5.	PDR Yemen	Hamraiya
6.	Saudi Arabia	Irzeiz
7.	Iraq	Zahidi, Halawy, Khadrawy, Sayer
8.	Iran	Ustaumran
9.	Libya	Bikraari, Taasfirt, Murzabad, Saidy
10.	Somalia	Sucotari, Succari
11.	Oman	Mabsaly, Fardh
12.	Mauritania	Tingerguel
13.	Morocco	Bufaguus, Medjool
14.	Sudan	Barakaavi, Mishrig-Khatib
15.	Tunisia	Fatumi, Deglet Noor
16.	Pakistan	Mozawati, Begam Jangi, Dhakki, Assi, Halawi
17.	India	Halawy, Khadrawy, Shamran, Zahidi, Barhee, Medjool. Zagloul and Khalas

13.12. References

Ahmed, H.S. and Jahjab, M.A. (1985) The pollination of date palm with pollen grains suspensions. *Date Palm J.*, **4**(1): 33-40.

Ahmed, M. and Ali, N. (1960) Effect of different pollens on the physical and chemical characters and ripening of date fruits. *Punjab Fruit J.*, **23**(80): 10-11.

Al-Salih, A.A. and Al-Jarrah, A. (1987) Chromosomes number of a date palm male: cultivars Ghannami, Akhdav. *Date Palm J.*, **5** (2): 128-137.

Anonymous, (2002) Annual Report 2001-2002. CIAH, Bikaner.

Beal, J.M. (1937) Cytological studies in the genus *Phoenix. Bot. Gax.*, **99**: 400-407.

Brown, R.M. and Perkins, E.G.V. (1972) Experiments with air craft methods of pollinating dates. *The Punjab Fruit J.*, **119**: 116-127.

Carpenter, J.B. (1979) Date growers Inst. Rpt. 54., pp 13-16.

Carpenter, J.B. and Ream, C.L. (1976) Date growers Inst. Rpt. 53., pp 25-33.

Chandler, W.H. (1958) Evergreen orchards. Lea and Fabiger, Philadelphia, USA.

Chevalier, A. (1952) Recherche surles *Phoenix* Africains, R.B.A., Mai-Juin.

Comelly, A. (1960) Le palmier dattier en Israel. *Fruits*, **15**: 223-231.

Dowson, V.H.W. (1982) Date production and protection with special reference to North Africa and the Near East. FAO Technical Bulletin No35, pp. 294.

Galeb, H.H., Mawlood, E.A., Abbas, M.J. and Abdelsalam, S. (1987) Effect of different pollinators on fruit set and yield of Sayer and Hallawy date palm cultivars under Basrah condition. *Date Palm J.*, **5** (2): 155-173.

Linne (1734) cited in Keaney, T.H. (1906) Date varieties and date culture in Tunis. Washington, USDA Bureau of Plant Industry, Bulletin No.92.

Manson, S.C. (1915) Botanical characters of the leaves of the date palms used in distinguishing cultivated varieties. USDA, Washington, D.C. Bulletin No. 223.

Milne, D. (1918) The date palm cultivation in the Punjab. Govt. Printing Press, Lahore.

Moore, H.E. (Jr) (1963) An annotated checklist of cultivated palms. *Principes,* **7**: 119-182.

Munier, P. (1973) Le palmier- dattier- techniques agricoles et productions tropicales, Maison Neuve et Larose, Paris, pp.217.

Nixon, R.W and Carpenter, J.B. (1978) Growing dates in the United States U.S. Dept. of Agriculture, Agric. Infor. Bull. No. 207: USDA Tech. Document, pp.63.

Nixon, R.W. (1966) Growing dates in the United States. *Agric. Inf. Bul.,* USDA Technical Document, pp.63.

Pareek, O.P. and Vishal Nath (1996) Coordinated Fruit Research in Indian Arid Zone – A Two Decade Profile. Published by NRC for Arid Horticulture, Bikaner.

Pareek, O.P. (1990) Date Palm. In: Fruits Tropical and Subtropical (Eds. T.K. Bose and S.K. Mitra), Naya Prokash Calcutta, pp. 667-689.

Pereau-Leroy, P. (1958) Le palmier dattier au maroc. IFAC, Paris, Minist. Agric., pp. 142.

Pliny, C. (1489) The elder. Trans. Histonia naturale, Book xiii, Cap.iii, 3 columns on the palmae. Translated in to Italian by Cristofore Landioro Fiorentino. Published by Bartolamino de Zani de Portesio.

Popenoe, P.B. (1973) The date palm. (Ed. Henry Field), Field Research Projects, Coconut, Miami, Florida, pp.247.

Popenoe, W. (1913) Date growing in the old and new world. *West India Gardens*. Altadena, California, pp. 316.

Popenoe, W. (1938) The date. In: Manual of Tropical and Subtropical Fruits. The Mc millan Company New York.

Saaidi, M., Toutain, G., Bannerot, H. and Louvet, J. (1981) *Fruits,* **35**:241-249.

Swingle, W.T. (1928) Metaxenia in date palm. *J. Heredity* **19**: 257-268.

Vanzyl, H.J. (1983) Date cultivation in South Africa. Information Bulletin No504, compiled by the Fruits and Fruit Technology Research Institute, Department of Agriculture, Stellenbosch, RSA pp.26.

Zaid, A. and Wetde, P.F. (1999) Origin, geographical distribution and nutritional values of date palm In: Date palm cultivation (Eds. A.Zaid and E.J. Arias-Jimenez). Publisher, FAO and Agriculture Organization of the United States.

14

Breeding of Litchi

Botanical name: *Litchi chinensis* Sonn.

Family: Sapindaceae

Chromosome number: 2n= 2x=30

Scientific classification

Kingdom	:	Plantae- Plants
Division	:	Magnoliophyta–Flowering plants
Class	:	Magnoliopsida – Dicotyledons
Subclass	:	Rosidae
Order	:	Sapindales
Family	:	Sapindaceae – Soapberry family
Genus	:	*Litchi Sonn. – lychee*
Species	:	*chinensis Sonn.*

Litchi is a delicious fruit and generally consumed as a table fruit. It is one of the most popular fruits of India which is commonly taken as fresh or in dried forms. Litchi is also recognized as "Queen of the fruits". Dried litchi commonly known as 'Litchi nut' is very popular among the Chinese living all over the world. This subtropical fruit cannot tolerate frost in winter and dry heat in summer. An interesting feature of the litchi plant is the symbiotic growth of Mycorrhiza fungi on roots of the tree. These fungi form nodules on the roots of litchi. The trees and the fungus live together to benefit each other. Litchi makes an excellent canned fruit and highly flavoured squash is prepared from inferior fruits. Besides, fruit, other parts of the plant like leaves, bark and roots are also used for various purposes. In China leaves are used for making poultices, bark and roots for making decoctions for throat gargle.

14.1. Centre of diversity

Litchi is indigenous to Southern China, particularly the provinces of Kwongtung and Fukien. It was introduced to Burma and India by the end of the 17th century (Goto,1960, Liang, 1981) and to the West Indies by the 18th century.

14.2. Distribution

Litchi is indigenous to southern China, particularly provinces of Kwangdung and Fukien. The litchi reached West Indies in 1775, South Africa in 1869, the Hawaii Islands by 1873 and Florida in 1883. Other countries, where it reached include Vietnam, Indonesia, Japan, Formosa, Australia, New Zealand, Brazil, etc. Litchi reached India through Burma and was first introduced in Bengal during the end of the 17th century and then spread to other parts of the country. Litchi, which was introduced in the country in the 18th Century has adapted well to the climate in Eastern India, i.e. Bihar, Jharkhand, West Bengal, Tripura, Uttar Pradesh, Uttarakhand, Chhattisgarh, Punjab and Himachal Pradesh. Due to its increasing demand, the area under cultivation has increased manifold.

Table 1: Varietal distribution of Litchi in different states in India

Sl.No.	State	Varieties
1.	Bihar and Jharkhand	Deshi, Ajhauli, Green Purbi, China, Kasba, Bedana, Dehrarose, Shahi, Mandraji, Longia, Trikolia, Kaselia and Swarnrupa
2.	Uttar Pradesh, Uttrakhand and Himachal Pradesh	Early Large Red, Bedana, Late Large Red, Rose scented, Calcuttia, Extra early, Gulabi, Pickling, Khatti, Dehradun, Piyazzi
3.	West Bengal and Assam	Bombai, Ellachi early, China, Deshi, Purbi and Kasba
4.	Haryana and Punjab	Calcutta, Early seedless, Late seedless, Sedless-1 and Seedless-2
5.	Chhattisgarh	Sarguja-1 and Sarguja-2

Table 2: Important Litchi cultivars in the world

Sl.No.	Country	Cultivars
1.	China	Sum Yee Hong, Souey Tung, Fay Zee Siu, Haak Yip, Kawi May, No Mai Chee, Tong Bok, Hong Pay, Bo Dy, Choo Mah Zee, Seong Sue Way, Bah Lup, Kwa Lok, Chong Yun Hong, TinNaan, Sai Kok Zee, Heong Lai Ah Neong Hai, Soot Wau Zee
2.	India	Calcutta Late, Dehradun Early Large Red, Early Seedless, Late Seedless, Swarnroopa, Saharanpur Selection, Sabour Madhu, Sabour Priya, Muzzafarpur, Rose Scented, Purbi, Bombai etc.
3.	Florida	Brewster, Mauritius, Sweet Cliff
4.	Hawaii	Kwai Mi, Hak Ip, Groft, Brewster, Charley Tong, Hilo Tree Nursery
5.	Thailand	Amboina, Peerless, Cheng, Erewhom, Hong Thai, Jim Jee, Kaloke Bai Yaow, Kom, Kom Hom Lam Chiak, Luk Lai, Sampao Kao

Contd.

6.	Australia	Taiso, Hakk Yip, Seong Sue Wai, Wai Chee, Souey Tung, Kwai May Red, Kwai May Pink, No Mai Chee Standard, No Mai Chee Late, Bengal, Brewster, Groft, Muzzafarpur
7.	Israel	Mauritius, Floridian, Bengali
8.	Vietnam	Thieuthauhha
9.	Nepal	Mujapuri, Raja Saheb, Dehradun, Calcuttia
10.	Bangladesh	Bombai, Muzaffarpur, Bedana, China 3
11.	Brazil	Bengal
12.	South Africa	Mauritius. McLean's red
13.	Phlipines	Sinco, Tai so, ULPB Red

14.3. Germplasm resources

The maintenance of germplasm is limited to field gene bank, little efforts have been made on development of in vitro conservation. Thirty-six accessions are maintained at Dehradun, Krishinagar, Muzaffarpur, Pusa (Bihar), and Saharanpur (Rathore, 1993). Further, 42 accessions of litchi collected from Bihar, West Bengal and Jharkhand are maintained in the field genebank at National Research Centre for Litchi, Muzaffarpur (Anon., 2002). Under AICRP on subtropical fruits, a total of 48 germplasm accessions and 13 superior seedlings were maintained at different centres i.e. 13 each germplasm accessions and superior seedlings at BCKV, Mohanpur, 21 at GBPUAT Pant Nagar and 15 at RAU, Pusa, Bihar including hybrid Sabour Madhu. List of institute/ organization/ university working on litchi are as under:

- Horticultural Research Station, Kahikuchi (Assam Agricultural University)
- Research Sub-station, Manikchak, Malda (Bidhan Chandra Krishi Vishwa Vidhyalaya)
- Horticulture Research Station, Jachh (Y S Parmar University of Horticulture and Forestry)
- Litchi and Mango Research Station, Nagvota.
- Regional Fruit Research Station, Dhaulakuan, Sirmur
- Fruit Research Station, Gandevi (Navsari Agricultural University)
- Regional Research Station, Ambikapur (Indira Gandhi Krishi Vishwa Vidhyalaya Krishinagar)
- Dr. Rajendra Prasad Central Agricultural University, Pusa, Bihar
- Agricultural Research Institute, Patna
- Bihar Agricultural College, Sabour

- National Research Centre for Litchi, Muzaffarpur
- HARP, Ranchi

14.4. Breeding objectives

- To develop precocious, prolific and regular bearer varieties.
- To have large size of fruits with high number of fruits per panicle.
- To develop attractive skin colour, with high aril content, small seed size with high sugar and pleasant aroma.
- Good keeping quality of the fruits with high sugar content and good flavour
- To evolve resistant to biotic (pest and diseases) and abiotic (fruit cracking) stresses.
- Dwarf tree stature.
- To have varieties with wider agro-ecological adaptability.

14.5. Botany

The Litchi (*Litchi chinensis* syns Euphoria litchi, Dimocarpus litchi and Sapindus litchi) is an evergreen tree belonging to the family Sapindaceae (Soap berry family) to which Rambutan (*Nephelium lappaceum*) Longan, Pulasan and Spanish lime also belong. This family consists of 125 genera and more than 1000 species. Genus Litchi has two species i.e. *Litchi chinensis* and Litchi philippinensis (wild type found in Philippines, can be used as rootstock) Litchi tree is medium to large, much branched, round topped, evergreen in nature, attains a height of 14 m, sometimes even more. The plants have alternate, pinnate and leathery leaves having 5-7 opposite or alternate leaflets. The leaves are glossy green on upper side and greyish green on the under surface. Haploid chromosome number in different litchi varieties, grown in India was found to be 14 as reported by Chaudhuri (1940). According to Liu (1954) Litchi species, have been derived from more than one wild progenitor and the number of haploid chromosomes may be 14 or 15or 16 or rarely 17. According to him "Mountain Litchi" is clearly different from the varieties commonly cultivated. It is also resistant to frost. According to Smith (1976) and Lu et al. (1987) chromosome number of *Litchi chinensis* is 2n=2x=30 whereas for Rambutan is 2n=2x=22 and for Longan 2n=2x=20 (IBPGR, 1986).

14.6. Floral biology and pollination

In Litchi, it has been reported that temperature has a direct influence on flower initiation. Records for 8 years from a Brewster litchi orchard suggest that a soil moisture deficit and recurring temperature between 45° F (7.2° C) and 32° F (0° C) in autumn and winter before flowering promote flowering and fruiting in Florida. (Young, 1970, Maity and Mitra, 1990). The time of flower initiation varies according to genotypes and environmental conditions e.g. in North India flower bud differentiation starts in December and completed by the end of January (Shukla and Bajpai, 1974). Subsequently, emergence of panicles starts by the end of January to end of February (Chadha and Rajput, 1969, Pandey and Sharma, 1989). However, in South India, flowering starts in December and fruiting in April – May (Pandey and Sharma, 1989). In China flowering occurs during March to May (Li and Li, 1948).

Inflorescence of litchi is determinate (Panicle) and composed of several multiple branched panicles on the current season wood (Menzel, 1983). The inflorescence is compound racemose type and flowers occur in cyme. Banerji and Chaudhuri (1944) classified cymes in six group based on the occurrence of different sex of the flower i.e. cymes having male flowers, cymes having female flowers, cymes terminating into male flowers with lateral flower female, cymes terminating in to the female flowers with lateral male flowers, cymes terminating into male flowers with the lateral flowers of different sexes and cymes terminating into female flowers with the lateral flowers of different sexes. Depending upon the development of male and female parts. Liu (1954) reported three groups i.e. Staminate, functionally pistillate and imperfectly hermaphrodite. Whereas Mustard et al. (1953) classified litchi inflorescence as staminate, hermaphrodite functioning as female and hermaphrodite functioning as male. Duration of flowering (anthesis to pollination) usually ranges from 26 to 35 days (Chadha and Rajput, 1969). Flowers are self-sterile, requiring insects for pollination (Scholefield, 1982). Inflorescence is a panicle with greenish white or greyish flowers. In some plants, high percentage of axillary panicles was noticed, but no varietal differences could be established. When terminal panicles develop on vigorous branching, they are generally in clusters that may include as many as ten panicles or more. In male flower, the pistil is pinkish white or greyish white and abortive. In hermaphrodite flowers, though there are anthers, they never develop and dehisce as in male. In pseudohermaphrodites further, the anthers are as in males but the ovary is neither ill developed as in male nor well developed as in hermaphrodite (Das and Choudhuri, 1958). Anthesis and pollen dehiscence continued throughout the day and night with peak opening early in the morning between 6.0 to 9.0 AM (Chadha and Rajput, 1969). As soon as the stigma starts dividing into lobes, it becomes receptive and remains so upto three days after anthesis (Chaturvedi and Saxena, 1965). Ovary consists of 2-4 carpels, each with its own stigmatic lobe (Scholefield, 1982).

14.7. Inheritance pattern

Genetic variability and heritability with respect to fruit characters in nine litchi cultivars was studied by Sarkar et al. (1991). Study revealed significant genetic divergence for different fruit characters. The coefficient of variability was highest for dry seed weight (44.6-49.8) and broad-sense heritability and genetic advance were 89.1 and 82.3%, respectively. High heritability and high genetic advance were recorded for fresh seed weight, fruit weight, fruit volume and fresh and dry pulp weight. High heritability (42.7-79.5%) in other fruit characters concludes that these characters can be improved by selection. It was also observed that fruit and seed weight had strong positive correlations with total sugar, ascorbic acid, protein and tryptophan content, but it had significant negative correlations with acidity and phenol content (Singh et al., 1987). Thus, the selection for two characters (fruit and seed weight) produce nutritionally superior genotypes. According to Huang and Qui (1987) path analysis and partial correlation analysis revealed a direct positive effect of the peel and seed coat and aril size. A negative partial correlation between embryo and aril and direct repressive effect of the former on the latter were confirmed. This may be taken into account while breeding a variety less prone to cracking. In nature, the extent of outcrossing in litchi varies from 65 to 87% depending upon the nearness to the pollen source. Stern et al. (1993) reported clear exhibition of inbreeding depression with respect to fruit and seed weight if selfing is done.

14.8. Breeding methods and achievements

14.8.1 Introduction

From China which is the home of litchi, there is much scope to introduce some cracking/pest resistant strains/genotypes.

14.8.2 Selection

Most of the cultivated varieties have been developed through selection. The important cultivars developed through selection are Groft, Brewster, Saharanpur Selection, Calcutta, Bedana, Dehradun, Rose scented, Early large red, Late seedless, Swarn Roopa etc.

14.8.3 Hybrdization

Intervarietal hybridisation at Bihar Agricultural College, Sabour resulted development of Sabour Madhu (H-105) (Purbi x Bedana) and Sabour Priya (H73) (Purbi X Bedana) hybrids and these hybrids have been released for commercial cultivation. Intergeneric hybridisation was also attempted between litchi and longan which did not produce improved hybrids (Mc Conchie et al., 1994). It was also observed that hybrid progeny could develop only when litchi was used as the female parents. Morphologically hybrid plants were similar to litchi but leaves were small.

Table 3: Litchi cultivars developed through selection and hybridization

S.No.	Cultivar	Major Characteristics
1.	Groff	It is a seedling selection from the Hak Ip variety. The fruit quality is superior to Hak Ip, a Chinese variety.
2.	Brewster	It is quite similar to Chinese cultivar Chenzi (Chen Family Purple). It requires a relatively severe winter (Maximum temperature < 70C) to initiate flowering. Fruits are medium in size (20-22 g), slightly fragrant and sweet. Seeds are small to mediumsized. Flesh recovery is 65 to 75 per cent
3.	Saharanpur Selection	This is a chance seedling selection. It is late maturing. Fruits ripen in the third week of June. Fruit TSS is around 19.8 per cent. Average fruit weight is 17.6 g. It has a very low percentage of fruit cracking (2% only) compared to other cultivars.
4.	Swarna Roopa	This is the outcome of the selection made at Ranchi from different collections of litchi cultivars. It has attractive deep-pink fruit colour, small seed and high TSS/acid ratio. Fruits are highly resistant to cracking. Fruits mature a week later than the late cultivar China.
5.	Sabour Madhu (H-105)	This hybrid resulted from Purbi x Bedana. It has higher number of fruits (24) per panicle and ripens 8 days later than another late maturing cultivar, Kasba. It has higher TSS and aril percentage than Purbi. Fruit shape resembles Purbi
6.	Sabour Priya (H-73)	This is a product of Purbi x Bedana. It has better fruit quality than Purbi in terms of higher aril percentage and TSS content. The fruit shape has combination of both the parents. The fruit weight is higher than the better parent (Purbi).

Table 4: Litchi cultivars based on ideotype

Plant yield characters	Wider adaptability	Shahi, China, Calcuttia, Bengal
	Early and Late maturity and high yielding	Early Bedana, Early Large red, Late Bedana and Late large red
	Large fruit types	Shahi, China, Rose scented, Muzaffarpur, Bombai, Calcuttia,
	Smooth skin and red colour	-
	Good edible quality fruits	Shahi, China, Rose scented, Muzaffarpur, Bombai, Calcuttia, Muzaffarpur and Gulabi
	High quality of seedless fruits	Early bedana, Late bedana
Suitability for processing	Dry fruits or nut naking	Not tried at commercial level
	Canning	Shahi, Rose scented, China, Early large red and Early bedana
Tolerant to abiotic stresses	Tolerant to hot and dry wind	Calcuttia
	Resistant to fruit cracking	Swarn roopa

14.8.4 Biotechnological tools

Attempts have been made with regards to regeneration of plants through in vitro culture of immature embryo (Kantharajah et al., 1992, Zuang et al., 1996), anther culture (Fu and Tang, 1983) and use of one to ten days old leaf tissue (Yu, 1991).

14.9. Important cultivars and varieties of Litchi

Groft - It is a variety of China developed through selection from HakIp. Fruit quality of this variety is superior to HakIp.

Brewster - It is selection from Chenzi, this cultivar requires severe winter to initiate flowering. Fruit is medium size; flesh recovery is 65 to 75%.

Swarn Roopa - It is local seedling selection, released from Horticultural and Agroforestry Research Programme, Ranchi. Fruit is round, TSS 19%, pulp content is 76.62% and fruit is less susceptible to cracking.

Sabour Madhu - Developed from the cross of Purbi x Bedana, at Bihar Agricultural University, Sabour. It is late maturing variety, medium fruit size with 68.6% pulp and TSS 21%.

Sabour Priya - Evolved at Bihar Agricultural College, Sabour from the cross of Purbi x Bedana. Fruit quality is better than Purbi with respect to aril content and TSS.

Saharanpur Selection - It is chance seedling selection, late in maturity. TSS is 20%, average fruit weight is 17g and there is less fruit cracking.

Muzaffarpur - It is the most important variety of Bihar. Fruits mature in the first week of May in eastern India, it is less prone to cracking. Average fruit weight is 20g and TSS is 18%.

Dehradun - It is grown in Uttar Pradesh and Uttaranchal. Plant is medium vigour, fruits are obliquely heart shaped to conical, sweet and susceptible to sun burn and cracking.

Calcuttia -This cultivar is comparatively more successful in hot and dry areas. It is heavy bearer and less susceptible to sunburn and cracking.

Rose Scented - It is one of the most important cultivars in India having a distinct rose scented fruit of heart shaped. It is moderately susceptible to sun burn and cracking.

Seedless Late - Fruits have shriveled seeds. Ripening of fruits starts in the third week of June, fruit weight is 29g and TSS is 18%.

China - It is excellent variety grown in West Bengal. Plant is semi dwarf, TSS 18%, fruit is less susceptible to sun burn and cracking.

Bombai - It is important variety of West Bengal. Tree is vigorous, TSS of fruits is17% and it is suitable for canning.

Longia - This cultivar is well distributed in North Bihar, and is preferred for late maturity. The tree is medium in size, leaves are small and light in colour and it has compact panicles. Fruits are medium in size and the aril has an excellent aroma. Due to shy bearing habit, there is a declining preference for this cultivar.

Kasba - This is one of the important cultivars of Bihar. Trees are medium in vigour attaining a height of 6.0 m and spread of 7.0m. Fruits ripen in the third week of May to first week of June. Fruit yield is high with 85-100 kg/ tree. Fruit weighs between 23-27g, but the number of fruit is less. Pulp is grayish-white, soft, juicy, TSS 16.80 brix and acidity 1.14 per cent. Seed is smooth, dark in colour, shining, mostly oblong shaped, average weight 3.5 g. The skin, seed and aril percentage are 17.6, 19.5 and 62.9, respectively. Fruits are less susceptible to sunburn and cracking. Interestingly, the cultivar performs better in marginal soils as it has the capacity to absorb more nutrients

Early Large Red - This cultivar has fruits, which are slightly more than 3.4 cm long, usually obliquely heart-shaped; crimson to carmine red in colour, with green interspaces. The skin is very rough, firm and leathery, adhering slightly to the flesh. Flesh is grayish-white, firm, sweet and flavored and is of very good quality. It is a moderate bearer and early maturing.

Early Bedana (Early Seedless) - It is a popular early cultivar in Bihar, Uttar Pradesh, Uttarakhand, Punjab and Bangladesh. Tree has medium canopy attaining an average height of 5.0 m and spread of 6.2 m. It is regular bearing and medium yielder (50-60 kg/tree) cultivar. Fruits are medium sized (15-18g), oval or heart-shaped, with rough, deep red skin at maturity. Over all fruit quality is good.

CHES-2 - It is a late maturing cultivar developed as clonal selection from Bombaia at Central Horticultural Experiment Station, Ranchi. It has inside canopy bearing habit, which helps in reducing the sunburn as well as fruit cracking. Fruits are free from sunburn and cracking. The fruits are deep red, conical shaped and appear in a cluster of about 15-20. The fruit has an average weight of 21.3 g containing 3.8 g seed and 16.1 g pulp. The fruits have 19.80 brix TSS and 0.20 per cent acidity. The skin: pulp: seed ratio is 18.0: 66.7: 15.3.

Deshi - It is an early cultivar, mainly grown in Bihar and West Bengal. Trees are of medium vigour and attain a height of 5.5 m and spread of 6.5 m. Maturity starts in the third week of May. Fruit yield is high (90-100 kg/ tree). Bearing is regular and profuse and fruits are generally heavy (22-24 g). Fruit shape is oval to oblong-conical, and the fruits are bright rose-pink at maturity. The fruit pulp is grayish-white, soft, and juicy. The TSS in pulp is 20.80 brix and acidity are 0.35 per cent. Seeds are smooth, dark chocolate colour, mostly oblong shaped and 3.7 g in weight. The skin, seed and aril percentage are 15.6, 16.7 and 67.7, respectively. It is less susceptible to sunburn and cracking. This cultivar is suitable for canning.

Elachi (Elaichi, Ellaichi) - It is an important cultivar in West Bengal and has bright prospects for commercialization. The tree is moderately vigorous, 5-6m high, 6-7m spread and mostly regular bearer. Fruit yield is 50-60 kg/tree. It matures in the mid-season, i.e. in the first week of June. Fruits are mostly conical, a mixture of nasturtium red and marigold orange in colour, weighing 12-15 g. Fruit pulp is creamy-white, sweet, soft, juicy with agreeable flavour. TSS is 18.0^0 brix, sugars 11.5 per cent, acidity 0.45 per cent, pulp: seed ratio 6.91:1. Seeds are relatively small, shining with average weight of 1.5-2.0 g. The fruits are less susceptible to sunburn.

Shahi - Identified through germplasm evaluation and released in 2001, Fruits are attractive deep red in colour with highly fragrant pulp (65%) and TSS 20.4^{o}B. Earliest maturity by second week of May. It is recommended for Bihar, Jharkhand, Uttarakhand, West Bengal and Odisha. Time of planting is July-August. Spacing is 10 m x 10 m. No. of plant/ha is 100. Average yield is 100-120 kg/tree.

Ambika Litchi-1

This strain has been developed at R.M.D. College of Agriculture and Research Station, Ambikapur through selection from local material at village Pratappur, district Surguja. It gives an average fruit yield of 31.41 kg which is 25.64% higher than standard varieties Sahi and Rose. The variety has less fruit cracking (6%) as against 28.62% in Sahi. It has desirable levels of sugar and T.S.S. It is tolerant to leaf minor, grey weevil, stink bug, leaf folder and bark eating caterpillar.

Indira Litchi -2

Parentage collected from Shri Triyambak Sharan Singh of Village Pratappur, District Surguja. It has minimum fruit cracking (05.91%) during high temperature. Recommended ecology is upland and upland bunded irrigated condition in Northen Hill Zone of Chhattisgarh. It has TSS -17.90%, Pulp-72.2%, average fruit weight-16.69g and average yield of 42.62 kg/tree.

Sabour Litchi

It is released in 2015. This hybrid matures in first week of June. Average fruit weight is 23 g. It has no fruit cracking problem.

14.10. References

Banerji, I. And Chaudhuri, K.L. (1944) A contribution of the life history of *Litchi chinensis* Sonn. Proc. *Indian Academy Sci Section* B – 19-7.

Chadha, K.L. and Rajput, M.S. (1969) Studies on floral biology, fruit set and its retention and quality of some litchi varieties. *Indian J. Hort.*, **26**:124-129.

Chaturvedi, R.B. and Saxena, G.K. (1965) Studies on the blossom biology of litchi (*Litchi chinensis* Sonn.). *Allahabad Farmer*, **39**:10-13.

Chaudhury, J.K. (1940) A note on the morphology and chromosome number of *Litchi chinensis* Sonn. *Curr Sci.*, **9**:416.

Das, C.S. and Choudhury, K.R. (1958) Floral biology of litchi (*Litchi chinensis* Sonn.). *South Indian Hort.*, **6**:17-22.

Fu, L.F. and Tang, D.Y. (1983) Induction of plantlets from pollen of *Litchi chinensis Sonn. Acta Genet. Sini.*, **10**: 369-374.

Goto, Y.B. (1960) Lychee (Litchi) and its processing. *Pacific Rim Food Conf.*,**1**:15-23.

Huang, H.B. and Qiu, Y.X. (1987) Growth correlations and assimilate partitioning in the arillate fruit of *Litchi chinensis* Sonn. *Australian J. Plant Physiol.*, **14**(2): 181-188.

IBPGR, (1986) Genetic resources of tropical and sub-tropical fruits and nuts (excluding Musa). Int. Board Plant Genetic Resources, Rome, pp.123-125.

Kantharajah, A.S., McConchie, C. A. and Dodd, W.A. (1992) *In vitro* embryo culture and induction of multiple shoots in lychee (*Litchi chinensis* Sonn). *Annals of Bot.*,**70:** 153-156.

Li, L.Y. and Li, C.S. (1948) Blooming of the orange, lychee and longan tree during the winter of 1948 in Foochow. *Fikien Agril. J.*, **10**: 163-168.

Liang, J. (1981) The litchi: A historical review of the origin, utilization and development of its culture. *J. Agric. Trad. Bot. Appl.*, **28**:259-270.

Liu, S.Y. (1954) Studies on *Litchi chinensis*, Sonn.. *Diss. Abst.*, **14**:1018-1019.

Lu, L.X., Cheu J.L. and Cheu, X.J. (1987) A studies on chromosome number and meiosis in pollen mother cell in *Litchi chinensis* Sonn. *J. Fugian Agril. College*, **16**:224-228.

Maity. S.C. and Mitra, S.K. (1990) Litchi. In: Fruits Tropical and Subtropical (Eds. T.B. Bose and S.K. Mitra), Naya Prokash Calcutta, pp. 436-437.

Menzel, C.M. (1983) The control of floral initiation in lychee: a review. *Scientia Hort*., **21**:210-215.

Mustard, M.J., Liu, S.Y. and Nelson, R.O. (1953) Observations of the floral biology and fruit setting in lychee varieties. *Proc. Fla. State Hortic. Soc*., **66**: 212-220.

Pandey, R.M. and Sharma, H.C. (1989) The Litchi. ICAR, Publication.

Rathore, D.S. (1993) Conservation of fruit genetic resources. In: Advances in Horticulture Vol. I (Eds. K.L. Chadha and O.P. Pareek), Malhotra Publishing House New Delhi, pp. 67-76.

Sarkar, T.K., Bandopadhyay, A. and Gayen, P. (1991) Studies on genetic variability and heritability of some fruit characters in litchi (*Litchi chinensis* Sonn.). *Haryana J. Hort. Sci*., **20** (1-2): 56-59.

Scholefield, P.B. (1982) A scanning electron microscope study of flowers of avocado, litchi, macademia and mango. *Scientia Hort*., **16**:263-272.

Shukla, R.K. and Bajpai, P.N. (1974) Blossom bud differentiation and autogeny in litchi (*Litchi chinensis* Sonn.). *Indian J. Hort.*, **31**: 226-228

Singh, A., Abidi, A.B., Srivastava, S. and Singh, I.S. (1987) Correlations among physical and biochemical traits in litchi (*Litchi chinensis* Sonn.). *Narendra Dev J. Agr. Res*., **2**(2): 160-163.

Smith, P.M. (1976) In: Evauation of crop plante (Ed. N.W. Simmonds), Longman London, pp. 301-324.

Stern, R.A., Gazit, S., El-Batsri, R. and Degani, C. (1993) Pollen parent effects on outcrossing rate, yield and fruit characteristics of Floridian and Mauritius lychee. *J. Am. Soc. Hortic. Sci.*, **118**:109-114.

Young, T.W. (1970) Some climatic effects on flowering and fruiting of Brewster lychees in Florida. *Proc. Annu. Meet. Fla. Sta. Hort. Soc*., **83**:362-367.

Yu, Y.B. (1991) Study on some factors in tissue culture of lychee (*Litchi chinensis* Sonn). *Fujian Agril. Sci. and Tech*., **5:**17-18.

Zuang, Z., Zhou, L., Ma, X., Cheu, J. and Cai, J. (1996) Studies on types of embryoid in tissue culture of Litchi chinensis Sonn. *J.Fruit Sci.,*(China) **13:**25-28.

15

Breeding of Coconut

Botanical name: *Cocos nucifera* L.

Family: Arecaceae or Palmae

Chromosome number: 2n = 2x = 40

Scientific classification

Kingdom	:	Plantae- Plants
Division	:	Magnoliophyta–Flowering plants
Class	:	Liliopsida – Monocotyledons
Subclass	:	Arecidae
Order	:	Arecales
Family	:	Arecaceae – Palm family
Genus	:	*Cocos L. – coconut palm*
Species	:	*nucifera L.*

Coconut is a monocotyledonous woody plant belonging to the family Arecaceae or Palmae. It is presumed that the generic name *Cocos* and popular name coconut are derieved from Spanish word 'Coco' meaning "Monkey face" a probable reference to the three scars on the base of the shell resembling two eyes and a nose of monkey face. Coconut is valued most for its nut, the other components are coconut water, kernel, shell and husk. Coconut water is one of the most refreshing drinks which is also known for its medicinal value. It increases blood circulation in kidneys, causes profuse diuresis and eliminates mineral poisoning. It is a good substitute for saline glucose under gastroenteritis conditions. The coconut has been regarded as Kalpavriksha (tree of heaven) because all the parts of the palm is useful to mankind in one way or the other.

The dried kernels commonly called as copra is richest source of vegetable fat i.e. 60 to 67% oil. Coconut is major source of edible oil and it is propagated through seeds. India is largest producer of coir and its product. The coconut plant has about 4-6 years of juvenile phase with 60 years of productive life.

15.1. Centre of diversity

Coconut is essentially tropical plant. Over 90% of the world's total area and production lies between 20° N and 20° S latitude. Coconut originated and domesticated in Malaysia where it was widely grown. Although the precise centre of origin of coconut is still a disputed issue with two distinct zones of diversity i.e. South Pacific Islands of Polynesia and Melanesia and South East Asia and South America. Its pan tropical distribution is limited to 23° N and 23° S of equator. The major coconut growing areas of the world are found in Asia, Oceania, West Indies, Central South America, East and West Africa. In India it is commercially cultivated in Kerala, Karnataka, Tamil Nadu, Andhra Pradesh, Gujarat, Orissa and West Bengal.

15.2. Germplasm resources

At present CPCRI Kasaragod is maintaining the world's largest collection of 41 indigenous and 86 exotics cultivars (Nair and Ratnambal, 1993). Germplasm collection are also maintained at Pilicode under Kerala Agricultural University and four coordinated centers in Tamil Nadu, Karnataka, Andhra Pradesh and Maharashtra to assess their regional adaptability and their evaluation for various characters (agronomic characters, yield, drought, biotic stress etc.) at nursery stage, juvenile stage and bearing stage.

15.3. Problems

- Unique floral biology.
- Long juvenile phase.
- Prolonged interval between generations, vigorous growth (tallness) which creates difficulty in studying genetic traits.
- Heterozygous in nature.
- Experimentation needs longer duration to get results.
- Lack of reproducible asexual methods of rapid multiplication.
- More area is required to conduct the experiment due to more spacing .

15.4. Objectives

- To develop biotic and abiotic resistant cultivars.
- To evolve the precocious and prolific bearer ideotypes.

- To develop varieties having better quality of copra and oil.
- To evolve the variety with wider agroclimatic adaptability.
- To breed variety for dwarf stature
- To exploit hybrid vigour.

15.5. Botany

In coconut mainly three types of genotypes are found:

15.5.1 Tall palms

It is referred as variety typical Nar. most commonly cultivated in all the coconut growing area of the world. Tall types palm generally grow upto the height of 25-30m which have 6-10 years pre bearing age. They are normally cross pollinated as there is usually no overlapping of male and female phases. Copra content is over 150g/nut and oil percentage varies from 65 to 70.

Examples: Laccadive Ordinary, West Coast Tall, East Coast Tall etc.

15.5.2 Dwarf palms

It is some times referred to as variety Nana (Griff.) Nar. are characterized by their short stature, precocious bearer, easy to harvest and short life span. But they have tendency of irregular bearing habit.

Examples: Chowghat Orange Dwarf, Chowghat Green Dwarf, Gangabondam and Malayan Yellow Dwarf.

15.5.3 Intermediate types

Coconut palm has only one growing point no visible trunk is formed until the palm is several years old. Internodes are very short and adventitious roots are produced at the lowest nodes. The stem maintains the slow and steady growth throughout its life span. If stem is damaged, wound cannot be repaired by new tissue due to lack of cambium tissue in coconut tree. Trunk is columnar, light greyish brown in colour, 20-40 cm in diameter, usually erect but often heads towards the light at the edge of the plantations and along the direction of the wind.

Examples: Laccadive Micro, Andaman Giant and Nadesa

Coconut tree borne as terminal radiating crown, usually not overlapping. The crown consists of 25-35 leaves and central bud which is the most vulnerable part of the tree with leaves on various stages of development. The phases of rapid elongation take place for 4-6 months and a fully mature leaf remain on the palm for 2.5 to 3 years. Leaves are large paripinnate 4.5 to 6m long with 10-15 kg weight. Petiole is stout with clasping sheath at the base which is attached to the trunk. Number of pinnae are about 200-250 with apex acute and stomata is confined to lower surface leaf. About 200 stomata can be observed per square metre. Coconut root has got a very typical water and nutrient absorption mechanism. The small length of epidermal layers on roots acts as an absorbing zone. Initially root is creamy white which later on turns to red colour without root hairs.

15.6. Floral biology and pollination

Coconut is monoecious plant with numerous male and female flowers on each spadix which borne in the axil of each leaf of a bearing palm. Flowering starts at 6-12 years of age. Developing spadices may abort, early in their growth due to drought and other reasons. Inflorescence length is about 1-2m with central axis and about 40 laterals branches called spikelets. Each spikelet bear about 200-300 male flowers at the top and 1 or 2 female flowers at the base. There are six stamens in male flower, female flowers are 2-3cm in diameter having a large ovary with three locules and a trifid stigma with three nectary glands. In tall variety female flowers do not become receptive until all male flowers in same spadix have shed their pollens. In dwarf and hybrid varieties interval between two phases i.e. pollen dehiscence and stigma receptivity are less thereby increasing the chances of self-pollination. Due to the fact that maturation of inflorescence is a progressive process, pollen discharge and anthesis is continued for about 18-20 days (Thangaraj and Muthuswami, 1990). Genetically, dwarf palms are considered as autogamous and tall palms as allogamous but hybrids and few dwarf types exhibit both the types of pollination. Pollination in coconut is carried out by wind and insects (bees are major pollinating agent). Fruit is fibrous drupe, weighing 1.2 to 2.0 kg and the nut of commercial value consists of seed and endocarp. It takes around 12 months to mature. The fruits consist of exocarp (outer skin), mesocarp (fibrous layer) and endocarp which is ovoid shell alongwith 3 ridges and seed (one) situated at calyx with a thin brown testa attached to the edible endosperm.

15.7. Breeding methods and achievements

The organized coconut breeding was started for the first time in the world in 1916, at coconut research station Neleshwar of Kerala. Important breeding methods used in improvement of coconut are as under

15.7.1 Introduction

The different cultivars of coconut from Ceylon, Indo-China, New Guenea, Java, Indonesia, Thialand, Philippines, Fiji, Laccadives etc. were introduced in India by Madras Agriculture Department. The Cochin Department of Agriculture had also introduced cultivars from Malaysia, Sea Island, and Philippines and recently cultivars from Solomon Islands, Borneo, Seychelles, Panama, East Africa, West Indies, etc.

Important varieties introduced from different countries are Java, New Guenea, Cochinchines, Philippines, Laccadiv Ordinary, Laccadiv Micro and West coast Tall etc.

15.7.2 Selection

Mass selection can be used as effective methods in improvement of coconut. In this method superior mother palms are selected. The success of this method depends on the ability of breeder and heritability of traits under consideration.

15.7.2.1 Selection of mother palm

Coconut palms are being selected on the basis of following characters (Pillai, 1993)

A. Yield

- It should be regular bearer
- Annual yield should not be less than 80 nuts per palm
- Copra content should be about 150g per nut

B. Age of the plant

It is advisable to select palms which have reached full bearing stage and giving higher yield for consecutively four years. Selection of very old tree (more than 60 years) should be avoided. Seed nut should be collected from established seed garden.

C. Nature and disposition of leaves

- A mother palm should have at least 30 fully opened leaves on the crown.
- The leaves should have short and strong petiole with wide leaf base firmly attached to the stem.
- The disposition of the leaves on the crown should be such that the leaf petioles of the lower whorl will provides adequate support to the developing branches thus reducing the possibility of bulking of branches and there by shedding of tender nuts in the immature stage itself.
- Tree with crown having leaf orientation as umbrella shape is ideally suited as mother palms.

- Tree having a drooping or upright position of leaves are generally avoided due to their poor yielding capacity.
- Nature and sequence of inflorescence initiation.
- Female flower distribution per spike determines the yield potential of tree.
- Generally, every leaf axil should be one inflorescence with several spikes.
- There should be 12 bunches on the crown.

D. Size and shape of nut

Size is variable factor and which varies from variety to variety. Generally, three shapes are common in coconut i.e. oblong, round and elliptical. Vigour of seedlings depend upon the size of the husked nut, thickness of kernel and size of embryo present in kernel. Palm producing barren nuts, irregular bearing should be avoided for selection as mother palm.

15.7.2.2 Collection of seed nuts

Mother palm selection would constitute an imperfect form of mass selection. Since the pollen donor is unknown and the female parent is heterozygous, it is not certain that generally high yielding palm would yield true to mother progeny.

15.7.2.3 Seedling selection

Since seed nut may not be genetically pure or uniform, rouging out inferior and unthrifty seedling in the nursery is as vital selection of mother palms and seed nuts. Nursery selection is based on characters attributed to yield of adult palm such as early sprouting, faster growth rate, splitting of unexpanded leaf into leaflets, seedling vigour in terms of girth at collar, height and number of leaves, besides, freedom from disease and pests (Menon and Pandalai, 1958).

15.7.2.4 Selection of prepotential palm

Harland (1957) has given the concept of prepotential palm and he defined prepotential palms as mother palms which, inspite of having indiscriminately pollinated by miscellaneous males, are sufficiently possessed of dominant yield factors to ensure that their offsprings are also high yielding. A prepotent palm one which is having stable gene combinations, particularly those controlling polygenic quantitative traits like yields which may be located on different chromosomes, tend to cohere but do not combine thus resulting in the 'enbloc' transmission of parental characters to the progeny even under random mating due to this coherence mechanism prepotent function as reservoirs for potential variability equilibrium between coherence and variation should be well established so that the organism is enabled to adjust to its dynamic environment (Clausen and Hiesey, 1960). This phenomenon would lead to functional homozygosity of a complex heterozygote like the tall variety of coconut constituting a prepotent palm (Iyer *et al.*, 1982).

Examples: Pratap from KVK, Dapoli, Lakshadweep Ordinary from CPCRI, Kasaragod, Kerala.

15.7.3 Hybridization

The possibility of making hybrids by controlled pollination when both male and female parents are known is designated as paired cross which formed the basis of a new method of tree breeding. As some of the high yielding paired crosses have given yield, more than twice the average, resowing of proved high yielding paired crosses, among the best of the paired crosses have given further increase in yield, so that with heavy manuring it reached a level about double that of the original unselected type. Process of identifying male transmitters can be speed up by the use of dwarf palm as female since they are largely self pollinated and are reasonably homogenous. The first coconut hybrid in the world was produced in 1934 by Dr J.S. Patel with West Coast Tall as female and Chowghat Green Dwarf as male parents (Anon., 1998).

The following cross combination have been attempted in coconut

- Dwarf x Dwarf
- Tall x Dwarf
- Dwarf x Tall
- Tall x Tall

15.7.3.1 Dwarf x Dwarf (DxD)

The crosses between DxD have not given satisfactory results with respect to yield potential traits.

15.7.3.2 Tall x Dwarf (T x D)

The hybrids possess desirable traits such as precocity in bearing, higher productivity than the parents. The important dwarf variety used as male parents are Dwarf Orange, Dwraf Green, Gangabondam and Malayan Dwarf yellow. It is reported that all the hybrids of Tall x Dwarf Green, Tall x Dwarf Orange and Tall x Gangabondam exhibited heterosis in the weight of nut, kernel content and nut water (Satyabalam *et al.*, 1970). Among the male parents Dwarf Orange and Gangabondam proved to be the best for production of economic hybrids with West Coast tall.

The superior performance of Gangabondam as male parents has been reported from Kesaragod, Mahwa and Coimbatore (Anon., 1981). In comparative trails involving West Coast Tall and LO as female parents and COD and Gangabondam as male parents the superior performance of LO x COD and CO x GB at Kasaragod has been reported (Anon., 1985). COD and Gangabondam are the desirable pollen parents and WCT and LO are desirable female parents.

Examples: Chandra Laksha (LO x COD), Laksha Ganga (LO x GB),VHC 1 (ECT x Dwarf Green), VHC2 (ECT x MYD), Vera Sankara (WCT x COD), Anandaganga (AO x GB), Kera Ganga (WCT x GB), Kera Sree (WCT x MYD)

15.7.3.3 Dwarf x Tall (D x T)

These hybrids have been found to be of higher production potential than T x D hybrids. The COD x WCT hybrid was found to have better nutrient utilization ability and responded well to lower levels of applied fertilizers (Khan *et al.*, 1986).

Example: Chandra Sankara (COD x WCT)

15.7.3.4 Tall x Tall (T x T)

The poor yield in tall combination may be due to the poor combining ability of the parents (Anon., 1992). Thus emphasis need to select promising tall parents based on combining ability traits.

15.8. Important coconut cultivars (Anon., 1998)

Cultivars	Growth habit	Copra content/ Nut (g)	Copra yield/ palm/ Year (kg)	Oil Content (%)
(A) Tall group				
West Coast Tall (WCT)	Tall	176	14.0	68
East Coast Tall (ECT)	Tall	125	8.7	64
Tiptur Tall (TPT)	Tall	178	15.3	68
Benaulim Tall (BENT) or Pratap	Tall	152	22.8	64
Lakshadweep Micro (LMT)	Tall	90	18	75
Lakshadweep Ordinary (LCT)	Tall	176	17.6	72
Andaman Ordinary (ADOT)	Tall	169	15.9	66
Philippines Ordinary (PHOT)	Tall	189	20.8	66
(B) Dwarf Group				
Chowghat Orange Dwarf (COD)	Dwarf	150	9.7	66
Chowghat Green Dwarf (CGD)	Dwarf	60	4.0	66
Gangabondam Dwarf (GBD)	Dwarf	153	10.2	67
Malayan Yellow Dwarf (MYD)	Dwarf	140	9.2	66

15.8.1 Hybrids of Coconut (Anon., 1998)

Cultivars	Growth habit	Copra content/ Nut (g)	Copra yield/ palm/ Year (kg)	Oil Content (%)
Chandrasankara (COD x WCT)	Semi tall	215	24.9	68
Kerasankara (WCT x COD)	Tall	187	20.2	68
Chandralaksha (LCT x COD)	Tall	195	21.3	69
Lakshaganga (LCT x GBD)	Tall	195	21.1	70
Keraganga (WCT x GBD)	Tall	201	20.1	69
Anandaganga (ADOT x GBD)	Tall	216	20.5	68
Kerasree (WCT x MYD)	Tall	216	28.1	66
Kerasowbhagya (WCT x SSAT)	Tall	196	22.7	65
VHC-1 (ECT x DG)	Tall	135	13.2	70
VHC-2 (ECT x MYD)	Tall	152	16.3	69
Godavariganga (ECT x GBD)	Semi tall	150	21.0	68

15.8.2 Coconut varieties released by CPCRI and SAUs for commercial cultivation

S.No.	Variety	Remarks	Area Recommended
1	Chandrakalpa	Drought tolerant High oil content-72%	Lakshadweep, Kerala, Tamil Nadu
2	Kerachandra	High Yield	All regions
3	Kalpa Pratibha	High nut, oil Yield, Tender nut, Drought tolerant	Kerala,Tamil Nadu
4	Kalpa Mitra	High nut, oil yield, Drought tolerant	Kerala,west Bengal,Andhra Pradesh
5	Kalpa Dhenu	High nut,oil yield, Drought tolerant	Andaman,Kerala, Tamil Nadu
6	Chowghat Orange Dwarf	Dwarf,tender nut	All regions
7	Gautamiganga	Dwarf,high Yield,Tender nut	Andhra Pradesh,Tamil Nadu
8	Kalparaksha	High nut, oil Yield in Rwd prevalent areas	Kerala
9	Kalpasree	Dwarf,Superior oil, High Yield in Rwd Prevalent areas	Kerala
10	Pratap	High Yield	Konkan Region, Kerala
11	VPM-3	High Yield, DroughtTolerant	Tamil Nadu
12	ALR 1	High Yield	Tamil Nadu
13	Kamrupa	High Yield	NE region-Assam
14	Kera Sagara	High Yield	Kerala
15	Kera Bastar	High Yield	Chattisgarh
16	Kera Keralam	High Yield	Tamil Nadu,Kerala
17	Kalyani Coconut	High Yield	West Bengal
18	Kalpatharu	Drought tolerant, Ball copra, High Yield	Kerala,Karnataka, TamilNadu

Kalpa Jyothi

It is released by Central ICAR-CPCRI, Kasaragod on 2014. It is suitable to Kerala, Karnataka and Assam with the yield of 20178 nuts/ ha/ year. Copra: 11-12 months and Tender nuts: 7-8 months. Dwarf palm. Yellow fruits suitable for tender nut and perennial.

Kalpa Surya

It is released by Central ICAR-CPCRI, Kasaragod. It is suitable to Kerala, Karnataka and Assam with the yield of 21771 nuts/ ha/ year. Copra: 11-12 months and Tender nuts: 7-8 months. Dwarf palm. Orange fruits suitable for tender nut and perennial.

Kalpa Haritha

It is released by Central ICAR-CPCRI, Kasaragod on 2014. Copra: 11-12 months Tender nuts: 7-8 months. Average yield is 20886 nuts/ ha/ year. It is suitable for Kerala and Karnataka. Tall Palms. Green fruits, tolerant to eriophyid mite.

Abhaya Ganga (GBGD x LCOT)

It is a dwarf x tall cross between Gangabondam green dwarf as female parent and Laccadive Ordinary tall as male parent. Semi Tall hybrid, early bearing comes to flowering in 38 to 40 months after planting. Highest oil content 72%. Recorded an increase in nut yield by 54, copra output by 95 and oil yield by 65 percent tender nut water content by 24 percent over local check (ECT) and 17, 10, 29 and 13.3 percent respectively over hybrid check (ECT X GBGD). The mean nut yield/palm/year was 136 with copra content of 170 g and oil content of 72 %. Moderately resistant to bud rot disease.

Gauthami Ganga

Dwarf stature (5.12 mts at 22 yrs) and early bearing comes to flowering in 36 months after planting. Higher quantity and quality of tender nut water and copra content ie., 59 and 26 % over East Coast Tall. Yields 85 to 94 nuts/ palm/year with copra content of 157 g /nut and oil content of 69 % with tender nut water of 447 ml with TSS-7.2 ^{0}Brix and Potassium content of 2035 ppm. It has good combining ability useful for crossing programmes for production of new hybrids.

Indira Nariyal – 1

Parentage Selection from IND-04 (material from CPCRI Kesargod, Kerla Maturity 84 months after planting Biotic stress Moderately susceptible to steam bleeding disease Abiotic stress Drought tolerant and low incidence of mite and rodents Recommended for Bastar Plateau Zone of Chhattisgarh Adopted under Irrigated condition of Chhattisgarh specially Jagdalpur, Kondagaon, Narayanpur, Dantewada, Sukma & Bijapur Average yield 15122 nut/ha/year (85.4 nut/palm/year with 2.46 t/ha copra)

Keramadhura

This variety is released in 2013. A superior genotype suitable for dual purpose (tender nut and copra) with more quantity of tender nut water (287 ml) with excellent quality. Number of nuts/plant 119; copra yield (196 g/nut).

Kalpa Samrudhi (MYDxWCT)

The mean annual yield is 117 nuts per palm. The copra yield is 4.38 t/ha and oil is 3.04 t/ha. The hybrid is suitable for tender nut purpose. This hybrid was recommended for cultivation in Kerala, and Assam.

Kalpa Sankara (CGDxWCT)

The mean annual yield is 85 nuts per palm. The copra yield is 2.5 t/ha and oil is 1.69 t/ha. This hybrid was recommended for cultivation in root (wilt) disease prevalent tracts of Kerala.

Kalpa Sreshta (MYD x TPT)

The mean yield is 167 nuts/palm/year, with estimated high copra out turn of 35.9 kg/palm/year or 6.28t/ha copra. The hybrid is suitable for tender nut purpose. This hybrid is recommended for cultivation in Kerala and Karnataka States.

Vynateya Ganga (PHOT x GBGD)

It is a tall x dwarf hybrid (Philippines Ordinary Tall x Gangabondam Green Dwarf). Semi tall hybrid, precocious comes to bearing in 48 months after planting. It is a dual purpose hybrid for yield (copra & oil) and tender nut water. Increased nut yield of 47 and 7, copra output of 119 and 22, oil yield of 120 and 17 and tender coconut water content of 23 and 17 percent over local check (ECT) and hybrid check (ECT X GBGD) respectively. Yields average nut yield of 118/palm/year with copra content of 190.50 g/nut and oil content of 66 %. High quantity of tender nut water (357 ml) with TSS-7.0 ^{0}Brix. Moderately resistant to Ganoderma, bud rot and stem bleeding diseases.

Godavari Ganga, Double Century, Kamrupa, Assam Green Tall, Kahikuchi Hybrid-1 are some other promising varieties of coconut.

15.9. References

Anonymous (1985) Proceeding of 7th workshop, All India Cordinated Coconut, Arecanut, Spices and Cashew Improvement projects Trivandrum, pp 182.

Anonymous (1992) Annual report for 1991-1992. CPCRI, Kasaragod Kerala.

Anonymous (1998) Promising coconut cultivars and hybrids. CPCRI, Kasaragod Kerala Technical Bulletin No.34 pp 1-10.

Anonymous, (1981) Progress Report for 1979-1980. AICRP on coconut and arecanut improvement, CPCRI, Kasaragod Kerala, pp45.

Clausen, J. and Hiesey, W.M. (1960) The balance between coherence and variation in evolution. *Proc. Natl. Acad. Sci.*(USA), **46**: 494-506.

Harland, S.C. (1957) The improvement of coconut palm by breeding and selection, cerculation paper No. 7/57. Coconut Res. Inst. Ceylon. Bull. No., 15 pp 14.

Iyer, R.D., Kuruvvinashetti, M.S., Mini, P.S., Prakash, P. and Raju, C.R. (1982) Some applications of tissue culture in coconut improvement. International Congress on Plant Tissue and Cell Culture, Tokyo, Japan.

Khan, H., Gopalasundaram, P., Joshi, O.P. and Nelliat, E.V. (1986) Effect of NPK fertilization on the mineral nutrition and yield of three coconut genotypes. *Fert. Res.*, **10**: 185-190.

Menon, K.P. V and Pandalai, K.M. (1958) The coconut palm- a monograph. Laden Central Coconut Committee, Ernakulum, pp384.

Nair, M.K. and Ratnambal, M.J. (1993) Genetic resources of coconut. In : Advances in Horticulture Vol. 9 (Eds. K.L. Chadha and P. Rethenam), Malhotra Publishing House, New delhi, pp51-63.

Pillai, R.V. (1994) Coconut seed production. In: Advances in Horticulture Vol.10 Part 2 (Eds. K. L. Chadha and P. Rethenum), Malhotra Publishing House, New Delhi, pp 655-670.

Satyabalan, K., Ratnam, T.C. and Kungan, P.V. (1970) Hybrid vigour in nut and copra characters of coconut hybrids. *Indian J. Agric. Sci.*, **40**:1088-1093.

Thangaraj, T. and Muthuswami, S. (1990) Coconut. In : Fruits Tropical and Subtropical (Eds. T.K. Bose and S.K. Mitra), Naya Prokash Calcutta, pp 336-385.

16

Breeding of Cashewnut

Botanical name: *Anacardium occidentale* L.

Family: Anacardiaceae

Chromosome number: 2n= 2x = 42

Scientific classification

Kingdom	:	Plantae- Plants
Division	:	Magnoliophyta–Flowering plants
Class	:	Magnoliopsida – Dicotyledons
Subclass	:	Rosidae
Order	:	Sapindales
Family	:	Anacardiaceae – Sumac family
Genus	:	*Anacardium L.*
Species	:	*occidentale L.*

Cashewnut is an important delicious and dollar earning cash crop of India. This crop was remained neglected due to its poor yield potential and prolonged harvesting season. Cashew was introduced mostly for soil conservation purposes. In the beginning of the 20th century cashew kernels gained importance in the international trade as edible nut and the crop began to be considered as having commercial value. Cashew is commercially grown for its kernels although cashewnut shell liquid (CNSL) and apples are also valuable by - products. It is considered to be the most important edible nut in the world trade (Ascenso, 1986). India is the largest producer and exporter of cashewnut in the world. Cashew kernel is the rich source of protein, carbohydrate, unsaturated fats and minerals like calcium, phosphorus and iron. Cashew proteins are complete with all essential amino acids. Fats of cashewnut are considered as complete, attractive, easily digestible and non fattening, it can be used by both olds and infants. The cashew apple juice is rich in vitamin C and it has some medicinal properties.

16.1. Centre of diversity

Cashew tree is believed to be a native of Brazil from where it was dispersed to many of tropical areas. It was introduced in the Malabar Coast probably served as a locus of dispersal to other countries in India and South East Asia (Decosta, 1578). It is presumed that about the same time it was also introduced in the East African countries. Bailey (1958) has described the *Anacardium* as a small genus with only eight species, however, 20 species of *Anacardium* are known to exist in Central and South America. Till date only *A. occidentale* was available in India but at National Research Centre on Cashewnut three other species of *Anacardium* viz. *A.microcarpum*, *A. pumilum* and *A. orthonianum* are also available. However only *A. occidantale* is commercially cultivated in all the countries for the edible kernels.

Table 1:

S.No.	Species	Centre of diversity
1.	*A. Brasiliense*	Brazil
2.	*A. curatellaefolium*	Brazil
3.	*A. encardium*	Malaysia
4.	*A. giganteum*	Brazil
5.	*A. humile*	Brazil
6.	*A. mediterraneum*	Brazil
7.	*A. nanum*	Brazil
8.	*A. occidentale*	Brazil
9.	*A. rhinocarpus*	Columbia, Venezuela
10.	*A. spruceanum*	Brazil
11.	*A. microsepalum*	Brazil
12.	*A. corymbosum*	Amazon region
13.	*A. excelsum*	Brazil
14.	*A. parvifolium*	Central America
15.	*A. amilcarianum*	Amazon region
16.	*A. kuhlmannianum*	Brazil
17.	*A. regrense*	Brazil
18.	*A. rondonianum*	Brazil
19.	*A. tenuefolium*	Brazil
20.	*A. microcarpum*	Amazon region

Source: Nambiar *et.al.* (1990)

16.2 Distribution

Cashewnut is distributed in India is mainly confined to the peninsular areas. It is grown in Kerala, Karnataka, Goa and Maharashtra along the west coast of the country and in Tamil Nadu, Andra Pradesh, Orissa and West Bengal along the east coast of the country. It is being cultivated in Chhattisgarh, North Eastern states (Assam, Manipur, Meghalaya, Tripura and Nagaland) and on Andaman and Nicobar to a limited extent.

16.3. Germplasm resources

The early attempts for germplasm collections were made during early fifties with the sanctioning of adhoc schemes in Composite Madras, Travancore, Cochin and Bombay. Research Stations at Kottarakkara (Kerala), Ullal (Karnataka), Vridhalum (Tamil Nadu), Bapatla (AP), Vengurla (Maharashtra) taken up the programme of collection of locally available elite type plants for further evaluation and selection.

Table 2: The number of accessions available at different centres are as under

S.No.	State	Research Station	No. of Accession	
			Seedling	Clonal
1.	Andhra Pradesh	Cashew Research Station, Bapatla	87	26
2.	Karnataka	CPCRI Regional Station, Vittal/Shantigodu	292	-
		NRC on Cashew, Puttur	-	300
		Agricultural Research Station, Ullal	111	-
		Agricultural Research Station, Chintamani	75	-
3.	Kerala	Cashew Research Station, Anakkayam	84	-
		Cashew Research Station, Madakkathara	93	68
4.	Maharashtra	Regional Fruit Research Station, Vengurla	80	77
5.	Orissa	Cashew Research Station, Bhubaneshwar	67	-
6.	Tamil Nadu	Agricultural Research Station, Vridhachalam	219	68
7.	West Bengal	Regional Research Station, Jhargram	81	08

Source: Bhaskar Rao and Swami (1994)

At ICAR Research Complex Goa, nine genotypes of cashewnut are being maintained for further evaluation. In-situ conservation of cashew germplasm is done only in the Amazon forests of Brazil. About one thousand collections including 69 exotic collections from Brazil, South Africa, Mexico, Tanzania, Malaysia, Nigeria and Srilanka are under evaluation for yield and associated characters and quality of nuts and apple at National Research Centre on Cashew, Puttur.

16.4. Objectives

- To evolve varieties having high yield potential with good quality nuts.
- To develop varieties with dwarf stature.
- To breed varieties resistant against to biotic stresses (tea mosquito, anthracnose).
- To evolve varieties suitable for export.
- To breed varieties with bold kernel.
- To develop suitable rootstock having dwarfness, stress resistance and easily asexually propagated.

16.5. Botany

Cashew belongs to the family Anacardiaceae which consists of 60 genera and 400 species. The genus *Anacardium* varies largely with respect to the size and shape of the peduncle, nut and leaves. Cashew is an evergreen perennial tree with extensive tap root system. Tree may grow upto the height of 15m. The branches rests on the ground a few meters away from the trunk and branches of older tree, which are grown without being disturbed may creep over soil to a considerable distance, sometimes rooting takes place where they touch the soil. Such growth habit makes the tree suitable for control of soil erosion. Leaves are glabrous, leathery oblong to obovate and rounded at the tip, simple entire, pinnately arranged, each leaf contain 20 pairs of veins arranged alternately. True fruit is kidney shaped nut whereas cashew apple is false fruit which is swollen pedicel. The economic life of cashew is about 30-40 years.

16.6. Floral biology and pollination

Cashew is polygamomonoecious plant which bears inflorescence having male and hermaphrodite flowers. Inflorescence is a terminal panicle either conical or pyramidal in shape. The time from first appearance of inflorescence to opening of first flower take place about 5-6 weeks. Panicle contains about 200-1600 flowers (Damodaran *et.al.*,1966). The anthesis takes place between 11.30 AM to 2.00 PM according to him 94% pollen fertility is reported in cashew. Stigma remain receptive for 48 hours from anthesis. Cashew is cross pollinated crop and pollination is carried out by wind, insect and ants (Smith, 1958). According to Ghosh (1988) there are 4 distinct flowering phases in cashewnut:

(1) First mixed phase with 5.04% male flowers and 13.87% hermaphrodite flowers.

(2) First male phase with 15.54% male flowers and 8.13% hermaphrodite flowers.

(3) Second mixed phase with 10.30% male flowers and 19.24% hermaphrodite flowers.

(4) Second male phase with 21.78% male flowers and 6.10% hermaphrodite flowers.

16.7. Inheritance pattern

Genetic or molecular basis of inheritance has yet to be studied in cashewnut. Attempts have been made to correlate yield with ratio of bisexual flowers, short and synchronized flowering phase, flowering intensity, number of fruits per panicle, nut size, shelling percentage and morphological characters. The yield of nuts from a plant is proportional to the number of fruit set and the total number of flowering shoots per unit area (Anon., 1975). It was observed that the weight of nut had a positive correlation with girth and internodal length. Damodaran *et.al.* (1979) reported that there is positive correlation between the proportion of bisexual flowers and gross yield of nuts on young trees. Kumaran (1985) observed highly significant correlation between seed and seedling characters and also between seedling and juvenile phase. Characters such as seed weight seed length, length from cotyledons to first leaf and canopy size had direct effect on yield. However he observed that there is significant negative correlation between nut yield and days taken for germination.

16.8. Breeding methods and achievements

16.8.1 Introduction

Cashew was introduced in to India in Goa region and Malabar Coast during 16th century. However, due to small quantity of seed/tree, thus there was a limited genetic base from which all the present-day cultivars in the country have been developed. This may be the reason for low variability in the cultivated cashew in India.

16.8.2 Selection

Survey was undertaken by various centres in India to identify and select elite cashew plant and their evaluation and multiplication for cultivation. Vengurla-1 was first released variety from KVK, Dapoli in 1974. Other important varieties of cashewnut are BLA-139-1, K-19-1 K-10-2, K22-1, K10-1, K-16-1, T-1, Murude ST-94, Seed Farm Selection-4, BPP-241, T-14 and Goa-1developed through selection.

16.8.3 Hybridization

It is one of the potential tools to evolve new variety. Cashew is highly heterozygous plant due to allogamous in nature.

16.8.3.1 Hybridization technique

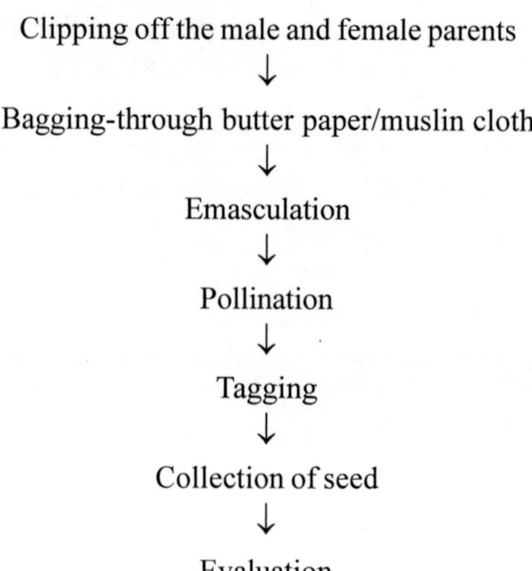

16.8.3.2 Achievements

S.No.	Hybrids	Parents
1.	V-3	V-1x Vetore56
2.	V-4	Midnapur red x Vetore56
3.	V-5	Ansur Early x Mysore Kotker
4.	V-6	Vetore56 x V-1
5.	BPP-1	Tree No.1 x Tree No.273
6.	BPP-2	Tree No.1 x Tree No.273
7.	BPP-8	Tree No.1 x Tree No.39
8.	Priyanka	BLA 139-1 x K30-1

16.8.4 Mutation

Occurance of natural bud sports in woody perennial tree is very rare. Induction of mutation with chemical or irradiation has not been practiced regularly. The work on mutation in cashew has been initiated at FRS, Vengurla in 1985. The irradiation of cashew bud stick with 1,2 and 3 kr. gama rays at BARC, Bombay. The soft wood grafting by using bud stick treated with 3kr gama rays causes 100% mortality. Whereas those subjected to 1 and 2kr dose gave 100% sprouting and variation in respect to phenotypic characters (Anon., 1985).

16.9. Brief description of important varieties

BPP-1

This cultivar was released from Cashew Research Station, Bapatla, Andhra Pradesh. It is a hybrid between Tree No. 1 and Tree No. 273. The mean yield is around 10.0 Kg/tree, having nut and kernel weight of 5.0 and 1.3 g respectively. Shelling percentage is 27.5 and kernel grade of W 400. The plant has extensive branching and open canopy. Flowering takes place between first fortnight of February to first fortnight of April and fruiting takes place between April-May.

BPP- 2

This cultivar is also a hybrid of same parentage i.e. Tree No.1 x Tree No. 273. The yield potential is about 11.0 Kg/tree. The nuts are smaller than BPP-1 and weighs around 4.0 g. Shelling percentage is 25.7 percent. Flowering takes place during second fortnight of April. Fruiting takes place during April-May.

BPP-3

It is collection from Andhra Pradesh and shelling is about 28%.

BPP-4

It is selection from Eprupalam. The younger leaves of this plant has light pink pigmentation and cashew apple is of yellow colour. Shelling is nearly 23%in this cultivar.

BPP-5

It is collection from Andhra Pradesh accession Tree No.1. The shelling percentage is around 24 and nut weight is 5.2g.

BPP-6

This is selection from Tree No.56 from A.P. The nut weight is 5.2g with 24% shelling percentage.

BPP-8

Released in early Nineties by Cashew Research Station, Bapatla, Andhra Pradesh. It is a hybrid between T.No.1 x T.No.39. It has bold size nut having nut weight. of about 8.2 g, kernel weight 2.3g, shelling percentage is 29.0. This hybrid is characterized by compact canopy and intensive branching. It flowers during February-April and fruiting takes place between April-May.

Dhana

This hybrid was released from Cashew Research Station of Kerala Agricultural University, Madkakathara, Kerala, during the year 1993. It is a hybrid between ALGD-1 x K30-1. The yield potential of this hybrid is 17.5 Kg/tree with nut and Kernel weight of 9.5 and 2.2g respectively, shelling percentage is 28.0 with kernel grade of W210. It is a vigorous cultivar having compact canopy and intensive branching. Flowering season is December-February, followed by fruiting from February to April.

Kanaka

This hybrid has also been derived from the same research station. It is a hybrid between BLA-139-1 x H3-13. The average yield is highest among all the hybrids developed so far. It has an yield potential of 19.0 Kg/tree. The nuts are medium sized with an average weight of 6.8g kernel weight 2.1g and shelling percentage being 31. It has an open canopy and intensive branching habit. Flowering takes place from November-January followed by fruiting in between January-March.

Priyanka

Evolved at Cashew Research Station, Madakkathra, (KAU), Kerala and recommended for release during the year 1995 under the name "Priyanka". It is a cross combination between BLA-139-1 x K30-1. The yield potential of this hybrid is 16.9 Kg/tree. This is a bold nut cultivar which has export grade of W 180 and nut weight of over 10g. Kernel weight is 2.8g and shelling percentage is 26.5. This cutivar is also recognized for its intensive branching with an open canopy. Flowering is late and takes place between December-March, fruiting period is extended up to May.

Amrutha

This is a hybrid with parentage BLA-139-1 x H3-13, same as Kanaka. The average yield potential of this cultivar is 18.4 Kg/tree and has medium sized nuts with a mean weight of 7.2g and kernel weight 2.2g. Shelling percentage is 31.6 with kernel grade of W210. Flowering and fruiting period is short, it takes place between November-December and January-February, respectively.

V-1

It is selection from germplasm collected from Ansur village in Vengurla. The plant has intensive branching habit with yellow apple colour. Percentage of perfect flower is 8%.

V-2

This is selection from germplasm collected from West Bengal. It has very short flowering and fruiting phase with 8% perfect flower.

V-3

Evolved at Regional Fruit Research Station, Vengrula, Maharashtra is a hybrid between Ansur-1 x Vetore 56. It has an yield potential of 14.4 Kg/tree. The mean nut and kernel weight is 9.1g and 2.4g, respectively. Owing to its better shelling percentage (27.0), kernel out put is higher. It has an export grade of W210. This cultivar also has same growth habit as Amrutha. However, flowering period in this cultivar is slightly earlier (October- December) and fruiting takes place between January - April.

V-4

Evolved from the same research station, Vengrula, Maharashtra was released for commercial cultivation during the year 1982. It is a cross between Midnapore Red x Vetore 56. The hybrid has a yield potential of 17.2 Kg/tree. Nuts are medium in size (7.7g) weighing about 140 nuts per kilogram. The kernel weight. is 2.4g. Shelling percentage is 31 with export kernel grade of W 210. Apple is red coloured having mean weight of 45g. The plant type is spreading with extensive branching habit. Flowering takes place between November-December and fruiting from February-May.

V-5

This is a hybrid with parentage Ansur Early x Mysore Kotekar 1/61. The cultivar has a compact canopy, with highest perfect flower content i.e. 50.5% per panicle. Flowering and fruiting takes place between October-December. and January-April, respectively. The average yield potential of this hybrid is 16.6 Kg/tree with export kernel grade of W 400. The nuts are smaller (<5.0) with kernel weight of 1.3g and shelling percentage is 30.0

V-6

It is a cross between Vetore56 x Ansur-1. It has an average yield potential of 13.8 Kg/tree. Nuts are medium in size (8.0g), kernel weight 2.2g, having shelling percentage of 28.0. It has export kernel grade of W210. Canopy type is compact, branching habit is intensive, flowering period is October-December and fruiting takes place between March-April.

V-7

A hybrid between Vengrula3 x M10/4 has an yield potential of 18.5 Kg/tree. The nuts are bold in size and the average weight is 10.0. Kernel weight is estimated to be 2.9g with shelling percentage of 30.5. The plant canopy is medium with intensive branching habit. Flowering occurs between December-February and fruiting between February-April.

VRI-1

It is selection from germplasm collected from Vazhisodhanipalayam in South Arcot District of Tamil Nadu. It contains 14.6% perfect flower.

VRI-2

This is the only variety released at national level based on the multilocational trial conducted at six coordinating centres. This national variety is a selection from the germplasm (T. No. 1668) collected from Kattupalli village of Minjur block of Changelpattu district of Tamil Nadu and released in 1985. This variety is found adaptable over wide range of soils and regions. The percentage of bisexual flowers is 10 with a setting of 5-8 fruits per panicle. The average yield is about 7.4 kg/tree. The nut size is small with 5.1g nut weight with shelling percentage of 28.3%. The kernel grade is W 320. The colour of the apple is pinkish yellow. This variety is withdrawn from recommendation for cultivation.

VRI-3

This is a selection from seedling progeny of a high yielding tree collected from a village Edayanchavadi in South Arcot District of Tamil Nadu and was released in 1991. It has 12.1% perfect flowers. The average yield of this variety is about 10 kg/tree, thus the increase over VRI-2 and VRI-1 being 35 to 39% respectively. The nut size is medium with 7.2g nut weight and shelling percentage of 29.1%. The kernel grade conforms to W 210 export grade. This variety is picking up fast among farmers of not only of Tamil Nadu but also of other states.

Bhubaneswar-1

It is a selection from seedling progeny of WBDC V (Vengurla 36/3), a collection from Regional Fruit Research Station, Vengurla and released in 1989. Flowering season is from January to March with medium duration of 70 days. It has cluster bearing habit with about 12 fruits per bunch. This variety has average yield of 10 kg/tree with small nut size (4.6g nut weight). The shelling percentage is high (32%) with kernel grade of W 320. It has been found suitable for cultivation in the sandy and laterite soils of the East Coast.

Goa-1 (Balli-2)

Goa-1 was developed and released from ICAR Research Complex, Goa in 1999. It is the first cashew variety released from the state of Goa. It is a selection from accession Balli-2 which is originated from a tree located in Balli village of Quepem taluk of Goa. The average yield of Goa-1 is 7.0 kg/tree with nut weight of 7.6 g (range: 7.3–7.9 g) and the shelling percentage of 30.0 (range : 28.9–31.0%). Kernel weight is 2.2 g. The kernel grade is W 210. Apple colour is yellow and with average weight of 66.7 g and with average juice content of 68.0%. It is recommended to the state of Goa.

Goa-2 (Tiswadi-3)

Goa-2 was developed and released from ICAR Research Complex, Goa in 2007. It is a selection from Ela village of Tiswadi taluk of North Goa District. The average yield of Goa-2 is 5.5 kg/tree with nut weight of 9.4 g (range: 9.2 – 9.6 g) and the shelling percentage of 29.25. Kernel weight is 2.3 g. The kernel grade is W 210. Apple colour is yellowish orange with cylindrical shape and with average weight of 105 g. Juice content ranges from 68.0 – 72.0 per cent. It is recommended to the state of Goa.

BLA 39-4 (Madakathara-1)

It is a selection from seedling progeny of Tree No. 39 of Bapatla. The variety was released in 1987. The flowering season is from November to January. The mean yield is 13.8 kg/tree. The nuts are medium sized with 6.2g nut weight. Shelling percentage is 26.8. The kernel weight is 1.6g and kernel quality conforms to W 280. Apple colour is yellow with a weight of 52g. Reducing sugar content is 10.5%.

NDR 2-1 (Madakathara-2)

This is a selection from germplasm collection made from Neduvellur in Kerala maintained at CRS, Anakkayam. This variety was also released in 1987. The mean yield is 17 kg/tree. The nuts are bold (7.3 g nut weight) with shelling percentage of 26.2%. Kernel weight is 2g having a count of W 240 export grade. Apple colour is red and with weight of apple 63.3g. Reducing sugar content is 7.8%.

Sulabha

It is selection released in 1996 with compact canopy and intensive branching. It is bold nut type with 9.8 g nut weight. The tree yields 21.9 kg of nuts with high shelling percentage (29.4%). The kernel weight is 2.88 g and grade is W 210. It bears light orange apples.

Dharashree

It is hybrid released during 1996 with a parentage of T 30 X Brazil-18. It has yield potential of 15.0 kg / tree with nut weight of 7.8 g. Shelling percentage is 30.5. The kernel weight is 2.4 g and grade is W 240. It bears yellowish pink apples.

Akshaya

It is hybrid released during 1998 with a parentage of H-4-7 X K -30-1. It has yield potential of 11.0 kg / tree with nut weight of 11.0 g. Shelling percentage is 28.4. The kernel weight is 3.12 g and grade is W 180. It bears yellow apples.

NRCC Selection-1

This variety was released in the year 1989. This is a selection from segregating progeny of germplasm 3/8 Simhachalam (VTH 107/3) originally a collection from AP. It is a late flowering type (December – February) with a flowering duration of 82 days. The number of fruits per bunch is 5. The yield, on an average, is 10 kg/tree. The nut weight is 7.6g. The shelling percentage is 28.8% and the kernel grade conforms to export grade of W 210. Apple colour is yellow. This variety is withdrawn from recommendation for cultivation.

NRCC Selection-1

This is a selection from the segregating seedling progeny of 2/9 Dicherla (VTH 40/1) originally a collection made from Andhra Pradesh. This variety was released in 1989. It has a mid-season flowering habit (November – January) with a flowering duration of 74 days. The number of fruits per bunch is 3. The average yield is 9 kg/tree. The nut weight is 9.2g. The shelling percentage is 28.6% and kernel grade conforms to export grade W 210. Colour of apple is pink.

Bhaskara

This variety was released during March 2006 for coastal region of Karnataka. This is having midseason flowering habit (Dec-Mar) with a flowering duration of 60 days and has potential to escape from the attack of the tea mosquito bug (TMB) under low to moderate outbreak situation. But the regular insecticidal spray against TMB is essential under severe out break situation. The number of fruits per panicle (bunch) ranged from 4 -13. The average yield on 13th year was 10.7 kg/tree with highest yield of 19 kg/tree. The nut and kernel weight are 7.4 g and 2.2 g respectively. The shelling percentage is 30.6 and kernel grade conforms to export grade W240. The apple colour is pinkish orange and juice content is 67.5%. This variety is very popular among the farmers of Dakshina Kannada District of Karnataka and also in neighbouring districts of Karnataka and Kerala.

Chintamani-1

It is a selection from 8/46 Taliparamba, a germplasm collection from Taliparamba in Kerala and released in 1993 from ARS, Chintamani. This variety is recommended for plain region of Karnataka. Its flowering period is from January to April with 2-4 nuts per panicle. The average yield of this variety is 7.2 kg/tree as against the 2 kg/tree of the local varieties. The nut weight is 6.9g with shelling percentage of 31%. The kernel grade is W 210.

Chintamani-1

It is a seedling selection from ME 4/4 of ARS, Ullal and released in 2007 from ARS, Chintamani. This variety is also recommended for plain region of Karnataka. The canopy type is compact and with intensive branching. Its flowering period is from December to January. The average yield of this variety is 12.4 kg/tree. The nut weight is 7.9g with shelling percentage of 30%. The kernel weight is 2.35 g. The kernel grade conforms to W 210. The colour of the apple is red purple with average weight of apple of 70g. Juice content is 60%.

Vengurla-9 (V-4 x M-10/4)

It is released by RFRS, Vengurle, Dr.BSKKV, Dapoli, Maharashtra on 2015. It is suitable for Konkan region of Maharashtra with 60-70 days crop duration. Yield is 11-11.5 q/ha with bunch bearing character.

Bidhan Jhargram-2

It is released by AICRP-Cashew, RRS, Jhargram BCKV, Kalyani on 2014. Cropping duration is mid- season (80 – 90 days) with 24-26 q/ha. It is suitable in Red and laterite zone of East coast with bold nut and shelling is 32%.

Anakkayam-1: This varicty is rclcascd in 1987. This is selected from Bapatla Collection - Tree 139-1.

BPP-10

High yielding, cluster bearing bold nut type (Nut Weight 8.1 gram) with highest shelling percentage (29.3%), Kernels show the export grade of W 210, Average nut yield per tree is 20.89 kg. This variety shows the medium in pest incidence and also less susceptible to foliage, flower and nut feeding pests. The percentage of hermaphrodite flowers 55.21 % with early bi sexual phase.

BPP-11

Early and regular bearing cluster type which escapes drought/moisture stress during flowering and fruiting phase. Suitable for high density planting due to compact panicle. Nut weight is medium (6.8 gram) with high shelling percentage (28.5%). The kernel count shows export grade of W240. Average nut yield per tree is 17.2 kg. This variety shows the medium in pest incidence and also less susceptible to foliage, flower and nut feeding pests.

Indira Kaju -1

It is clonal selection from natural pollinated seedling. Matures 4 year after planting (1460 days). Nut weight is 10.50g. It has field tolerance to tea mosquito bug and cashew stem borer. Shelling percent is 28.65% and colour of mature nut is grey. Average yield is 15.53 kg/tree. Tolerant to low temperature and recommended for Bastar Plateau Zone, Jashpur and Raigarh (Zone VII of Indian Agroclimatic Zone).

Madakkathara-1: This variety is released in 1987. This is selected from Bapatla Collection - Tree-39-4.

Madakkathara-2: This variety is released in 1990. This is selected from NDR-2-1.

Mrudula: This variety is released in 1996. Selected from PTR-1-1.

Poornima: It is released in 2006. It is hybrid between BLA 139-1 x K-30-1. A high yielding (14.08kgnuts/tree/year), compact, intensively branching, mid season, cashew hybrid with export grade nuts, high shelling percentage and kernal weight.

Sree: It is released in 2010. This variety is tolerant to tea mosquito bug.

Table 3: Cashew cultivars recommended for different states (Awasthi *et al.*, 2004)

State	Cultivars	Parentage
Andhra Pradesh	BPP-4	9/8 Epurupalen
	BPP-6	T.No.56
	BPP-8	T.No.1xT.No.39
	VRI-2	M44/3
	Madakathara-1	BLA 39-4
Andaman and Nicobar Islands	Ullal-1	8/46, Taliparamba
	VRI-2	M 44/3
	Vegrula-1	Ansur-1
	Vengrula-4	Mid Red x Vetore 56
Chhattisgarh	Vegrula-4	MidRed x Vetore 56
	Madakathara-1	BLA 39-4
Goa	Vengrula-1	Ansur-1
	Vengrula-4	MidRed x Vetore56
	Vengrula-6	Vetore 56 x Ansur-1
	VRI-2	M 44/3
	Goa-1	Balli-2
Karnataka	NRCC Sel-1	VTH 107/3
	NRCC Sel-2	VTH 40/1
	Ullal-1	8/46 Taliparamba
	Ullal-2	3/67, Guntur
	VRI-1	M10/4
	VRI-2	M44/3
	Vegrula-1	Ansur-1
	Vengrula-4	MidRedxVetore 56
Kerala	Madakathara-1	BLA 39-4

Contd.

	Madakathara-2	NDR 2-1
	K-22-1	Kottarakkara 22
	Dhana	ALGD 1 x K-30-1
Maharashtra	Vengrula-1	Ansur-1
	Vengrula-4	MidRed x Vetore-56
	Vengrula-6	Vetor 56 x Ansur-1
	VRI-2	M44/3
Manipur	VRI-2	M44/3
	Madakathara-1	BLA 39/4
Orissa	Bhubaneshwar-1	Vengrula 36/3
	VRI-2	M44/3
	BPP-2	T.No.1xT.No.273
Tamil Nadu	VRI-1	M10/4
	VRI-2	M44/3
	VRI-3	M26/2
West Bengal	Jhargram-1	T.No. 16 of Bapatla
	Madakathara-1	BLA 39/4

16.10. References

Anonymous (1975) Prog. Report All India Coordinated Spice and cashewnut Improvement Project, CPCRI, Kasaragod.

Anonymous (1985) Annual Report of Kokan Krishi Vidyapeeth, Dapoli, pp.57.

Ascenso, J.C. (1 986) Potential of cashew crop. *J. Agric Intl.*, **38**:324-327.

Bailey, L.H. (1958) The Standard Encyclopedia of Horticulture Vol1 (17 Edn.), Mac Millan, New York pp.1200.

Bhaskar Rao, E.V.V. and Swami, K.R.M. (1994) Genetic Resources of Cashew In: Advances in Horticulture Vol. 9 part1 (Ed K.L.Chadha and P. Rethinum), Malhotra Publishing House New Delhi, pp.79-97.

Damodaran, V.K., Abraham, J and Alexander, K.M. (1966) The morphology and biology of cashew flower (*A occidantale*) 1 anthesis, dehiscence, receptivity of stigma, pollination, fruit set and fruit development. *Agri.Res. J. Kerala* **4**:78-84.

Damodaran, V.K., Vilaschandran, T. and Vasalkumari, P.K. (1979) Research on cashewnut in India, Kerala Agri. Univ., Trichur pp.79.

Decosta, C. (1578) Tratodo das progase Medicinas das Indias Orientais 1964 Edn. Junta de investigacoes do uttramar, Lisbon pp356.

Ghosh S.N. (1988) *The Cashew* **2**:14-16.

Kumaran, P.M. (1985) PhD thesis, IARI, New Delhi.

Nambiar, M.C., Bhasker Rao,E.V.V. and Pillai, P.K.T. (1990) Cashew. In: Fruits Tropical and Subtropical (Ed. T.K.Bose and S.K.Mitra), Naya Prokash Calcutta, pp.386-419.

Smith, F.G. (1958) Bee keeping operation in Tonganyka. 1940-57. *Bee World*, **39**: 29-36.

17

Breeding of Pineapple

Botanical name: *Ananas comosus* L.

Family: Bromeliaceae

Chromosome number: 2n= 2x = 50

Scientific classification

Kingdom	:	Plantae- Plants
Division	:	Magnoliophyta–Flowering plants
Class	:	Liliopsida – Monocotyledons
Subclass	:	Zingiberidae
Order	:	Bromeliales
Family	:	Bromeliaceae – Bromeliad family
Genus	:	Ananas Mill.
Species	:	*Comosus* (L.) Merr.

Pineapple is one of the most important commercial fruits of the world. It is good source of vitamin A, B and fairly rich in vitamin C, minerals like calcium phosphorus and iron. Fresh fruit of pineapple is used as dissert fruit in the form of slices. Fruit is also used for juice, squash, jam, mixed jam and in canned form, fruit cone is used for preparing candy. Some cultivars of pineapple are used for extraction of fibre. The important states of pineapple growing are Assam, West Bengal, Tripura, Kerala, Andhra Pradesh, Bihar, Karnataka and Madhya Pradesh (Bal, 1997).

17.1. Centre of diversity

Pineapple is believed to be originated in Brazil. The generic name Ananas is derived from the Indian "nana". The wild Brazilian pineapple (*Ananas microstachys* Lindle) is considered as ancestor of cultivated pineapple. It reached India during 1548.

17.2. Germplasm resources

In India much attention has not been given in the development of field gene banks. However, commercial cultivars are maintained at BCKV, Kalayani, Agriculture Research Station Kovvur (APAU), Department of Horticulture, College of Agriculture Jorhat, Regional Research Station, Diphu (Assam Agril. University), Department of Horticulture, College of Horticulture Navsari (GAU), Banana and pineapple Research Station, Kannara (KAU), College of Horticulture at Vallanikkara and Trichur etc. A list of 135 varieties was published by Johnson (1935), although some of them were synonymous.

17.3. Objectives

- To develop high yielding, early maturing varieties with wider geographical adaptability.
- Plant should be hardy, vigorous, capacity to produce good ratoon crop, leaves should be spineless.
- Fruit stalk should be short and strong.
- There should be flat eyes and small cones.
- Development of varieties resistant to biotic and abiotic stresses (e.g. multiple crown, fasciation, wilt, heart rot, root rot and nematode.)

17.4. Botany

Pineapple (*Ananas comosus* L. merr.) belongs to the family Bromeliaceae. Plant of pineapple just like small shrubs, basal leaves are long, slender and spiny. Roots are ariel, covered by relamen (loose mass of tissue at the tip of the root) which helps in absorption. Lower roots enter in the ground and upper are clasped with stem. Plants are poorly anchored in the ground. There are two genera in this family i.e. Ananas in which syncarp bears conspicuous coma of foliaceous bracts and plants produce slips but never produce stolon. Whereas in Pseudananas syncarp at maturity bears a minute inconspicuous coma of bracts and produces elongated stolons and no slips. Both the genera having basic chromosome number x=25, which is common throughout the Bromeliaceae but Ananas is typical diploid (2n=2x=50) while Pseudananas is tetrapolid (2n=4x=100). Some triploid genotypes with 2n=3x=75 (Cayenne BR59, Spanish GU75-2 and Dos Indios BR47) have also been reported (Dujardin,1991). Genera Ananas has five species i.e. *Ananas bracteatus*, *A. ananassoides*, *A. fritzmuelleri*, *A. comosus* and *A. erectifolius*. However, the genera Pseudananas is monotypic and has one species i.e. *P. sagenarius*. It is incompatible with (*A. comosus*) cultivated pineapple.

17.5. Floral biology

Pineapple plant generally flowers after attainment of certain vegetative growth and it is attained in11-12 months after planting and formation of at least 40 leaves. Pineapple plants produce one fruit in its life. Flowering in pineapple can be induced with the use of growth regulators. Among the auxins, NAA and Planofix have been reported to be very effective for induction of flowering.

Ananas comosus is the only self-incompatible species in genus, whereas other wild species and Pseudananas are partially self-fertile. According to Brewbaker and Gorrez (1967) pineapple showed gametophytic S-allele types of incompatibility and induction of polyploidy does not breakdown the incompatibility system (Pickers gill, 1976). After completion of the vegetative growth, terminal bud is terminated to peduncle which bears inflorescence. Many reddish-purple sessile flowers are arranged spirally on the axis, each subtended by a pointed bract. This stage can be termed as red bud stage. Inflorescence is panicle raceme often terminal spike with highly coloured bract. Flowers are regular, bisexual and inflorescence may consist of 100 to 200 florets. Each floret may consist of 06 perianth in two series, segments are free sometimes united, stamens six mostly remain attached to the base or adnate attached to whole length, ovary with three celled and three stigmatic lobes.

In pineapple, fruit develop parthenocarpically. It is not known whether pollination is required to initiate the fruit development. Self-pollination does not occur because pollen tube development ceases before reaching the ovary. Fruit with seed is harvested occasionally, but it is not known whether they result from cross pollination by insects (humming birds, honey bees, pineapple beetles) or an environmentally induced change in physiology which breaches the normal barrier to self-pollination. The chance of getting seeds with artificial crossing is fair because all diploid cultivars produce functional pollens and ovules. The ovules are borne in two rows near the top of each locule. Number of ovules per locule may be 14 to 20. Further, 2000 to 3000 seeds can be obtained from a fruit after hand pollination (Collins, 1960).

17.5.1 Classification of Pineapple

The horticultural classification of pineapple varieties of Hume and Miller (1904) is currently followed. They divided cultivated varieties of pineapple into 3 main groups viz.,

i) Cayenne

ii) Queen

iii) Spanish

Cayenne group is by far the most important group. Most of the varieties in India may be accommodated into anyone of the 3 groups. For example, Kew or Giant Kew, synonymous with Smooth Cayenne, grown most extensively in India, represents Cayenne group, and Queen another popular variety belongs to Queen group. Recently, Py et.al. (1987) classified cultivars grown throughout the world into 5 distinct groups, the additional two being 'Pernambkuco' and 'Mordilonus-Perolera- Maipure'. The varieties of Cayenne and Spanish group are dual-purpose ones whereas varieties of Queen group are grown exclusively for fresh-fruit markets, as they are not suitable for canning; owing to deep eyes.

17.6. Inheritance pattern

According to Collins, 1960 spineless form in which spines are restricted to the top few inches of the leaf are dominant to spiny wild type. Further, many commercial cultivars, possess spiny leaves. *Ananas ananassoides* contributes a distinctive flavour, disease resistance and ability to grow in cool, moist conditions. *Ananas bracteatus* is also a source of disease resistance. The cultivar Pernambuco is donar for good flavour and aroma, tender non-fibrous juicy fruits, early fruiting, resistant to heart and root rot. Queen can be donor for crisp, non-fibrous, deep yellow flesh, early ripening, and good fruit shape. Red Spanish is good source of vigour, resistance to wilt, heart and root rot. Singapore Spanish can be good donor for square-shouldered fruits with golden yellow flesh and the wild species are good source for vigour, resistance to various diseases and pests. Spiny tip and spiny characters are the phenotypic expressions of a single pair of alleles, with spiny tip being dominant (Collins and Kerns, 1946). Homozygous SS and heterozygous Ss produce spiny tips, recessive ss may give rise to spiny plants progeny (Collins and Kerns, 1946). The piping and non-piping characters are controlled by another non linked pair of alleles with the gene P (piping) being epistatic to S and s. The homozygous PP genotype produces pronounced piping than Pp genotypes. Frequent mutations of S to s occur. (Sen, 1996). *Ananas erectifolius* is having Se gene of smooth tip leaves (Collins, 1960).

17.7. Breeding methods and achievements

17.7.1 Selection

Most of the improvement of the pineapple has taken place by simple selection of mutant clones within cultivars and by hybridization between cultivars followed by selection of the highly heterozygous progeny. Selection in the Singapore Spanish population in Malaysia had led to a new cultivar, Masmerah. It is more vigorous, possessing more leaves, which stand more erect and bears heavier fruit than the parent cultivar (Wee, 1974). Several such selections have been made in different pineapple-growing areas.

17.7.2 Hybridization

A cross between Red Spanish and Cayenne has led to the development of a new hybrid PR-1-67 in Puerto Rico (Remirez, 1970). This hybrid shows better plant vigour and resistance to wilt disease. However, self-fertile somatic mutants obtained from cultivar Cayenne show a loss of vigour on selfing and heterosis on crossing. A hybrid (H-7) has been produced by crossing Valera Monendi x Kew. This hybrid produces large fruits, individually weighing on an average 3.0-3.5 kg.

17.7.3 Mutation

Utilization of mutation induction in this crop seems to be quite feasible. However, due to wide natural variation, limited attempts have been made for induced mutations. In Kerala irradiation of plants of cultivar Kew and Mauritius led to growth retardation and premature suckers. Merz (1964) reported the induction of self-fertile mutants by X-irradiation of pollen during meiosis. Several morphological mutations were found when 1.0 to 1.5-month-old detached slips were treated with chemical mutagens like Ethylenimine (EI), N-Nitroso-N-Methyl Urethane (NMU) and Diethyl Sulphate (DES). One mutant produced spineless plants from cultivar Queen and was economically significant (Singh and Iyer, 1977).

17.7.4 Biotechnological tools

Attempts have been made for rapid multiplication of the plants through micropropagation by using different kinds of explant i.e. leaf base, shoot base, excised lateral buds, meristem tips from crown etc. (Ma and Wang, 1977, Mathew and Rangan, 1979, Drew, 1980, Liu *et al.*, 1987). In the crosses where fertilization fails due to incompatibility, embryo culture technique can help to rescue the hybrid. The genetic transformation of pineapple clones has been attempted with the objective to acquire ability to introduce desirable genes (Thomson *et al.*, 1998).

17.8. Major group and varieties of Pineapple

Important group	Cultivars	Important Characteristics
Abacaxi (Brazilian)	Abakka, Amarella, Papelon, PinaValera, Sugar Loaf, Venezolana, Vermelho, Yupi.	Fruit conical in shape, weighing about 1.2 to 1.5 Kg, yellow rind, pale yellow or white flesh, sweet, tender, and juicy, leaves spiny, disease resistant, fruit neither processed nor survine export well, grown for fresh domestic consumption.
Cayenne	Boron, Baraonne de Rothschild, Champaka, Cayenne Lisse, Emeralda, Gautemalan, Giant Kew, Hilo Cayenne, Kew, Rothschild, Smooth, St. Michael, Smooth Cayenne, Typhone	Fruit shape is cylindrical with a slight upward tapering and flat eyes, most suitable for canning, fruit weight is about 1.8 to 3.0 Kg. Colour of rind is dark orange and flesh is pale yellow to yellow, sweet, mildly acid with low fibre and a tender juicy texture, leaves smooth with few spines near the tip, highly susceptible to mealy bug wilt, suitable for export.
Maipure	Bumanguesa, Legrija, Maipure, Marquita, Monte Lirio, Perolera, Piamba de, Rondon.	It is sweeter than the Cayenne, aromatic, fibrous but tender and very juicy, leaves completely smooth, grown for fresh consumption, fruit ovoid to cylindrical in shape, fruit weight is about 0.8-2.9 Kg, rind colour is yellow to dark orange or red, flesh is white or deep yellow.
Queen	Alexandria, Bakhat, James, Jhaldhup, MacGregor, Mauritius, Netal, Queen, Ripley,Victoria, Z. Queen.	Fruit shape is conical, weighing about 0.5 to 1.1 Kg, rind is yellow, flesh is deep yellow, less acidic than Cayenne, sweet, low in fibre, spiny leaves, more disease resistant than the Cayenne.
Spanish	Betek, Cabezona, Castilla, Espanola Roja, Gandol, GreenSelangor, Masmerah, Nangka, PR1-67, PR1-56, Red Spanish, Singapore Spanish.	Fruit shape is globose, fruit weight is 0.9-1.8Kg, rind deep reddish, flesh pale yellow to white with spicy acid taste, fibrous texture, leaves spiny, resistant to mealy bug wilt, susceptible to gummosis, suitable for export and fresh consumption.

17.9. References

Bal, J.S. (1997) Fruit growing. Kalyani Publishers, pp 106-114.

Brewbaker, J.L. and Gorrez, D.D. (1967) Genetics of self-incompatibility in the monocot genera, Ananas (pineapple) and Gasteria. *Amer. J. Bot.*, **54.**

Collins, J.L. (1960) The Pineapple, Leonard Hill Ltd. London, pp.294

Collins, J.L. and Kerns, K.R. (1946) Inheritance of three leaf types in the pineapple. *J.Hered,* **37**: 123-128.

Drew, R.A. (1980) Pineapple tissue culture unequalled for rapid multiplication. *Queensland Agric. J.,* **106**: 447-451.

Dujardin, M. (1991) Cytogenetics of pineapple. *Fruits* (Paris), **46**:376-379.

Johnson, M.O. (1935) The pineapple. Paradise of the Pacific Press, Honululu, Hawaii.

Liu, L.J., Rosa-Marquez, E. and Lizardi (1987) In vitro propagation of spineless Red Spanish pineapple. *Phytopathology*, **77**(12): 1711.

Ma, S. and Wang, S. (1977) Tissue culture propagation of pineapple. *J. Hortic. Soc.China*, 23 **(3)**: 107-113.

Mathews, H.V. and Rangan, T.S. (1979) Multiple plantlets in lateral buds and leaf explants in vitro culture of pineapple. *Sci. Hort.*, **11**: 319-328.

Merz, G.S. (1964) Study of self-incompatibility in the pineapple. *Agril. Res. Pretoria,* **2-6:** 561-562.

Pickersgill, B. (1976) Pineapple. In: Evolution of crops plants (Ed. N.W. Simmonds) Longman, London, pp 14-18.

Py, C. Lacoeuilhe, J.J. and Teisson, C. (1987) The pineapple: cultivation and uses. G. P. Masonneuve and Larose. 15, Rue –Victor, Paris. 568pp.

Remirez, O.D. (1970) Pineapple PR1-67, *Rev. Café, Puerto Rico*, **25**:117-20.

Sen, S.K. (1996) Pineapple. In: Fruit Tropical and Subtropical (Eds. T.K. Bose and S.K. Mitra), Naya Proskash Calcutta, pp. 252-279.

Singh, R. and Iyer, C.P.A. (1977) Standardization of dosimetry and Technique for inducing mutations in pineapple. In: Fruit Breeding in India (Ed. G.S. Nijjar) Oxford and IBH Publishing Co, New Delhi.

Thomson, K.G., Thomas, J.E. and Dietzgen, R.G. (1998) Retrotransposon-like sequences integrated in to the genome of pineapple (*Ananas comosus*). *Plant Molecular Biology*, **38**(3) :461-465.

Wee, Y.C. (1974) The Masmerah pineapple. *World Crops*, **26**: 64-67.

18

Breeding of Underutilized Fruits

18.1. Bael

Botanical Name -*Aegle marmelos* C.

Family – Rutaceae

Chromosome number 2n=2x=18

Scientific classification

Kingdom	:	Plantae
Division	:	Magnoliophyta
Class	:	Dicotyledonae
Subclass	:	Polypetalae
Order	:	Sapindales
Family	:	Rutaceae
Genus	:	Aegle
Species	:	*marmelos* C.

Bael is one of the oldest and nutritional fruits of India (John and Stevenson, 1979). The tree is very hardy and thrives in almost all types of soil and climate. The ripe fruits of bael are rich in protein, fat, minerals, riboflavin, carotene, thiamine, niacin and vitamin C. The ripe fruit is lucrative while unripe fruit is prescribed for diarrhoea and dysentery (Mitra 1999, Roy, 1996, Singh and Ali, 1992, Singh, 2000). The marmelos in is most probably the therapeutically active principle of bael fruits.

18.1.1 Centre of diversity

The Bael tree has its origin from Eastern Ghats and central India. Its indigenous to Indian subcontinent and mainly found in tropical and sub-tropical regions. Even though it is native to India, it is found throughout Southeast Asia (Neeraj *et al.*, 2017). The tree is also found as a wild tree, in lower ranges of Himalayas up to an elevation of 500 meters. Bael is also found growing along foothills of Himalayas, Uttaranchal, Jharkhand, Madhya Pradesh and the Deccan Plateau and along the east coast (Sharma *et al.*, 2007).

18.1.2 Distribution

Bael is widely distributed and found in India, Ceylon, China, Nepal, Sri Lanka, Myanmar, Pakistan, Bangladesh, Nepal, Vietnam, Laos, Cambodia, Thailand, Indonesia, Malaysia, Tibet, Sri Lanka, Java, Philippines and Fiji. It grows well in the dry forests on hilly and plain areas (Neeraj *et al.*, 2017).

18.1.3 Germplasm resources

Field gene banks of Bael is maintained at Department of Horticulture, NDUAT Faizabad, CISH, Lucknow, Department of Horticulture, GBPUAT Pantnagar, CIAH, Bikaner etc. which can be utilized in improvement programme.

18.1.4 Objective

- Dwarf plant stature
- Medium size of fruits (about 750 g to 1250 g)
- Thin shell of fruit
- Less number of seeds
- Low mucilage content
- High sugar content
- Low phenolics
- Pleasant flavour
- Less prone to fruit drop and fruit cracking
- Resistant to biotic and abiotic stresses

18.1.5 Botany

The genus *Aegle* belongs to the family Rutaceae and consists of 2 to 3 species. The generic name is of Greek origin and the species *marmelos* is of Portuguese origin. The tree is deciduous, 6-8 m in height, with trifoliate aromatic leaves. The leaves are divided into three leaflets in a pair and a terminal one, the terminal one is usually (but not always) the largest (Allen 1969). The branches are usual with long, straight spines. The bark is shallowly furrowed and corky. The bisexual flowers are nearly 2 cm wide and borne in clusters, sweet scented and greenish white in colour. The calyx is shallow with fine short, broad teeth, pubescent outside. There are five petals (rarely four), which are oblong oval, blunt, thick, pale greenish white and dotted with glands. Stems are numerous, sometimes coherent in bundles. The ovary is oblong ovoid, slightly tapering, the axis being wide, and are numerous 8-20, small and arranged in a circle with so many ovules in each cell. The fruit is usually globose with a pericarp nearly smooth, greyish yellow, about 3mm thick, hard and filled with soft, yellow and orange very fragrant and pleasantly flavoured pulp. The number of cells (seed cavity) in the fruit arranged in a circle is equal to the number of cells (seed cavity) in a ovary. Seeds are numerous, compressed and arranged in closely packed tiers all surrounded by a very tenacious, slimy, transparent mucilage which becomes hard when dry. The testa is white with wooly hairs and the embryo has large cotyledons (Singh and Roy, 1984).

18.1.6 Breeding of underutilized fruits

Till date the improvement work on bael is confined to exploration and evaluation of germplasm and selection of suitable ideotype. Hybridization, mutation and polyploidy breeding etc. are not attempted so far in this crop. A few selections have been made in U.P., Bihar and West Bengal during 1961-71. Some popular types are Mirzapur, Darogaji Ojha, Rampuri, Kaghji Gonda, Kaghji Etawah and Deoria Large. An extensive survey in U.P. and Bihar has been made and 18 genotypes were collected at NDUAT, Faizabad. Out of 18 genotypes NB4, NB5, NB7 and NB9 are most promising (Pathak and Pathak 1993, Srivastava *et. al.* 1998, Singh and Kumar, 1991). Further, a collaborative programme of the NBPGR, New Delhi with NDUAT, Faizabad has been under taken. The 16 germplasms including 6 in fruiting stage are being maintained at NBPGR, New Delhi. Three promising lines viz. Basti collection–1, Basti collection–2 and Basti collection–4 have been identified superior in physico-chemical characters.

18.1.7 Important varieties and cultivars of Bael

Thar divya – It is released by CIAH, Bikaner. It is early maturing (second forth night of Feb), precocious bearer and suitable for growing under dryland areas. Yield is 50-55q/ha.

Goma Yashi – It is released for cultivation in semi-arid and arid parts of the country. Produces very good quality fruit with weight of 1.25 kg per fruit. Fruits are ovate in shape, greenish yellow. Flesh colour is straw. Fruit shell weight is 180. Number of locules is cross section/ fruit is 17.

NB 5 - Prolific bearer and fruit size is medium, round in shape having smooth surface at maturity, low mucilage, moderately fibrous and have soft flesh with excellent taste.

NB 6 - Fruit size is medium, round with smooth surface, and have thin rind, few seeds, soft flesh, low mucilage and mild acidic.

NB-7 - Fruits are very large in size, flattened round, yellowish green in colour

NB 16 - Elliptical round, pulp yellow, Av. weight 1.3 kg, TSS 31%, medium seed and low fibre content.

NB 17 - Fruits oblong, big sized, fruit quality excellent, seed content less.

Pant Shivani - Mid season cultivar with ovoid oblong shape, size 2 kg, colour lemon yellow on ripening, fiber and mucilage content medium, rind medium thick, pulp light yellow with very good taste and pleasant flavor (Gupta 1999).

Pant Bael-10 -This variety was found highest yield (Jaiswal 1997).

Pant Aparna - Dwarf tree with drooping foliage, almost thorn less, heavy bearer. Fruits average weight 1.0 kg. Late cultivar with small fruit size (0.6 - 0.8 kg), globose shape, and seed, mucilage, fiber and acidity are low. Its flesh is yellow, sweet, tasty and having good flavor.

Pant Sujata - Trees are medium dwarf, heavy bearer. Fruit weight varied from 1 to 1.5 kg.

Pant Urvashi- Pant Urvashi were found to be best on yield and yield characteristics (Gupta 1999).

CISH Bael-1- It is a mid-season variety which matures during April-May. Precocious and heavy bearer. Average fruit wt. =1.0 kg. Suitable for canning and slices preparation

CISH Bael-2 – Trees are dwarf with medium spreading habit. Suitable for processing with pleasantly aromatic pulp. It is released from CISH, Lucknow for cultivation.

'Narendra Bael-5' and 'Narendra Bael-9' – It produced fruits of moderate size with better quality in terms of thin skull, less fiber content, low number of seed/fruit and high total soluble solids, ascorbic acid and total sugars in the pulp (Srivastava and Singh, 2004)

Sabour Bael-1

High yielding (700-800 fruits/tree), Fruit weight 0.90- 1.20 kg per fruit. Pleasant aroma, thin peel (2.35 mm), Pulp contains less mucilage & seeds (91), TSS-43.6%, Acidity- 0.58%; Vit C 34 mg/100g.

18.1.8 References

Allen, B.M. (1969) Malayam fruits. Doland Moore Press Ltd., Singapore.

John, L. and V. Stevenson (1979) The complete book of fruit. Angus and Robertson Publishers, Sydney.

Mitra, S.K. (1999). Other fruits. In: Tropical Horticulture Vol. I. (T.K. Bose, S.K. Mitra, A.A. Farooqi, M.K. Sadhu), Naya Prokash Calcutta, pp390-399.

Neeraj, Vinita Bisht and Vishal Johar (2017). Bael (*Aegle marmelos*) Extraordinary Species of India: A Review. *Int. J. Curr. Microbiol. App. Sci.* **6**(3): 1870-1887.

Roy, S.K. (1996). Bael. In: Fruits Tropical and Subtropical (T.K. Bose and S.K. Mitra), Naya Prakash Calcutta, 740-746.

Singh, I.S. and W. Ali (1992) Bael; A Panicea. *Indian Hort.* **37**;35.

Singh, I.S. (2000). Bael: Bulletin, Published by Department of Horticulture, NDUAT, Faizabad.

Singh, R.N. and Roy, S.K. (1984). The Bael. ICAR Publication, New Delhi.

Pathak, R.K. and Pathak, R.A. (1993) Improvement of minor fruits. In: Advances in Horticulture Vol I (Eds. K.L. Chadha and O.P. Pareek), Malhotra Publishing House New Delhi, pp. 407-421.

Singh, I.S. and S. Kumar (1991). Improvement of Bael (*Aegle marmelos Correa*) tree through selection from existing germplasms necessary. *New Agriculturist,* **2**: 191-94.

Sharma, P.C., V. Bhatia, N. Bansal, Sharma (2007). A Review on Bael tree. *Ind. J. Nat. Prod. Resour.,* 6(2): 171-178.

Srivastava, A.K., Singh, H.K. and Singh, I.S. (1998). Genetic diversity in Bael (*Aegle Marmelos* correa). *Indian Horticulture* Campaign. Jan – March.

Srivastava, K.K. and H.K. Singh (2004) Physico chemical quality of Bael (*Aegle marmolous Correa*) cultivars. *Agri. Sci. Digest,* **24**(1): 65-66.

18.2. Jamun

Botanical Name - *Syzygium cumini* Skeels

Family – Myrtaceae

Chromosome number – 2n=66

Scientific classification

Kingdom	:	Plantae- Plants
Division	:	Magnoliophyta–Flowering plants
Class	:	Magnoliopsida – Dicotyledons
Subclass	:	Rosidae
Order	:	Myrtales
Family	:	Myrtaceae – Myrtle family
Genus	:	Syzygium P. Br. Ex Gaertn
Species	:	*cumini* (L.)

Jamun (*Syzygium cumini* Skeels) belongs to the family Myrtaceae. It is native to India. The fruit is highly tolerant to drought and best suited to wasteland. Genus *Syzygium* contains about 1000 species. It is evergreen tree. It is having small dark purple fruit with prominent elongated seeds. The colour of flesh is also purple and the fruit is astringent even when ripe. *S. jambos* commonly known as rose apple or gulab jamun, produces fruits yellow in colour insipid in taste with high pectin content. *S. uniflora* (Surinam cherry cv. Pitanga) is found in south India and produces deep crimson fruits with spicy taste (Pathak and Pathak, 1993). *S zelanica* is found in the western ghat of India. The choice of seedlings has been for multipurpose trees which promote socio- economic development, raising rural income and those that could be managed by small farming communities on a sustainable basis. Owing to crosspollination and seed propagation, there is a enormous variability in its fruit morphology, fruit quality, maturity and productivity. A number of seedling strains with a lot of variation in fruit shape and size, pulp colour, T55, acidity and earliness, particularly in Uttar Pradesh, Gujarat and Maharashtra provide a good scope for selection of better varieties.

18.2.1 Centre of diversity

The jamun is native to India, Burma, Ceylon and to the Andaman Islands (Zeven and de Wet, 1982) and available throughout Indian plains as well as in Kumaon hills up to 1,600 m. It is found grown as a wild and semi-wild in tropical and subtropical parts of India viz.,Punjab, Haryana, Uttar Pradesh, Maharashtra, Rajasthan, Gujarat, Madhya Pradesh, Bihar, Chhattisgarh, Jharkhand, Karnataka, Kerala, Tamil Nadu and Andhra Pradesh (Singh *et al.* 2011). It is widely distributed in Sri Lanka, Malaysia, Thailand, Philippines and Australia.

18.2.2 Distribution

In India, jamun has a lot of diversity which are grown in a wide range of ecological situations, under tropics, subtropics and temperate. It is well-known that genetic resources are building blocks of new varieties as the common heritage of mankind. It is widely distributed in Sri Lanka, Malaysia, Thailand, Philippines and Australia (Mishra *et al.*, 2018). Common throughout Upper Gangetic Plain, Bihar, Orissa, especially in moist localities, planted in West Bengal, Deccan and Konkan region, and all forest districts of South India, both in the plains and in the hills up to 1800 m, usually along river banks, coast ghats (Rakesh Shukla, 2013). The native diversity, in particular, occurs largely in small farms/orchards, in backyards, roadside avenues and marginal lands across the country, which has predominantly arisen from seedling origin.

18.2.3 Botany

It is a large evergreen and densely foliaceous tree with greyish-brown thick bark, exfoliating in woody scales. The wood is reddish-grey, close grained and durable. The leaves are leathery, oblong-ovate to elliptic or obovate-elliptic with 6 to 12 centimeters long (extremely variable in shape, smooth and shining with numerous nerves uniting within the margin), the tip being broad and less acuminate. The panicles are borne mostly from the branchlets below the leaves, often being axillary or terminal, and are 4 to 6 cm long. Flowers are scented, greenish-white, in clusters of just a few or 10 to 40 and are round or oblong in shape and found in dichotomous paniculate cymes. The fruits are berries and are often obviously oblong, 1.5 to 3.5 centimeters long, dark-purple or nearly black, luscious, fleshy, and edible; it contains a single large seed (Stephen, 2012).

18.2.4 Floral biology

The trees flowers once in a year. New vegetative shoots in the jamun emerged in 2 distinct flushes from February to May and from August to October, the number of inflorescences per shoot was maximum in February flush and least in those produced in October (Mishra and Bajpai, 1975). The inflorescences in jamun are generally borne in the axils of leaves on 5 months to one-year old branchlets. The floral buds assumed a club shaped appearance by the first week of March, and the flowering started from the last week of March and continued up to the last week of April. The trees were in full bloom during the second to third week of April and each inflorescence carried about 35 to 45 flowers. However, at Lucknow the period for panicle emergence was recorded from 3rd week of February and continued till 2nd week of March. The flowering period lasted for 14-19 days, full bloom occurring 5-8 days after initial anthesis in individual plants. Further fruit ripening period lasted for 67-70 days. The total flowering phase culminating in fruit set and ripening lasted for 119 to 126 days with long phase of flower bud initiation, lasted 45-50 days. Flowers are hermaphrodite and light yellow, borne in the axils of leaves on branchlet; calyx lobes 4, calyx tube not extending beyond summit of ovary; petals 4, white, spreading, stamens many, ovary inferior and 2 celled (Bailey and Bailey, 1978). The flowers are sessile, small (7-12 mm), white in color and with thin membranous petals. They are arranged mostly in threes and appear usually from the scars of the fallen leaves (Warrier *et al.*, 1996). The flower length varied from 9.25-13.21 mm. The stamen and style length varied from 5.20-9.00 mm and 7.25-8.90 mm respectively (Singh *et al.*, 2007).

18.2.5 Objectives

- Early bearing, high quality fruits with high yield and early synchronized ripening
- High pulp percentage, large size fruit with small seed, sweet aromatic taste and enhanced shelf life
- Large sized fruits of deep purple or bluish black colour with high juice content
- Rapid juvenile growth
- Dwarf tree stature for high density planting
- Resistance against pests like fruit borer and fruit fly and diseases like anthracnose and leaf spot

18.2.6 Conservation and Breeding

There are no standard varieties available in India, however some of common types grown in north India are known as Raja mum which produces big size, oblong, deep purple fruits having purple pink, juicy and sweet pulp and small seeds. A type having large-sized fruits known as 'paras' in Gujarat. A selection known as Narendra Jamun-6 has been identified with desirable traits at Faizabad, however it has not been popularized. In recent years, a total of 43 superior genotypes were identified in situ on the basis of quality attributes form Lucknow, Barabanki, Unnao, Raebareilly, Faizabad, Saharanpur, Mujaffarnagar, Badot,

Gautam Budha Nagar and Mathura districts in Uttar Pradesh; Jhajjar, Sonipat, Karnal and Faridabad districts in Haryana; Vadodhara, Godhra and Anand districts in Gujarat, Vengurla and Belgaum, Gokak and Arbhavi areas in Karnataka.

18.2.7 Breeding of underutilized fruits

18.2.7.1 Selection

Jamun is an open pollinated fruit crop and predominantly propagated by seed to raise new plantations (Srivastava *et al.*, 2012). Selection from these seedlings can be used to obtain superior genotypes with respect to tree morphology, fruit yield and quality. Several attempts have been made to select superior strains from seedling population (Kumar *et al.*, 1993, Kumar *et al.*, 2006, Shakya *et al.*, 2010, Ghojage *et al.*, 2011).

Considerable variability in jamun is found in Maharashtra and Uttar Pradesh which provides good scope for better variety (Pareek and Sharma, 1993). Jamun germplasm has been widely collected from throughout India and is being maintained at few places in the field gene banks, important field gene banks are CISH, Lucknow, CHES (CIAH), CIAH, Bikaner (Rajasthan), Godhra and MPKV, Rahuri. In Gujarat, Paras a seedling selection is reported to yield sweet fruits.

18.2.7.2 Morphological and Molecular Methods

The use of biochemical and molecular techniques along with morphometric methods provides clues to demarcate in situ conservation sites, and for better understanding of diversity. The molecular analysis has been initiated by RAPD primers. Good differentiation and diverse genetic bases of Indian accessions was demonstrated. As J-28, J-30 and J-33 are selections from Karnataka (Arvhavi), these have shown close affinities among each other. This demonstrates agroecological adaptation to be important for clustering. The dendogram analysis clearly grouped the accessions as per their different geographical regions. The seedling selection form the same geographical regions where they have been traditionally grown over the years normally group together.

18.2.7.3 Hybridization

Lancaster and Bose (1965) observed that F1 hybrid of interspecific hybridization between Alba variety of *S. javanicum* (wax apple) and *S. jambos* (rose apple) possessed high fruiting habit with fragrance and sweetness of rose apple. This hybrid was found to be prolific fruiting habit.

18.2.8 Important varieties and cultivars of Jamun

Goma Priyanka (GJ - 2) - This variety was released from CHES, Godhra (CIAH-ICAR), Gujarat (Singh and Singh, 2012). It is semi-dwarf, spreading growth habit, dense foliage and drooping branches, early, precocious bearer (starts flowering in 4th year) and suitable for high density planting. It starts flowering in the month of March, ripens in the fourth week of May and recorded 19.86 g average fruit weight, with 85.06 per cent pulp. Fruits are good in taste having 16.80oBrix TSS, 0.38% titratable acidity, 12.10% total sugar, 6.11% reducing sugar, 45.44 mg/100 g vitamin C.

GJ – 8 - Selection made at CHES, Godhra. Peak period of flowering is in the month of March. It ripens in the second week of June and recorded 17.00 g average fruit weight, 81.82 per cent pulp, 14.20 per cent TSS, 0.39 per cent acidity, per cent 11.35 per cent total sugar and 45.10mg/100g vitamin C.

GJ – 19 - Selection made at CIAH, Bikaner. Peak period of flowering is in the first week of March. It ripens in the first week of June and recorded 20.50 g average fruit weight, 82.86% pulp, 11.100 Brix TSS, 0.36% acidity, 8.93% total sugar and 44.00 mg/100g vitamin C.

GJ – 23 - Selection made at CIAH, Bikaner. It ripens in the middle of June with an average fruit weight of 20.00 g, 84.45% pulp, 12.000 Brix TSS, 0.32% acidity, 9.03% total sugar and 44.00 mg/100 g pulp vitamin C.

GJ – 40 - Selection made at CIAH, Bikaner. It ripens in the fourth week of June and recorded 16.00 g average fruit weight, 82.0% pulp, 14.00% TSS, 0.39% acidity, 11.00% total sugar and 46.10mg /100g pulp vitamin C.

Katha Jamun - Selection made at CIAH, Bikaner. It ripens in the middle of June and recorded 8.00 g average fruit weight, 70.0% pulp, 9.00 Brix TSS, 0.42% acidity, 34.00 mg/100 g pulp vitamin C, such types are preferred for processing purpose.

Narendra Jamun 6 - Selection made at NDUAT, Faizabad. Improved selection with better fruit qualities (Shakya *et al*., 2010). The fruit is oblong type, higher fruit weight and higher pulp: seed ratio.

Krian Duat - It has larger, thicked leaves and red inner bark. Fruits with purple flesh are more astringent than the white fleshed types.

Paras - This was released from GAU. It is seedling selection from Gujrat known for its large sized sweet and juicy fruits.

Konkan Bahdoli - Selection made at KKV, Dapoli. It has heavy and cluster bearing habit with bold fruits, small seeds, high pulp to seed ratio, better table and processing qualities.

Rajendra Jamun – 1 - This was released from Bihar Agricultural College, Bhagalpur, Bihar. It is early (May -June), high yielding (450 kg/ tree), TSS 18.200 Brix, acidity 0.31%, fruit weight 12.86 g with 88.40% pulp.

Rajamun -This is a common cultivar grown under north Indian conditions (Singh *et al.*, 2007). It produces fruits of big size having fruit length of 2.5 to 3.5 cm and diameter 1.5 to 2.0 cm. Fruits are oblong with deep purple or bluish black colour at full ripe stage. Fruit is juicy with small seed size and ripens in the month of June-July.

CISH J – 37 - Selection made from land race of Lucknow district of U.P. by CISH, Lucknow. A mid-season type, fruiting during second week of June. This seedling selection has superiority due to its bold large sized fruits with attractive deep purple colour. It has high pulp seed ratio (90-92%) and TSS (16-170 Brix). Fruits are rich in ascorbic acid (49.88 mg/100g) and total antioxidant value (38.30 mg AEAC/g). The fruit is oblong and has average weight of 24.05 g, length 3.90 cm, diameter 3.03 cm. This is a mid-season variety and fruits mature during second week of June. Yield potential is 200–300 kg/ plant.

CISHJ – 40 - It is promising genotype collected at CISH, Lucknow. It ripens in the month of June and recorded 17.23 g fruit weight, 91.00%) pulp, 12.400 Brix TSS, 0.76% acidity, 5.02% total sugar and 29.54 mg/100g vitamin C.

CISH J – 42 - Selection made at CISH, Lucknow. Seedless accession obtained from a land race of Chandauli district of UP. Its uniqueness lies in seed lessness (rudimentary seed) with high pulp content (97-98%). The fruit is round shaped and has average weight 6.87 g with a length of 2.57 cm. Fruits are good in test having 14-16 TSS 0 Brix. Fruits are rich in Vitamin C (34.14 mg/100g) and total antioxidant value 15.54 mg AEAC/g of fresh weight and shelf life of 5-7 days at ambient temperature. Yield potential is 180-250 kg/ tree. Fruits mature during second week of June (Singh *et al.*, 2012).

Selection 1- Selection made at Department of Horticulture, Institute of Agricultural Sciences, Banaras Hindu University, Varanasi. It was found most promising with regard to fruit weight (14.55g), minimum seed weight (1.73g), higher pulp content (90.05), higher total soluble solid (21.23%) and total sugar (20.24%) (Prakash, *et al.*, 2010).

AJG – 85 - It has been released by K.R.C College of Horticulture, Arabhavi, Karnataka, for its desirable traits.

18.2.9 References

Bailey, L.H. and E.J. Baily (1978) Hortus Third- A concise dictionary of plants cultivated in the United States and Canada. MacMillan Publishing Co., Newyork, PP. 457-58.

Ghojage, A.H., G.S.K. Swamy, V.C. Kanamadi, R.C. Jagdeesh, P. Kumar, C.P. Patil, and B.S. Reddy (2011) Studies on variability among best selected genotypes of jamun (*Syzygium cumini* skeels.). *Acta Hort.,* **890**: 255-260.

Kumar, R., J. Singh and H.P. Singh (2006) Release of new Jamun variety-Rajendra Jamun 1. *In Proceedings of the national symposium on production, utilization and export of underutilized fruits with commercial potentialities,* Kalyani, Nadia, West Bengal, India, 22-24 November, 2006, pp. 50-53.

Kumar, S., I.S. Singh and S.M. Ashraf (1993) Improvement of jamun (*Syzygium cuminii* (L.) Skeels) through selection. *Haryana J. Hort. Sci.,* **22**(1): 17-21.

Lancaster, P.S. and T.K. Bose (1965) Studies on hybrids of *Syzygium. Indian J. Hort.,* 22: 87-88.

Mishra, R.S. and Bajpai, P.N. (1975) Studies on the floral biology of the jamun (*S.comini* L. Skeels). *Indian J. Hort.,***32**: 15-24.

Pareek, O.P. and Sharma, O.P. (1993) Genetic resources of under-exploited fruits. In: Advances in Horticulture Vol I (Eds. K.L. Chadha and O.P. Pareek), Malhotra Publishing House New delhi, pp. 189-225.

Pathak, R.K. and Pathak, R.A. (1993) Improvement of minor fruits. In: Advances in Horticulture Vol I (Eds. K.L. Chadha and O.P. Pareek), Malhotra Publishing House New Delhi, pp. 407-421.

Prakash, J., A.N. Maurya, and S.P. Singh (2010) Studies on variability in fruit characters of jamun. *Indian J. Hort.,* **67**: 63-66.

Rakesh R Shukla (2013) Jambu (*Syzygium cumini*) -a fruit of gods. Ayurpharm Int J Ayur Alli Sci., 2(4): 86 – 91

Warrier, P., V. Nambiar and C. Ramankutty (1996) Indian Medical Plants. 1st ed. Hyderabad: Orient Longman Ltd., **5**: 225-228.

Shakya, R., S. A. Siddiqui, N. Srivatawa, and A. Bajpai (2010) Molecular Characterization of Jamun (*Syzygium cumini* L. Skeels) Genetic Resources. *International Journal of Fruit Science*, **10**: 29–39

Singh, S., A.K. Singh, H.P. Singh, B.G. Bagle, and T.A. More (2011) Jamun. ICAR. pp.1-46.

Srivastava, V., P. Kumar and P.N. Rai (2012) Correlation study for physic-chemical characters in jamun. *Hort Flora Res.Spec.,*1(1): 83-85.

Singh, Sanjay, H.K. Joshi, A.K. Singh, V. Lenin, B.G. Bagle, and D.G. Dhandar (2007)
Reproductive biology of jamun (*Syzygium cuminii* Skeels) under semi-arid tropics of western India. *The Horticultural Journal*, **20**(2):76-80.

Stephan, A (2012) *Syzygium cumini* (L.) Skeels: A multipurpose tree, its phytotherapic and pharmacological uses. *Journal of Phytotherapy and Pharmacology*, **1**(4): 22-32.

Zeven, A.C and J.M.J. de Wet (1982) Dictionary of cultivated plants and their regions of diversity: excluding most ornamentals, forest trees and lower plants. Centre for Agricultural Publishing and Documentation (Pudoc), Wageningen, I-II.

18.3 Karonda

Botanical name: *Carissa carandas* L.

Family: Apocynaceae

Chromosome number: 2n=2x=12

Scientific classification

Kingdom	:	Plantae
Division	:	Magnoliophyta
Class	:	Magnoliopsida – Dicotyledons
Subclass	:	Asteridae
Order	:	Gentianales
Family	:	Apocynaceae
Genus	:	Carissa
Species	:	*carandas* L.

Karonda (*Carissa carandas* L.) belongs to the family Apocynaceae. Genus *Carissa* has about 30 species. *C. inermis* Vahl. Syn. *C. macrophylla, C. paucinervia* D.C., *C. spinarum* L. Syn. *C. diffusa* grow wild in India (Pareek and Sharma, 1993, Singh and Arora, 1978). Plant of karonda is an evergreen spiny shrub. Karonda is well known protective hedge plant due to its strong axillary spines. It is very hardy, drought tolerant and widely grown in India (Singh and Singh, 2000). It does well in tropical and subtropical climatic conditions. This plant produces berry like fruit which are very attractive in colour. Fruits are sour and astringent in taste. It is one of the richest sources of iron and it also contains fair amount of vitamin C. Fruits of karonda are very useful for the cure of anaemia and have antiscorbutic properties.

18.3.1 Centre of diversity

Karonda (*C. congesta* Wt. Syn. *C. carandas* L.) is native to India. However, *C. grandiflora* (Natal plum) came to India from South Africa and *C. edulis* Vahl. has been introduced from the USA (Singh *et al*., 1982, Pareek and Sharma, 1993). Diversity of karaonda occurs in the North West India particularly in Mount Abu, Ghate and Khandala in Maharashtra (Mehra and Arora, 1982, Pareek and Sharma, 1993). Since karonda is mostly propagated through seeds which leads to considerable heterozygosity and variation in the fruit characters.

18.3.2 Distribution

Karonda is suitable for growing throughout sub-tropical and tropical climatic zones of India. The plant is found to be distributed in the Himalayas at elevations of 300-1800 m, in the Siwalik Hills, the Western Ghats, in Nepal, Afghanistan, India, Sri Lanka, Java, Malaysia, Myanmar, Pakistan, Australia, and South Africa. In India it is cultivated in the states of Maharashtra, Bihar, West Bengal, Chhattisgarh, Orissa, Gujarat, Madhya Pradesh, Rajasthan, and in the Western Ghats. In Maharashtra, the major area under this crop is scattered in sub mountain area like Kolhapur, Ratnagiri, and Pune district (Sawant *et al.*, 2002)

18.3.3 Botany

It is a very hardy, drought tolerant plant that thrives well in a wide range of soils. Karonda is an evergreen deciduous, generally 2-4 m tall shrub. Its stem is rich in white latex, having sharp spines on branches. The spines are straight and 1-3 cm long. The leaves are oblong and conical, 4-6-inch-long and 2-3-inch-wide, green on the top and brown below. They are green with shine above and a dull green shade below. Old leaves keep shedding throughout the year. Karonda wood is hard, straight-grained and use for firewood purpose. The green plant serves as a protective hedge around agricultural holdings. The plant produces white colored flowers, measuring 3-5 cm in diameter. The fruit is a berry, which is formed in clusters of 3-10 fruits, with 5-1 hard angles curving upwards, glabrous with five to seven wings, woody, and fibrous. The fruit shape is globose to broad ovoid consisting of several seeds. Young fruits are pinkish white, while ripe fruit became red to dark purple. Ripe fruit color varies from white, green and pinkish red depending on the genotype. Seed 3-5 per fruit, blackish brown, flat, elliptical, and light in weight. Fruits are generally harvested at the immature stage for vegetable purpose, fully ripen fruits are consumed fresh or processed.

18.3.4 Floral biology

Flowering starts in the month of January - February and fruits mature in May June. Flower is pentamerous, white coloured star shaped. Inflorescence terminal corymbose cyme. Calyx has pink coloured sepals, 5 almost fused from base around the corolla tube showing separation at the tip. Corolla is white in colour with 5 petals fused at the base forming about 1.5 cm long pink coloured corolla tube. Anthers are very small and 5 in number, bilobed, filaments fused with corolla tube all along its length. Single style of about 1 cm long, stigma just below the anthers favours self-pollination. Variable percentage of anthesis was registered in different genotypes. Peak period of anthesis was recorded from 3-6 P.M. or 8 to 10 P.M. depending on the locality, cultivar and atmospheric temperature. Dehiscence of anthers takes place before anthesis between 12 noon and 4 P.M. The dehiscence takes place in the longitudinal fashion and complete within an hour. Atmospheric temperature and humidity influence the time and duration of dehiscence. The maximum receptivity of stigma is found two days before anthesis and two days after anthesis. The initial fruit set by self-pollination was 73.06% which reduced to 24.24% at last. The initial fruit set was highest in plants with most number of umbel segments. Hence number of umbel segments could therefore be used selection criteria for higher fruit yield.

18.3.5 Objectives

- Early bearing, high quality fruits with high yield potential
- Big size fruit with attractive skin colour
- High TSS with low in acidity and high in ascorbic acid
- Good quality and higher pulp content with a smaller number of seeds
- Cultivar with dwarf structure with less spine intensity
- Resistance against drought, pest and diseases

18.3.6 Breeding of Karonda

18.3.6.1 Selection

Karonda is such an underutilized plant that not much research work has been done on varietal improvement. There are no well-established varieties of karonda. Based on colour of the fruit genotypes, karonda are grouped in to four types i.e. Grape green, Green with purple blush, white with pink blush and Maroon (Kumar and Singh, 1993, Singh *et al.*, 1999, Singh and Singh, 2000). Average fruit weight ranged from 1.6 to 4.7g and average number of seeds per fruit from 5 to 11. Wide variation was also observed in the biochemical composition of the fruit, with total soluble solids ranging from 3 to 4.5%, ascorbic acid from 10.26 to 17.94 mg/100g, reducing sugars from 0.93 to 2.4% and non-reducing sugars from 0.57 to 1.33%. Big sized fruit with attractive colour and good quality are the selection criteria for karonda. At Rahuri some promising types No.3, No.12, No.13, and No.16 have been identified (Karale *et al.*, 1989).

18.3.6.2 Hybridization

Crossing may be done to develop high yielding variety having large size, bright red coloured fruits between two desirable parents. There is genetic diversity with respect to morphology and qualitative characters which can be exploited through clonal selection and hybridization.

18.3.7 Important varieties and cultivars of Karonda

Pant Suvarna: It was developed by selection and released from G.B. Pant University of Agriculture and Technology, Pant Nagar in 1986. It has leaf size of 2.41 x 1.41cm, leaf area 2.8 cm2, 45 leaves/shoot. Plants are upright growing and sparse. Fruit size 2.25 cm x 1.6 cm, colour dark brown blush on green background. Average fruit weight 3.62gs, seeds 5.89/ fruit, flesh 88.27%, dry weight 12.39%, TSS 3.836, total titratable acidity 2.30% and yield 22 kg/plant. On ripening, fruit colour changes to dark brown.

Pant Manohar: It was developed by selection and released from G.B. Pant University of Agriculture and Technology, Pant Nagar in 1991. Plants are medium-sized dense bushed with leaf size, 2.35x1.36 cm, having fruits of 2.13 x 1.69 cm, dark pink blush on white background. The fruit length is 2.14 cm, fruit diameter 1.70 cm, fruit weight 3.95 g with 3.95 seeds / fruit. It contains 88.28% flesh, TSS 3.92%, total titrable acidity 1.82% and yield 27 kg per plant.

Pant Sudarshan: It was developed by selection and released from G.B. Pant University of Agriculture and Technology, Pant Nagar in 1991. It has leaf size of 2.475 cm x 1.447 cm, leaf area 2.84 cm2, 41 leaves/shoot, size 2.16 cm x 1.69 cm, pink blush on white background. On ripening fruits become dark brown. Average fruit weigh 3.46 g, seeds 4.68/fruit, flesh 88.47%, dry weight 11.83%, TSS 3.45%, total titratable acidity 1.89% and yield 29 Kg/ Plant.

Thar Kamal: This variety was released by ICAR-CIAH, Bikaner for hot semi-arid regions. It flowered during March -April with fruit yield up to 13kg/plant and having high fruit weight (5g) and TSS (9.56^{0}Brix). It is suitable for candy and jelly making with the yield of 10kg/plant.

Konkan Bold: It was developed by Fruit Research Station, Vengurla, Maharastra. It is sweet type variety. It has large fruit size weighing 16.23g, better pulp content (92%) and low acidity. (Salvi *et al.*, 2006).

CISH Kr-11: It is a superior genotype selected from CISH, Lucknow. Average Fruit weight is 6.0g, length 2.30 cm, pulp 4.60 g, seed 1.40 g, TSS 6.10 ^{0}Brix, ascorbic acid 16.8 mg/100g and antioxidant 189.89µg/100ml AEAC unit.

Maroon Coloured: It is identified as a promising selection from NDUAT, Faizabad, it has an attractive maroon coloured fruits. Average fruit weight is 3.2g, TSS 7.1%, acidity 4.1 %, vitamin C 17.9 mg/100g anthocynin (OD) 0.7 iron 3.8mg 100g-1. Average yield/bush is 3.7 kg (Singh and Singh, 2000).

White-Pink blush: This promising type of karonda has identified at NDUAT, Faizabad. Fruit weight is comparatively lesser than maroon coloured (2.9g). Average number of seeds/fruit is 7.0, TSS 6.6%, acidity 4.7% vitamin C 15.4mg/

100g, anthocyanin (OD) 0.2 and iron 4.15mg/100g. the average yield/bush is 2.8 kg (Singh and Singh, 2000).

CHES-K – 2: It is promising genotype developed at Central Horticultural Experiment Station, Godhra. Peak period of flowering was recorded in the month of March. It recorded 10.00 kg fruit yield/plant, fruit weight 5.10 g and TSS 6.100 Brix at the time of maturity.

Maru Gaurav: It is a selection, released by Central Arid Zone Research Institute, Jodhpur, Rajasthan (Anon., 2011). Fruit size-21.79x16.35mm, elongated ovoid, green with purple blush coloured fruit, mean fruit weight-3.74g, TSS-9.4%, acidity-2.82% flesh dry matter 12.85%, vit. C-35.88mg 100-1g.

CZK – 2022: It is a selection, to be released by Central Arid Zone Research Institute, Jodhpur, Rajasthan (Anon., 2011). Fruit size-22.47x19.14mm, ovoid white with pink coloured fruit, mean fruit weight-4.18g, TSS-8.5%, acidity-2.14%, flesh dry matter-12.77%, vit. C-37.80mg 100-1g.

CZK – 2031: It is a selection, to be released by Central Arid Zone Research Institute, Jodhpur, Rajasthan (Anon., 2011). Fruit size is 22.66 × 19.57mm, ovoid half white half pink coloured fruit, mean fruit weight-5.01g, TSS-8.7%, acidity-2.95% flesh dry matter-12.96%, vit. C-37.4 mg 100-1g.

At Rahuri some promising types No. 3, No.12, No.13 and No.16 have been identified (Karale *et al.*, 1989).

18.3.8 References

Karale, A.R., Keskar, B.G., Shete, M.B., Dhawale, B.C., Kale, P.N. and Choudhari, K.G. (1989) Seedling selection in Karonda (*Carissa carandas* L.). *Maharashtra J. Hort.*,**4** : 125-129.

Kumar, S. and Singh, I.S. (1993) Variation in quality traits of karonda (*Carissa carandas*L.) germplasm. *South Indian Hort.*,**41**:1018-110.

Mehra, K.L. and Arora, R.K. (1982) Plant Genetic Resources of India – Their Diversity and Conservation. NBPGR, New Delhi.

Pareek, O.P. and Sharma, O.P. (1993) Genetic resources of under-exploited fruits. In: Advances in Horticulture Vol I (Eds. K.L. Chadha and O.P. Pareek), Malhotra Publishing House, New Delhi, pp. 189-225.

Singh, H.B. and Arora, R.K. (1978) Wild edible plants of India. ICAR, New Delhi.

Singh, R., Chopra, D.P. and Gupta, A.K. (1982) Some *Carissa* for Western Rajasthan. *Indian Hort.*, **27** :10-11.

Singh,I.S. and Singh, V. (2000) Karonda. Bulletin Department of Horticulture, N.D. University of Agril. and Technology, Kumarganj, Faizabad.

Singh, I.S., Srivastava, A.K. and Singh, V. (1999) Improvement of some underutilized fruits through selection. *J. Appl. Hort.*, **1**: 34 - 37.

Sawant, B.R., Desai., U.T., Ranpise, S.A., More, T.A. and Sawant, S.V (2002) Genotypic and Phenotypic variability in Karonda (*Carissa carandas* L.,). J. Maharashtra Agric. Univ., **27**(3): 266-268.

18.4. Phalsa

Botanical name: *Grewia subinaequalis* L.

Family: Tiliaceae

Chromosome number: 2n=2x =36

Scientific classification

Kingdom	:	Plantae- Plants
Division	:	Magnoliophyta–Flowering plants
Class	:	Rosopsida
Subclass	:	Dilleniidae
Order	:	Malvales
Family	:	Tiliaceae
Genus	:	*Grewia*
Species	:	*subinaequalis L.*

Phalsa is a bushy plant and capable to grow under adverse soils and climatic conditions (Meghwal, 1997). This plant can tolerate even a temperature of 450 C and freezing temperature for few days (Singh *et al.*, 2000). It bears many small fruits of deep reddish brown colour. Ripe fruits are sub acidic in taste and relished in its fresh form as well as in form of excellent soft drink. It is fair source of vitamin A, vitamin C and minerals such as phosphorus and iron. Phalsa fruits have cooling effect, it helps in curing inflammation, heart and blood disorders. The leaves are applied on skin eruptions and they are known to have antibiotic action. The fresh leaves are valued as animal fodder. Fruit can be processed into juices and candies. The ripe phalsa fruits are consumed fresh, in desserts, or processed into refreshing fruit and soft drinks enjoyed during summer months in India (Salunkhe and Desai, 1984).

18.4.1 Centre of diversity

Phalsa is native to India, probably Vadodara, Gujarat. Since it is mostly propagated through seeds therefore, a lot of variability exists among the populations. Maximum variability of phalsa exists in Central India, Rajasthan, Bihar and drier parts of South India, where it is essential to conduct the exploration to collect the promising types. At of the early part of the 20 th century, the fruit was introduced to Indonesia and East Indies, where it has been naturalized very well. It was introduced into the Philippines before 1914 and is naturalized at low elevations in dry zones of the island of Luzon (Morton, 1987).

18.4.2 Distribution

It is widely cultivated, in tropical and subtropical India, Pakistan and Bangladesh. In India it is mainly grown in the Himalayan regions of India, and thrives at elevations up to 3,000 feet. The major areas in India cultivating the fruit commercially are Punjab, Uttar Pradesh, Madhya Pradesh, Haryana and Rajasthan and on small scale it is also grown in Maharashtra, Gujarat, Bihar, Karnataka, Andhra Pradesh and West Bengal.

18.4.3 Botany

Phalsa (*Grewia subinaequalis* L) belongs to order malvales and family Tiliaceae which includes 18 genera and 350 species. The genera Grewia has about 140 species, out of which about 40 species occur in India (Pareek and Sharma, 1993). It is a large shrub or small tree which grows to height of 4.5 m or more, the phalsa has long, slender, drooping branches, the young branchlets densely coated with hairs. The alternate, deciduous, widely spaced leaves are broadly heart shaped or ovate, pointed at the apex, oblique at the base, up to 20 cm long and 6.25 cm wide, and coarsely toothed, with a light, whitish bloom on the underside. Small, orange yellow flowers are borne in dense cymes in the leaf axils. The round fruits, on 2.5 cm peduncles are produced in great numbers in open, branched clusters. Largest fruits are 1.25 to 1.6 cm wide. The skin turns from green to purplish red and finally dark purple or nearly black. It is covered with a thin, whitish bloom and is thin, soft and tender. The soft, fibrous flesh is greenish white stained with purplish red near the skin and becoming suffused with this colour as it progresses to over ripeness. The flavour is pleasantly acid, somewhat grapelike. Large fruits have 2 hemispherical, hard, buff colored seeds 3/16 in (5 mm) wide. Small fruits are single- seeded.

18.4.4 Floral biology

Plants are small tree or shrubs, leaves are alternate, simple and serrated, inflorescence is complex cymose. Flowers are bisexual, complete, actinomorphic and ovary is hypogynous. Sepals are five polysepalous or cornate and basally velvety. Petals are again five in number and pale yellow to greenish yellow in colour. Stamens are numerous, free, anthers two celled and dehiscence occur longitudinally. Ovary is bicarpellary, syncarpus and superior. Pollination is carried out by the insect. It is also observed that flower bud development in phalsa took about 29 days (Pathak and Pathak, 1993), whereas, Narayanaswamy *et al.* (1986) observed that 16.7 days were required for complete bud development. Anthesis started at 6.00 hrs. and continued up to 15.00 hrs. with peak at 10.00 hrs. Parmar (1976) observed more than 70% fruit set with open pollination and that honey bee was the chief agent for pollination. Phalsa pollen had high viability, Narayanaswamy *et al.* (1986) observed that stainability was to the extent of 88-89% by acetocarmine.

18.4.5 Objectives

- Development of synchronous ripening and keeping good quality varieties
- High yielding varieties with good blend of sugar and acid
- Early bearing with good quality fruits
- Resistance against to biotic and abiotic stresses

18.4.6 Breeding of Phalsa

Not much attention has been given for the improvement of phalsa. There are no established varieties in phalsa. However, two distinct type i.e. Tall and Dwarf, have been identified. Dwarf type is very productive and commonly grown. Fruits of dwarf type contain 60-70% juice, 2-3% citric acid, and 18-20% TSS, 35-40mg/100g Vitamin C, 700-750 IU Vitamin A and 0.4 to 0.6% minerals with caloric value of 72/100g.

18.4.7 Varieties and cultivars

Some local selections have been named as Local, Sharbati, Tall and Dwarf etc (Pareek and Panwar, 1981, Anon., 1998).

Table 1: Characteristics of tall and dwarf types of Grewia

Characteristics	Tall	Dwarf
Plant height (m)	4.5	3.4
Internodal length (cm)	101.6	97.6
Leaf size (cm)	20x10	18x15
Lower leaf surface	Light green	Greenish white
Fruit yield (Kg)	5.2	3.5
Fruit size (cm)	2.07	2.26
Fruit weight (g)	0.478	0.544
Pulp (%)	81.5	60.3
Juice (%)	5.4	34.6
TSS (OB	14	12.1
Acidity (%)	4.64	3.63

Dhawan *et al.*, (1993)

Total 19 genotypes have been collected by CIAH, Bikaner. From that six genotypes from Chiraigaon, Rustampur, Bariyasanpur and Chunadih villages of Varanasi and two genotypes from Lawarpur, Gandhinagar, Gujarat were collected. Maximum fruit weight was recorded in CHESP-13 (2.00 g), closely followed by CHESP-14 and CHESP-15. Maximum TSS was also recorded in CHESP-13 (20.10 Degree Brix). Vitamin C was also noted highest in CHESP13 (CIAH Annual report, 2014-15).

18.4.8 References

Anonymous, (1998) Biennial Report, All India Cordinated Research Project on Arid Zone Fruits. NRC for Arid Horticulture, Bikaner. CIAH (2014-15) Annual report.

Dhawan, K., Malhotra, S., Dhawan, S.S., Singh, D. and Dhindsa, K.D (1993) Nutrient composition and electrophoretic pattern of protein in two distinct types of phalsa (*Grewia subinequalis* DC). Plant Food Hum. Nutr., **44**: 255-260.

Meghwal, P.R. (1997) Exploiting underutilized fruits of arid zone. *Indian Hort.*, **42** (3) : 26-27.

Morton, J.F (1987) Phalsa. In: Fruits of warm climates. Julia Morton, Miami, FL, Pp. 276-277.

Narayanaswamy, P., Mokashi, A.N., Mohan kumar, G.N., Nalawadi, U.G., Prabhakar, N. and Deivar, K.V. (1986) Floral biology of Phalsa (*Grewia asiatica* L). *South Indian Hort.*, 34 : 115-116.

Pareek, O.P. and Panwar, H.S. (1981) Vegetative, floral and fruit characteristics of two phalsa (*Grewia subinaequalis* L) types. *Annals of Arid Zone*, **20**: 281-290.

Pareek, O.P. and Sharma, O.P. (1993) Genetic resources of under-exploited fruits. In: Advances in Horticulture Vol I (Eds. K.L. Chadha and O.P. Pareek), Malhotra Publishing House, New Delhi, pp. 189-225.

Parmar, C. (1976) Pollination and fruit set in phalsa (*Grewia asiatica*L). *Agric.Agro. Indust. J.*, **9**: 12-14.

Pathak, R.K. and Pathak, R.A. (1993) Improvement of minor fruits. In: Advances in Horticulture Vol I (Eds. K.L. Chadha and O.P. Pareek), Malhotra Publishing House New Delhi, pp. 407-421.

Salunkhe, D.K. and Desai, B.B. (1984) Phalsa. In: Salunkhe and Desai (eds.), Postharvest Biotechnology of Fruits. CRC Press, Boca Raton, FL. **2**: 129.

Singh, I.S., Singh, V. and Singh, H.K. (2000) Phalsa. Bulletin of Department of Horticulture, NDUAT, Faizabad.

18.5. Khirni/Rayan

Botanical Name: *Manilkara hexandra (Roxb.) Dubard*

Family: Sapotaceae

Chromosome number: 2n=24

Scientific classification

Kingdom	:	Plantae- Plants
Division	:	Magnoliophyta–Flowering plants
Class	:	Dicotyledonae
Order	:	Ebenales
Family	:	Sapotaceae
Genus	:	Manilkara
Species	:	*Hexandra*

Manilkara hexandra (Roxb.) Dubard commonly known as Rayan/Khirni belongs to sapotaceae family. It includes about 70 genera and 800 species. The characteristic feature of member of this family is presence of reddish-brown hairs on the leaf undersides and other plant surfaces). It is originated in Indian sub-continent and well adapted to semi-arid regions and tolerant to drought conditions. It is suitable for sandy wastelands of semiarid regions. Well drained loamy soils and medium black soils are suitable for cultivation. However, it is commonly grown in laterite and sandstone soils. Fruits and bark of this tree species have economical value as mature fresh fruits are very sweet and eaten raw as well as after drying and bark is used for several medicinal purposes. The seeds contain approximately 25 per cent oil which is used for cooking purposes. The fruit is good source of iron, sugars, minerals, protein and carbohydrate etc. It is commercially used as a rootstock for vegetative propagation of sapota in different parts of the country. In the tribal area of Rajasthan, Gujarat and Madhya Pradesh this tree plays very important role in the socio-economy and livelihood security of small and landless farmers. Owing to its economical importance as fruit tree, this species is cultivated in backyards, home gardens, parks, farmers' fields and community lands.

18.5.1 Centre of diversity

It is indigenous to Srilanka, Deccan peninsula, Bengal, South East Asiatic mainland and Hainan and it is cultivated throughout the central India and Deccan peninsula (Anonymous 1962: Hammer 2001).

18.5.2 Distribution

M. hexandra is indigenous to India, found wild in the forests of South India, Northcentral India, parts of Gujarat and Rajasthan. It is found in dry evergreen forest and deciduous forests of western and central India. In India this species is generally cultivated near villages, backyards and homestead gardens in the parts of Madhya Pradesh, Gujarat, Rajasthan and Vidarbha region of Maharashtra and also found as natural wild populations (Malik *et al.*, 2010). They also located near Pritam Pura in Ratlam, Chanderi in Ashok Nagar districts, Neemach and Dhar districts of Madhya Pradesh, Panchmahal, Bharuch, Dahod and Sambarkanta districts of Rajasthan (Keerthika *et al.*, 2017).

18.5.3 Botany and floral biology

The tree is medium to large size ever green tree with spreading growth habit and erect trunk, 15-20 m tall with blackish grey and deeply furrowed bark with conical degenerated branchlets. Leaves are elliptic, ovate or oblong with 7-10cm long and 3.5-6.0cm broad, apex obtuse or emarginated, margin entire, glabrous, dark green in colour. Petiole is up to 2.5cm long. The flowers solitary white or pale yellow. Flowers are bisexual, in axillary fascicles of 2-6, pedicles stout, calyx 6 lobbed, ciliolate, corolla 16 or 24 lobed in 2 or 3 series, outer one linear, stamens 6 and anthers acute as long as filaments, staminodes 6 or 8, alternate with stamens. Ovary is 12 celled and hairy, Fruit ellipsoid 1.5 - 2.0 cm long. Berry type fruits and ovoid shape. Seed 1or rarely 2, reddish brown and shining. Seeds are endospermic, oily with hard and glossy light brown to blackish seed coat.

18.5.4 Breeding of Khirni/Rayan

There are no or less improved cultivars but there is some variation from tree to tree for characteristics viz., size, shape, colour, pulp content and taste of fruits in local grown trees. Germplasm accessions are collected and being evaluated at ICAR-Central Horticultural Research Station (CHES), Godhra, Gujarat and ICAR- Central Institute of Subtropical Horticulture (CISH), Lucknow. A wide range of variation was observed in fruit and seed characters which show prevalence of variability in natural populations (Malik *et al.*, 2012).

Table 2: Germplasm accessions assembled in different institutions

S.No	Germplasm accessions	Location	Characterization	Conservation	References
1	30	ICAR-CIAH, Regional Station, CHES, Godhra	Promising genotypes: CHESK-1, CHESK-11 and CHESK-10	Field gene bank	http://www.ciah.ernet.in/germplasm_developed.php Sanjay Singh, 2015
2	99	ICAR-NBPGR, Delhi	Morphological characterization has been done for 47 accessions. 23 accessions genetically evaluated using RAPD markers revealing 78% polymorphism	Cryogene bank	Malik et al., 2012
3	28	ICAR-CIAH, Lucknow	To identify accessions having bold fruits with low seed content	Field gene bank	Annual report, 2009-2010 (CSIH)

18.5.5 Varieties and cultivars

Thar Rituraj: This variety is released by ICAR-CIAH, Bikaner for semi-arid region. The yield per hectare is 13 quintals. It is suitable for table purpose and processing.

CISH K-10: a superior genotype established in field gene bank of CISH, Lucknow, recorded average fruit weight of 4.44 g, length 2.83 cm 100/g and antioxidant 189.89 g/ml AEAC unit. It is propagated vegetatively.

GK-10: it was collected form Parwadi village of Panchamahal district in Gujarat. The peak period of flowering is in December. It ripens in third week of May and gives fruits weighing 5.2g each 25.50 % TSS, and 26.90mg.100g vitamin C.

18.5.6 References

Anonymous (1962) The wealth of India: Raw materials, Vol 6. Publications and Information Directorate, CSIR, New Delhi: 298-300.

Annual report, 2009-10, CISH.

Hammer, K. (2001) Sapotaceae. (In) Hanelt P, Institute of plant and genetics and crop plant research (Eds.) Mansfeld's encyclopedia of agricultural and horticultural crops, vol 3. Springer, Berlin, Pp. 1647-1667.

Keerthika, A., Gupta, D.K., Noor mohamed, M.B., Vikas Khandelwal, Shukla, A.K., Jangid, B.L and Subbulakshmi,V. (2017) Rayan/Khirnis. In: Horticultural crops of high nutritive values (Eds.) K.V.Peter. Brillion publishing, New Delhi. Pp. 327-337. Malik, S.K., Chaudhary, R., Dhaliwal, O.P. and Bhandari, D.C (2010) In: Genetic resources of Tropical underutilized fruits in India. NBPGR, New Delhi, P.168.

Malik, S.K., Kumar, S., Choudhary, R., Kole, P.R., Chaudhary, R and Bhatt, K.V. (2013) Assessment of genetic diversity in Khirni (*Manilkara hexandra* (Roxb.) Dubard): An important under-utilized fruit species of India using RAPD markers. *Indian J.Hort.*, **70**(1): 18-25.

18.6. Jackfruit

Botanical name : *Artocarpus heterophyllus* L.

Family : Moraceae

Chromosome number: 2n=4x=56

Scientific classification

Kingdom	:	Plantae- Plants
Division	:	Magnoliophyta–Flowering plants
Class	:	Magnoliopsida – Dicotyledons
Subclass	:	Hamamelidae
Order	:	Urticales
Family	:	Moraceae - mulberries
Genus	:	*Artocarpus*
Species	:	*heterophyllus* Lam

Jack fruit is a important fruit crop which is used for both culinary as well as table purpose. Genus *Artocarpus* contains about 50 species and most of them are monoecious in nature. Jackfruit (*Artocarpus heterophyllus* Lam) produces more yield than any other tree species, and bear the largest known edible fruit (up to 35 kg). The jackfruit tree has several uses. Flakes of ripe fruits are high in nutritive value; every 100 g of ripe flakes contains 287-323 mg potassium, 30.0-73.2 mg calcium and 11-19 g carbohydrates. In Bangladesh, it is commonly referred to as "poor man's food" as it is cheap and plentiful during the season. The nutritious seeds are boiled or roasted and eaten like chestnuts, added to flour for baking, or cooked in dishes. The tree is also known for its durable timber, which ages to an orange or reddish brown color, with anti-termite properties (Arung *et al*., 2006).

18.6.1 Center of diversity and distribution

Jackfruit has been cultivated since prehistoric times and has naturalized itself in many parts of the tropics, particularly in Southeast Asia, where it is today an important crop of India, Myanmar, China, Sri Lanka, Malaysia, Indonesia, Thailand and Philippines. It is also grown in parts of Africa, Brazil, Surinam, Caribbean, Florida and Australia. It has been introduced to many Pacific islands since post European contact and is of particular importance in Fiji, where there is a large population of Indian descent (Om Prakash *et al*., 2009)

18.6.2 Botany

Jackfruit is a medium size, evergreen tree that typically attains a height of 8–25 m (26–82 ft) and a stem diameter of 30–80 cm (12–32 in). The canopy shape is usually conical or pyramidal in young trees and becomes spreading and domed in older trees. The tree casts a very dense shade. This species is monoecious, having male and female inflorescences (or "spikes") on the same tree. Male and female spikes are borne separately on short, stout stems that sprout from older branches and the trunk. Leaves are dark green, alternate, entire, simple, glossy, leathery, stiff, large (up to 16 cm [6 in] in length), and elliptic to oval in form. Leaves are often deeply lobed when juvenile and on young shoots. Jackfruit has a compound or multiple fruit (syncarp) with a green to yellow brown exterior rind that is composed of hexagonal, bluntly conical carpel apices that cover a thick, rubbery, whitish to yellowish wall. Seeds are light brown, rounded, 2–3 cm (0.8–1.2 in) in length by 1–1.5 cm (0.4–0.6 in) in diameter, and enclosed in a thin, whitish membrane (Om Prakash *et al.*, 2009).

18.6.3 Floral biology

Individual flowers borne on an elongated axis and forming a racemoid inflorescence; male spikes produced singly, elongated, whitish-green or dark green with smooth skin, becoming yellowish and rough when mature, oblong, cylindrical, clavate, ellipsoidal or barrel shaped, distal end with a 1.5-2.5 mm wide annular ring, 3-10 x 1-5 cm, slightly hairy. Hanging or drooping peduncle 1.5-3.5 cm long and 4-5 mm thick, many densely crowded sterile or fertile flowers; sterile flower has a solid perianth and the fertile one is tubular and bi-lobed. Flowers are tiny, pale green when young, turning darker with age. Female flowers are larger, elliptic or rounded, with a tubular calyx. The flowers are reportedly pollinated by insects and wind, with a high percentage of crosspollination (Orwa *et al.* 2009).

Artocarpus heterophyllus L. is tetraploid and having separate male and female flowers. Cauliflory (emergence of flower from the main trunk) is also common in this fruit crop. Pollination in jack fruit is carried out by the wind. Under open pollination, the fruit set is about 75% which can be improved by hand pollination (Samaddar, 1985). Being a monocious plant the female sex is located separately which reduces the job of emasculation. For the crossing selected female spikes are covered with a bag before anthesis. When anthesis of the small flowers of female spike starts, hand pollination is carried out with the desired pollen source.

18.6.4 Objectives

- Early bearing, high quality fruits and early synchronized ripening
- Large sized fruits with high yield
- Rapid juvenile growth
- Dwarf tree stature
- Resistance against drought, pest and diseases

18.6.5 Breeding of Jackfruit

Varietal improvement is limited to the selection of the suitable ideotypes through exploration. Srinivasan (1970) described a variety Muttam Varikka which produced fruits of about 7Kg. Maximum variability in jack fruit is observed in evergreen forest of Western Ghats, Gorakhpur, Deoria and Allahabad districts of Uttar Pradesh.

As a result of survey, 16 types were collected from Faizabad of which NTJ 1, NTJ2, NTJ3, and NTJ4 had larger fruits of excellent quality and suitable for table purpose. However, NJC1, NJC2, NJC3 and NJC4 had small to medium size fruits with thin rind and soft flesh which is suitable for culinary purpose (Pareek and Sharma, 1993).

18.6.6 Important varieties and cultivars of Jackfruit

Swarna Manohar: It was identified through selection from the land races collected from Jharkhand, Bihar, West Bengal, U.P., Odisha. Fruit has attractive skin colour, uniform in shape, high quality fragrant pulp, sweet and firm flake. Highly nutritive with natural sugars and fragrance, TSS–18.0°B, Titratable acidity – 0.09% and total sugar – 8.33%. It is suitable for table purpose and recommended for Jharkhand and adjoining areas. Early fruit maturity. Average fruit weight is 13.7 kg and average yield is 548 kg/tree.

Swarna Poorti: It was selected from the land races collected from Jharkhand, Bihar, West Bengal, U.P., Odisha. Fruit has attractive light green colour, roundish uniform shape. It is highly nutritive, TSS-18.0°B, titratable acidity0.14% and total sugar – 7.46%. It is suitable for vegetable purpose. Average fruit weight is 4.20 kg and average yield is 120.0 kg/tree. Late fruit maturity. It has recommended for Jharkhand and adjoining areas.

Sindoor

It is released in 2015. Jack for table purpose, with attractive sunset orange flakes having; bear fruits twice/ year; medium fruits 11-12 kg, 52 cm long; 25 fruits/ tree / year; distinct aroma, taste and sweetness.

18.6.7 References

Arung, E.T., K. Shimizu and R. Kondo (2006) Inhibitory effect of artocarpanone from *Artocarpus heterophyllus* on melanin biosynthesis. *Biol. Pharm. Bull.* **29**: 1966-1969.

Om Prakash, Rajesh Kumar, Anurag Mishra, Rajiv Gupta (2009) *Artocarpus heterophyllus* (Jackfruit): An Overview. Phcog. Rev., **3**(6): 353-358.

Orwa,C., A. Mutua, Kindt, R., Jamnadass, R., Anthony, S. (2009) Agroforestree Database: A tree reference and selection guide version 4.0 (http://www.worldagroforestry.org/sites/treedbs/treedatabases.asp)

Pareek, O.P. and Sharma, O.P. (1993) Genetic resources of under-exploited fruits. In: Advances in Horticulture Vol I (Eds. K.L. Chadha and O.P. Pareek), Malhotra Publishing House, New Delhi, pp. 189-225.

Samaddar, H.N. (1985) Jack fruit. In Fruits of India Tropical and Subtropical (Ed. T.K. Bose), Naya Prokash Calcutta, pp. 487-497.

Srinivasan, K. (1970) Muttam Varikka – A promising Jackfruit variety. *Agric. Res. J. Kerala,* **8**: 51-52.

19

Breeding of Temperate Fruits

19.1 Apple

Botanical name: *Malus x domestica Borkh.*

Family: Rosaceae

Scientific classification

Kingdom	:	Plantae- Plants
Division	:	Magnoliophyta–Flowering plants
Class	:	Magnoliopsida – Dicotyledons
Subclass	:	Rosidae
Order	:	Rosales
Family	:	Rosaceae – Rose family
Genus	:	*Malus*
Species	:	*Malus x domestica* Borkh

Apple (*Malus x domestica* Borkh) is the most widely grown temperate fruit of the world considered as king of temperate fruit. It is one of the most delicious fruit rich in carbohydrates, minerals like calcium, phosphorus, iron, potassium, vitamin B6 and low in calories. Apples are extremely rich in important antioxidants, flavanoids, and dietary fiber. Apple is recommended to reduce the incidence of dental caries, and the phytonutrients and antioxidants may help reduce the risk of developing cancer, hypertension, diabetes and heart disease. Apple is chiefly used as table fruit. The various processed product like juices, jelly, preserve, canned slices, wine, concentrated juice, cider and powder are prepared from apple fruit.

19.1.1 Center of origin and distribution

The primary center of species richness and diversity is in southwest China and Central Asia, with several species ranging eastwards to Manchuria and Japan, and others ranging westwards to Europe (Ignatov and Bodishevskaya, 2011; Luby, 2003). A secondary centre is in North America, with four native species (Hancock *et al.* 2008; Luby 2003; Rieger 2006; Robinson *et al.* 2001). It is widely cultivated in Afghanistan, Armenia, Azerbaijan, Bhutan, China, India, Indonesia, Iraq, Iran, Israel, Japan, Jordan, Kazakhstan, Korea (Republic of), Kyrgyzstan, Lebanon, Myanmar, Nepal, Pakistan, Philippines, Syria, Tajikistan, Thailand, Turkey, Turkmenistan, Uzbekistan, Vietnam, Yemen (CABI 2012; Hancock *et al.* 2008) . Production in tropical Asia is limited to high altitudes, in India this includes the temperate North Western Hills region and to a lesser extent the North Eastern Hills region (Papademetriou and Herath 1999).

Apple was introduced into the country by the British in the Kullu Valley of the Himalayan State of H.P. as far back as 1865, while the colored 'Delicious' cultivars of apple were introduced to Shimla hills of the same State in 1917. The apple cultivar 'Ambri', is considered to be indigenous to Kashmir and had been grown long before Western introductions. It is predominantly growing in temperate zone of India viz., Jammu and Kashmir, Himachal Pradesh and hills of Uttar Pradesh which is accounting 90% of the total production. Its cultivation has also extended to Arunachal Pradesh, Sikkim, Nagaland, Meghalaya in north eastern region and Nilgiri hills in Tamil Nadu.

Table 1: Apple growing belts in India

S.No.	State	Growing belts
1.	Jammu & Kashmir	Srinagar, Budgam, Pulwama, Anatnag, Baramullah, Kupwara
2.	Himachal Pradesh	Shimla, Kullu, Sirmour, Mandi, Chamba, Kinnaur
3.	Uttarakhand	Almora, Pithoragarh, Tehri Garhwal, Uttarkashi, Chamoli, Dehradun, Nainital
4.	Arunachal Pradesh	Tawang, West Kanneng, Lower Subansiri

19.1.2 Germplasm resources

The cultivated apple is likely the result of interspecific hybridization and at present the binomial *Malus* ×*domestica* has been generally accepted as the appropriate scientific name (Korban and Skirvin 1984). Cultivated apple Malus X domestica are diploids (2n =34) and few are triploids (2n=51) and ancestor of it is Malus sylvestris. It is a deciduous fruit tree mostly grown in temperate region of the world. Despite high genetic variability, the thousands of cultivars distributed throughout the world and the world-wide breeding programs, mainly based on organoleptic traits, aesthetic standards and disease resistance, the size of the genetic resources used by breeders has been limited and reduced to a few varieties such as 'Cox's Orange Pippin,' 'Golden Delicious,' 'Jonathan,' 'Red Delicious,' and 'McIntosh' (Noiton and Alspach, 1996). Though the germplasm is placed at the bottom of pyramid of any breeding but genetically diverse germplasm of crop is only the sources of new traits for any new cultivars. In India apple collection has around 6734 total accessions, of which 5226 are in the fieldgene bank and 1508 are seedlots. Non spur and spur type are two important categories of apple germplasm. Good collection of Germplasm has been collected at NBPGR Phagli, CITH Srinagar (Jammu and Kashmir), IARI regional station Shimla, Fruit Research centre Shalimar, Hill Fruit Research centre Chaubattia and Chakrata , Dr. YSPUHF Solan, SUKAST Srinagar.

19.1.3 Objective

19.1.3.1 For Scion

- To evolve varieties, red in colour with early maturity
- High yielding, unique flavors, exceptional crispness, enhanced storage and shelf life
- Resistant to diseases (scab) and pests
- To obtain tree habits those allow high productivity and regular bearing.
- To develop cultivars adapted to extreme climatic conditions.

19.1.3.2 For rootstocks

- To focussed on productiveness and tree size control
- To develop woolly apple aphid resistant rootstock
- To evolve rootstock capable of surviving under a wide range of environmental conditions especially winter hardy and drought tolerance
- Inducing precocity, productivity and fruit quality into scion.

19.1.4 Botany

Small to medium sized tree rarely shrubs, deciduous rarely evergreen. Tree size and shape is heavily dependent on rootstock and training system. Buds void with several imbricate scale. Leaves oval or elliptic to broad ovate, bluntly serrated, glabrous and more or less glossy above and light pubescence on underside. Initiation of floral buds occurs during the summer before blossom period. Most cultivars are self-unfruitful. Cross-incompatibility is rare, so most cultivars that bloom at the same time will serve as pollinizers, including crab apples. A few cultivars are pollen-sterile. Honey bees and Mason bees are the most effective pollinators. The fruit is pome, the fleshy edible portion derives from the hypanthium or floral cup, not the ovary that is why it is considered as false fruit. Fruiting begins 3-5 years after planting, although a few fruit may be produced in the 2nd year, varies with rootstock and cultural practices. Fruit are usually thinned to one per spur, with spurs spaced 4-6 inches apart for attainment of marketable size. The apple contains 5 seed cavities with generally 2 seeds each. Seeds are relatively small and black, and mildly poisonous. Most apples reach maturity about 120-150 days after bloom, with a few cultivars maturing in as short as 70 days, and others as long as 180.

19.1.5 Floral biology

The apple trees require chilling to break rest period which varied from 1000-1600 hrs depending on cultivars. Flowering occurs in early spring when white to deep pink flowers develop in a cyme-like inflorescence of 4-6 flowers. The center flower which opens first is often referred to as the "king bloom". Most flowers are borne terminally on spurs and less frequently, laterally on long shoots. Flowers borne on short spurs begin the transformation from vegetative buds to flower buds 4-6 weeks before lateral buds (Jackson 2003). The flowers are hermaphroditic with the ovary embedded in the floral cup and flower parts located above the ovary. A normal flower contains five carpels each with two ovules, five sepals, petals and styles and usually 20 stamens (Dennis 2003; Rieger 2006). The ovary being enclosed by non-ovarian tissue that remains attached to the ovary giving rise to false fruit or pome. Most apple cultivars requires cross pollination duet to self-incompatibility and triploidy nature of cultivars. Pollen fertility of most of the cultivars is cent percent but is reduced in some cultivars such as McIntosh by unknown factors and in Jonagold by triploidy. Denis (2003) reported the flower development takes around ten months from vegetative to reproductive (late June in Northern hemisphere) and ending with anthesis (late April to the beginning of June) in the subsequent year. Sharma *et al*, 2005 recorded full bloom in fourth week of April and flowering period was 16-19 days. Stigma attained receptivity two days prior to anthesis and remain receptive up to two days after anthesis with peak receptivity on the day of anthesis.

19.1.6 Inheritance pattern

Knowledge on the genetic control of important characteristics is essential way to improve breeding efficiency. It is also helpful in proper parent selection for crosses in breeding programmes and correct selection of progenies. Most of the fruit and tree characteristics investigated so far proved to be controlled by one or a few genes (monogenic traits). However, many traits of agronomic importance like fruit quality, size and flavour, resistance against certain diseases and tree architecture are under polygenic control (Vissen *et al.* 1968; Brown and Harvey 1971; Kellerhals and Furrer 1994; Conner *et al.* 1998; King *et al.*

2000). Apple ripening date showed polygenetic inheritance. Compact growth habit and cold hardiness is controlled by several genes polygenic. Fruit colour is controlled by three independent genes A, B, and C. Yellow (any dominant gene), red (any two dominant genes) and green (recessive gene). Most commonly introduced gene for apple scab resistance is Vf. PL1 and PL2 (M. Zumi and M. robusta) are single major genes of powdery mildew.

In apple, the thicker and the shorter the leader, the shorter the juvenile period (Denardi *et al.* 2010). The bromcresol-grecn-test for presence of the malic acid gene in apples (Nybom, 1959 and Brown, 1975). The degree or time of leaf fall (colour) is a possible pre-selection criterion of the time of picking (Visser, 1967). Lateral branching has high heritability and is correlated with topological traits (Segura *et al.*, 2006). Nyborn (1959) found a correlation between the pH of the leaf sap and the pH of the fruit juice. Another correlation that is often claimed is between red fruit color and anthocyanin pigmentation of other parts of the tree such as 1 year old shoots or leaf petioles.

19.1.7 Breeding methods and achievements

19.1.7.1 Introduction and selection

There are about 10,000 documented cultivars of apple. Most of the commercial cultivars of apple are the chance seedling selections. The Delicious apple (discovered by chance in Iowa, 1872) and Golden Delicious (selection at west Virginia, 1905) still dominates the apple production in the world.

19.1.8 Important Varieties and cultivars

Apple varieties should have climatic adaptability, attractive fruit size, shape, colour, good dessert quality, long shelf life, resistance to pest and diseases and drought along with high productivity.

19.1.8.1 Promising cultivars in major production region of India

Jammu and Kashmir: Benoni, Irish Peach, Cox's Orange Pippin, Ambri, White dotted Red, American Apirouge, Red Delicious, Golden Delicious

Himachal Pradesh: Tydeman's Early, Mollies Delicious, Starkrimson, Starking Delicious, Red delicious, Richared, Granny-Smith, Red Spur, Top Red, Red Chief, Oregon Spur, Golden Spur, Michal, Schlomit

Uttar Pradesh: Early Shanburry, Chaubattia Princess, Fanny Benoni, Red Delicious, Starking Delicious, Rymer, Buckingham

19.1.8.2 Varieties based on bearing habit (Ted, 2002)

Tip bearer : Ballyfatter, Bullock, Cornish Gilliflower, Cortland, Cox stone, Dougherty, Dulect, Empire, Fuji, Fuji Lynd-Spur, Golden Russet, Granny Smith, Irish Peach, Kerry Pippin, Nova Spy, Paula Red, Priam, Pristine, Redfree, Rome Red, Scarlet, St. Edmunds Pippin, Summer Red, Swe*et al*ford, Sweet Caroline, Twenty Ounce Pippin.

Partial tip bearer: Adam's Pearmin, Blenheim Orange, Bramley's Seedling, Cornish Gilliflower, Discovery, Ginger Gold, Golden Noble, Gravenstein, Greensleeves, King of Tompkin's County, Lodi, Lord Lambourne, Northern Spy, Paula Red, Pink Lady, Tydeman's Early Worcester, Vista Bella, Wealthy, Worcester Pearmin, Yellow Newton Pippin.

Spur type cultivar - 12 varieties were introduced through NBPGR New Delhi at regional research station , Mashobra (H.P), in 1980. Amongst some important are Red spur delicious, Golden spur delicious, Oregon spur, Red chief, Miller, Sturdee spur, white spur , Stark crimson.

Colour spots – Royal Red, Vance delicious. Top red, Skyline supreme, Red Delicious

Low chilling variety – Sufficient attempt have not been made to introduce low chilling apple cultivars in Indian. From early introductions cultivars such as Tropical beauty and Perlin beauty were found to be best w.r.t. productivity and fruit quality. Some cultivars of Sub–Tropical apples are Maayan, Michael, Schlomit and Vared where introduction from Israel and are under evaluation.

*Scab resistant var*iety – Prima, Priscilla , Liberty, Florina, Macfree, Freedom, Coop-12

19.1.8.3 Pollinizing cultivars (Verma, 2013)

Early Bloomer: Tydeman's Early Worcester, Manchurian, Everest, Malus floribunda. Mid

Bloomer: Snow Drift, Gloster, Red Gold, Chestnut, Spartan, Cox Organe Pippin, Yellow Transparent, York Imperial.

Late Bloomer: Golden Delicious, Golden Spur, Rome Beauty, Granny Smith, Worcester Pearmain, Golden Hornet, Starkspur Golden.

Most adopted apple cultivars for commercial cultivation includes Oregon Spur, Red Chief, Well Spur, Vance Delicious, Hardy Spur, Gold Spur, Ruby Red, Hardy Bright, Ultra Red, Supreme, Wagon Spur, Red Spur, Millers Sturdy Spur, Silver Spur, Top Red, Hardyman, Bright-n-Early and Red Fuji.

19.1.9 References

Brown, A.G. and Harvey, D.M. 1971The nature and inheritance of sweetness and acidity in the cultivated apple. *Euphytica*,**10**: 68-80.

CABI. 2012. Crop protection compendium. CAB International, Wallingford, UK.

Conner, P.J., Brown, S.K. and Weeden, N.F. 1998. Molecular marker analysis of quantitative traits for growth and development in juvenile apple trees. *Theor. Appl. Genet*., **96:** 1027-1035.

Hancock, J. F., Luby, J. J., Brown, S. K. and Lobos, G. A. 2008. Apples. J. F. Hancock, ed. Temperate Fruit Crop Breeding: Germplasm to Genomics. Springer Science+Business Media B.V., New York, NY (New York): 1-37

Ignatov, A. and Bodishevskaya, A. 2011. Malus. C. Kole, ed. Wild crop relatives: Genomic and breeding resources Temperate fruits. Springer-Verlag, Berlin, Heidelberg: 45-64

Janick, J., Cummins, J.N., Brown, S.K. and Hemmat, M. 1996. Apples. In: Janick J, Moore JN (eds) Fruit breeding, vol I, Tree and tropical fruits. Wiley, New York, USA: 1–77

Juniper, B.E., Watkins, R. and Harris, S.A. The origin of the apple. *Acta Horticulturae.* 484.

Kellerhals, M. and Furrer, B. 1994. Approaches for breeding apples with durable disease resistance. *Euphytica*,**77**: 31-35.

King, G.J., Maliepaard, C., Lynn, J.R., Alston, F.H., Durel, C.E., Evans, K.M., Griffon, B., Laurens, F., Manganaris, A.G., Schrevens, E., Tartarini, S. and Verhaegh, J. 2000. Quantitative genetic analysis and comparison of physical and sensory descriptors relating to fruit flesh firmness in apple (*Malus pumila* Mill.). *Theor. Appl. Genet.*,**100**: 1074-1084

Luby, J. J. 2003. Taxonomic classification and brief history. D. C. Ferree, I. J. Warrington, eds. Apples: Botany, production and uses. CABI International, Cambridge, UK: 1-14

Noiton, D. A. M., and Alspach, P. A. 1996. Founding clones, inbreeding, coancestry, and status number of modern apple cultivars. *J. Am. Soc. Hortic. Sci.* **121**: 773–782.

Papademetriou, M. K. and Herath, E. M., (eds.). 1999. Deciduous fruit production in Asia and the Pacific. Food and Agriculture Organization of the United Nations: Regional Office for Asia and the Pacific, Bangkok.

Rieger, M. 2006. Introduction to fruit crops. Food Products Press, Binghamton.

Robinson, J. P., Harris, S. A. and Juniper, B. E. 2001. Taxonomy of the genus Malus Mill. (Rosaceae) with emphasis on the cultivated apple, Malus domestica Borkh. *Plant Systematics and Evolution*, **226**: 35-58.

Ted L. Swenson. 2002. Tip or Partial Tip Bearers ID & Special Pruning Techniques, Pome News, Tigard, Oregon, U.S.A., Spring: 77 – 79.

Verma, M. K., Awasthi, O.P. and Giri, R.K. 2013. Prospects of Temperate Fruit Production in subtropical climate. In: Precision Farming in Horticulture (Eds. J. Singh, Jain, S.K., Dashora, L.K. and Chundawat, B.S.). New Indian Publishing Agency, New Delhi, India.

Vissen, T., Schaap, A.A. and de Vries, D.P. 1968. Acidity and sweetness in apple and pear. *Euphytica,* **17**: 153-167.

19.2. Pear

Botanical name: *Pyrus communis L.*

Family: Rosaceae

Scientific classification

Kingdom	:	Plantae- Plants
Division	:	Magnoliophyta–Flowering plants
Class	:	Magnoliopsida – Dicotyledons
Subclass	:	Rosidae
Order	:	Rosales
Family	:	Rosaceae – Rose family
Genus	:	*Pyrus*
Species	:	*communis L.*

Pyrus communis L. is one of the important fruit crops which belongs to the family Rosaceae. It is a deciduous small to medium-sized tree to 10 m tall (normally 3-5 m in cultivation), with a pyramidally shaped crown. The conical erect trunk bears small, reddish-brown, narrow-angled branches. The grey-brown bark has shallow furrows and flat-topped scaly ridges. Leaves alternate, simple, elliptic/ovate with a finely serrated margin, obtuse tips, 2.5-10 cm long and 3-5 cm wide, shiny green above, paler and dull below, glabrous. The petioles are stipulate and the buds are involute, with imbricate scales. Flower corymbose inflorescences, 5-7.5 cm wide, containing 5-7 showy white, 2.5-3.5 cm wide flowers, borne from terminal, mixed buds of short spurs, appearing before or with the leaves. The spurs are very short and lateral branches. The ovary is epigynous, or inferior, with the 5-carpellate ovary embedded in receptacle tissue, containing up to 10 ovules (2 per carpel); peduncle thin, 2.5-5 cm long. Fruit a pyriform (pear-shaped) pome with persistent or deciduous calyx, 4- 12 cm long, greenish colored, dry and gritty. Seed blackish, 8.4 by 4.8 mm, each with a thin layer of endosperm.

19.2.1 Center of origin and distribution

The primary center of origin is China and Asia minor to the Middle East in the Caucasus mountains. Secondary center is in Central Asia (Vavilov, 1992). In world scenario, pears are mainly cultivated in China, USA, Argentina, Italy Turkey, Spain, India, South Africa, Japan, Belgium, etc. Pear is grown under a wide variety of climatic regimes, ranging from cold dry temperate hilly conditions to warm humid subtropical conditions on the plains of northern India. In India Pears are widely cultivated in Punjab, Himachal Pradesh, Jammu and Kashmir, Uttar Pradesh, Uttarkhand, etc.(Milind Parle and Arzoo, 2016)

19.2.2 Varieties and Cultivars

In higher altitude conditions, high chilling requirement varieties (like Bartlett) are mainly grown. In more recent years, red-color strains of pear like Max Red Bartlett, Red Bartlett and Starking are replacing yellow colored cultivars. In warmer sub-mountainous areas of H.P. and sub-tropical Punjab oriental pear cultivars like Baghugosha, Kieffer, China and sand pear Patharnakh are cultivated commercially both for table and processing purpose.

Cultivars grown in Jammu and Kashmir: William, Kashmir Nakh, Vicar of Wakefield, Beuree D. Amanlis, Goshbagu, Beurre Hardy

Cultivars grown in Himachal Pradesh: China, Bartlett, Max Red Bartlett, Flemish Beauty, Devoe, Manning's Elizabeth, Winter Nellis

Cultivars grown in Uttar Pradesh: Thumb Pear, Doyene du Comice, Victoria, William's Bartlett, Beurre Hardy, Flemish Beauty

19.2.3 References

Milind Parle and Arzoo. 2016. Why is pear so dear. *Int. J. Res. Ayurveda Pharm.*, **7**(1): 108-113.

Vavilov, N.I. 1992. Origin and geography of cultivated plants, Cambridge University Press, Cambridge, UK, Vol. 15.

19.3. Peach

Botanical name: *Prunus persica (L.) Batsch*

Family: Rosaceae

Scientific classification

Kingdom	:	Plantae- Plants
Division	:	Magnoliophyta–Flowering plants
Class	:	Magnoliopsida – Dicotyledons
Subclass	:	Rosidae
Order	:	Rosales
Family	:	Rosaceae – Rose family
Genus	:	*Prunus L. – plum*
Species	:	*persica (L.) Batsch*

Peach (*Prunus persica* (L.) Batsch) is one of the most important fruit in the Rosaceae family grown in warm temperate zones of the world. Fruits are rich in protein, sugar, minerals and vitamins. Peaches are highly valued as table fruit for their attractive colour and palatability. Canned peaches, dried frozen preserves, jams, nectar are principal processed products of peach in many regions. Prunacin is the principal glycoside present in pulp while amygdalin is present in the seed of peach.

19.3.1 Centre of origin and distribution

China is the center of origin for peach and was domesticated there 4-5000 years ago (Hedrick, 1917; Wang and Zhuang, 2001; Aranzana *et al.*, 2011). *P. kansuensis* Rehd. and *P. davidiana* (Carr.) Franch. are native to north western China and north China respectively. Whereas, native of *P. mira* Koehne is Tibetan plateau and areas neighbouring Nepal and India. *P. ferganensis* (Kost. and Rjab.) Kov. and Kost. is native to western China (Scorza and Sherman, 1996). China is also place of origin for Chinese wild peach (*P. consociiflora* Schneid.). Flat peaches (*P. persica* var. platycarpa) also originated in China. It is believed that nectarines might have originated in Europe, which was introduced to China. However, it is also considered that nectarines are also native to China. The main routes of peach movement from China to west were sea through India and mid-east as well as silk route through Persia (Janick, 2003).

It is one of the most important temperate fruit crops grown mainly in Jammu and Kashmir, Himachal Pradesh, Punjab, Uttarakhand, Nilgiri hills, Jharkhand and North Eastern States (Josan *et al.*, 2009) valued for its fresh and canned fruits. Now days peach has become pride to poor and marginal hilly farmers of sub mountainous regions, plains of northern India and Southern hills and also in irrigated arid and plateau ecosystem. It is relatively performed well at an altitude ranging between 600-1000m from mean sea level.

19.3.2 Germplasm resources

China being the centre of origin of peach, has the richness of genetically diverse germplasm (Li *et al.*, 2008; Yoon *et al.*, 2006). Chinese peach cultivars have more genetic diversity than has been reported for other peach germplasm collections (Zhang *et al*, 2006). Peach germplasm collection occurred in Shanghai, Dalian and Shanxi. More than one thousand germplasm accessions have been collected and maintained to safeguard against genetic erosion. In India, a rich diversity of seedling population of peach landraces is found in Kashmir (Jammu & Kashmir) and Kinnaur (Himachal Pradesh). At present Central Institute of Temperate Horticulture (CITH) has vast germplasm collection of peaches and recently identified some unique genotypes viz. Cresthven, Glohaven, Red Globe, Fantasia for different uses. Field gene bank of peaches are maintained at following stations

Station	Germplasm	References
China		Wang *et al*, 2002
Nanjing	560	
Zhenzhoui	280	
Beijing	650	
India		Dhillon and Rana 2004
NBPGR Shimla	52	
NBPGR Bhowali	19	
NBPGR Akola	10	
Dr. YSPUHF Solan	46	
GBPUA&T Pantnagr	12	
SKUAST Srinagar	39	
CITH Srinagar	32	Annual Report CITH

19.3.3 Objectives

19.3.3.1 For Scion

- To develop cultivars with low chilling requirement
- Extension of season of maturity and early maturing variety harvest before rain in sub tropical areas

- Tolerance to high summer temperature
- Maturity between 60-7 0 days after full bloom
- Attractive colour, firm flesh, freedom of stone from pulp and non browning of flesh

19.3.3.2 For rootstocks

- Resistant to root knot nematode, iron chlorosis and water logging
- Control of tree vigour for HDP
- Ease of propagation and stock-scion interaction with higher graft compatibility
- Wide range of soil and climatic adaptation

19.3.4 Botany

The peach tree is a deciduous in nature, vigorous growing, but small tree with a spreading canopy, usually 6-10 feet. Trees are short-lived, generally living only 15-20 years as productive life. Leaves are linear, lanceolate with acute tips, finely serrate margins, folded slightly along the midrib, sickle-shaped in profile, pinnately veined, 2-63 in length. The flowers are produced in early spring before the leaves. They are solitary or paired; Light pink to carmine to purplish with five petals, 1-1.53 in diameter. Ovary is perigynous, surrounded by a hypanthium. Self-pollinating, normally grown without pollinizers. The fruit is drupe, yellow or whitish flesh, and a skin that is velvet hairy in different cultivars. The flesh is the mesocarp, the skin the exocarp. The fleshy mesocarp may adhere to the stone (clingstone) or may separate from it (freestone). The flesh is very delicate and easily bruised in some cultivars, but is fairly firm in some commercial varieties, especially when green. The bony endocarp (pit) surrounds a single, large, ovate seed approximately 1.3-2.0 cm long. Trees are very precocious, producing some fruit in the 2nd year after planting. Peaches require extensive thinning (80-95% of flowers) for proper size development. Usually, thinning is done by hand 30-45 days after full bloom, leaving about 1 fruit per 6 inches of 1-year-old shoot length.

19.3.5 Floral biology

The peach trees require chilling to break rest period which varied from 500-800 hrs depending on cultivars. The flower buds develop in leaf axil on current season's growth which are reddish-green and turn dark grey when old. Generally, each node bears three buds, two lateral flowers buds and one vegetative bud in middle. The flower is perigynous, hermaphrodite, pale to dark pink in colour, however white to dark red is also found. There are five sepals, five petals arranged alternately and thirty or more stamens. There are two shapes of corolla i.e. showy and non-showy. Showy with large petals hidden anthers until the petals separated while non showy with smaller petals, the anthers protrude beyond the corolla. Pistil become receptive one or two days after anthesis and remain receptive for several days. Flower anthesis usually as the day warm up. The pistil has two ovules but only one develop into seed. Most of the peach cultivars re-self fruitful and self pollinated.

19.3.6 Breeding methods and achievements

Systematic breeding work on improvement of peaches was initially started in China which has highest and diverse germplasm. Planned breeding programme for improvement of peaches was started in 1957 at Saharanpur. Some work was also initiated at Pantnagar where peach germplasm collected and evaluated but no variety was found suitable for commercial cultivation. Latter on rigorous breeding work was carried out at Dr. YSPUHF Solan and Punjab Agricultural University Ludhiana

19.3.6.1 Introduction

A large number of low chilling varieties e.g. Flordasun, Sunred and Sun Gold and some other selections were introduced successfully. Flordared and Flordabelle were introduced at PAU Ludhiana during late sixties from Florida and California. Of these introductions Flordasun, Sunred, Flordared and 16-33 (Latter named Shan-e-Punjab) became very popular in India. TA 170 popularly known as Pratap identified as seven days early than Flordasun is also an introduction from USA. Latter introduction from Florida, Flordaprince and Earligrand have been recommended for commercial cultivation for the plains of Punjab and adjoining areas. Flordaprince is an early ripening, whereas Earligrand is a mid-season variety.

Table 1: Some of popular varieties of peach Introduced in India (*Source*: Kanwar *et al.*, 2002)

Country	Name of introduced varieties	Place of Introduction
USA	Kanto 5, Sumerset, Dawne	H.P.
Japan	Shimizu, Hakuto, Okubo	H.P.
Florida USA	Florda Sun, Sun Red, 16-33, Florda Red, Florda Balle, Early Amber	PAU Punjab
California USA	Bonita, Rochan, Vantura	PAU Ludhiana
USA	Flordagrande, Flordaprince, Earligrande, Valleygrande, Tropical Beauty, Tropical Snow, Tropical Sweet	

19.3.6.2 Selection

Availability of great genetic variability in peach warrants the development of diverse cultivars through selection for wider adaptability. A large number of low chilling varieties e.g. Sun Red, Sun Gold etc developed through selection are popular in Indian conditions. Sharbati is a chance seedling selection selected at Saharanpur.

19.3.6.3 Hybridization

Out-crossing, inbreeding and back crossing are the principal breeding system in peach. Peach does not exhibit inbreeding depression clearly due to its high selfing nature. Backcrossing has been found essential in improving fruit quality characters as well incorporating resistance into commercial cultivars. Intensive hybridization programme was carried out in USA, Canada, Italy, USSR, South Africa and Australia. Redhar is a cross between 'Halehaven' and 'Kalhaven bred' at USA. Domesticated peach can be readily hybridized with native populations of native P. persica and all the wild species of peach. Nemagard, a hybrid between *P. persica* x *P. davididasa* is a widely used root-knot nematode resistant rootstock, which is immune to Meloidogyne incognita. Successful hybrids have also been produced between peach and almond, apricot, plum and sour cherry. In most cases, these wide hybrids are largely sterile. Planned hybridization work on peach was started in 1957 at Saharanpur, India. Peach Saharanpur Prabhat (Sharbati x Flordasun) was released.

19.3.6.4 Mutation Breeding

Two varieties have been released through gamma irradiation 'Magnof 135' (1968) for fruit size and 'Plovdiv 6' (1981) for better yield. Bud-sports Early ripening mutant from 'Redhaven' have been identified are 'Garnet Beauty' and 'Stark EaliGlo'

Table 2:

Cultivars	Source	Features
Early Red goldza	Natural mutation of cv. Stark Redgold	Flesh yellow to yellow orange, mid season
Yuhua 2	Natural mutation of 63-9-22 (sims x Phillips)	Mid season variety, flesh white, hard melting and sweet
Magnif 135	Mutant of Magnif 43 by gamma radiation	Good fruit size
Plovdiv 6	Mutant of dupnishka by gamma radiation	Better yield
Lady Nancy	Mutant of Jersey queen	Ripens about four weeks after Redhaven

Table 3: Important varieties of peach

Variety	Features
Arctic Supreme	Clingstone, red-over cream skin; white flesh, Midseason
August Pride	freestone large, round peach with red-blushed yellow skin; yellow flesh, Midseason
Babcock	semi-freestone, white flesh, Early harvest
Bonita	Freestone, medium to large, light yellow skin with yellow flesh, Midseason
Dessert Gold	Semi-clingstone, round fruit with red-blushed yellow skin, yellow flesh, Early
Dixi Red	semi-freestone, red skin , yellow flesh, Early midseason
Early Amber	Freestone, dark red-blushed yellow skin, orange-yellow flesh, Early
Elberta	Freestone, large, blushed-red skin, yellow, Midseason.
Flordasun	Semi cling stone, large fruit, deep yellow skin, grooves present on stone, vigorous tree.
Flordaprince	Semi-freestone, medium to large, red-blushed yellow skin, yellow flesh , from Florida. Very early.
Golden Gem	Large, yellow skin and flesh, Early.
Golden Glory	very large fruit, golden skin, yellow flesh, Late
Halehaven	Freestone, medium to large fruit, red over greenish yellow skin, yellowflesh,
J.H. Hale	Freestone, very large, yellow skin and flesh, Midseaon
July Elberta	Freestone, medium round fruit , greenish yellow skin, small pit, midseason.
Redhaven	semi-freestone, medium fruit, brilliant red over yellow skin; yellow flesh ,early to midseason
Rio Oso Gem	Freestone, large fruit bright red skin, yellow flesh ,midseason.
Sun Red	semi-freestone, small fruit, bright red skin, yellow flesh, excellent quality for dessert
Sharbati	Clingstone, medium fruit, greenish yellow skin, white flesh
Pratap	Freestone, fruit medium, red blush skin, semi vigorous tree, midseason,
Prabhat	Clingstone, fruit medium, skin yellow with green tinge, semi vigorous tree, Early,
Shan-e Punjab	Semi Clingstone, large fruit, skin yellow with red spot, vigorous tree, Late.

Some other important varieties of peach in different peach growing countries are May Pride, Snow Beauty, Summerset, Suncrest, Sunhaven, Tropic Snow, Ventura, Veteran, Loring, Empress, Eva's Pride, Madison, Nectar, Q 1-8, Rio Grande, Roza, Rubidoux, Santa Barbara, Tropi-berta.

19.3.7 References

Anonymous. 2003. Package of practices for fruit crops. Directorate of Extension Education, Dr. YSP University of Hort.& Forestry, Solan. H.P., India.

Aranzana, M.J., Abbassi, EL-K., Howad, W and Arus, P. 2010. Genetic variation, population structure and linkage disequilibrium peach commercial varieties. *Bio. Med. Cen. Genet.*, **11**(69): 1-12.

Das, B, Sharma, A.K, Srivastava, K.K and Mir, M.S. 2004. Studies on phenology and fruit characters of some commonly grown apricot cultivars/landraces in Ladakh. *The Hort. J.*, **17**(3): 205-211.

Hedrick, U.P. 1917. The peaches of New York. Rep New York Agric. Exp. Sta.

Huxley, A. 1992. New RHS dictionary of gardening. Macmillan.

Janick, J. 2003. English title: History of Asian horticultural technology. *Acta Hort*., **620**: 19-32.

Josan, J.S, Thind, S.K, Arora, P.K and Kumar, A. 2009. Performance of some low chilling peach cultivars under north Indian conditions. *Environment and Ecology*. **27**(4B): 1923-1926.

Scorza, R and Sherman, W.B. 1996. Peaches. In Janick J, Moore JN (eds.) Fruit Breeding: Vol. 1- Tree and Tropical Fruits, John Wiley & Sons, Inc., New York, pp. 325-440.

Sharma, R.L. 1993. Genetic resources of temperate fruits. In Chadha KL, Pareek OP (ed.) Advances in Horticulture Vol. 1-Fruit crops Part 1, Malhotra Publishing House, New Delhi, India, pp. 244-263.

Wang, Z.H and Zhuang, E.J. 2001. China fruit monograph- peach flora. *China Forestry Press*, Beijing, pp. 42-51.

19.4. Plum

Botanical name: *Prunus salicina L.*

Family: Rosaceae

Scientific classification

Kingdom	:	Plantae- Plants
Division	:	Magnoliophyta–Flowering plants
Class	:	Magnoliopsida – Dicotyledons
Subclass	:	Rosidae
Order	:	Rosales
Family	:	Rosaceae – Rose family
Genus	:	*Prunus L. – plum*
Species	:	*salicina L.)*

Prunus salicina L. is commonly called as Japanese plum belongs to the family Rosaceae. It is an economically important temperate fruit which is used as fresh and dry form for table purposes. The total area under plum cultivation is 22000 ha with annual production of 82000 tonnes and productivity is 3.72 t ha^{-1} (Dinesh *et al.*, 2018). The processed products include candy, frozen fruit, jams, jelly products and traditional Serbian plum for alcoholic beverages (Milosevic *et al.* 2010). The ripe fruits are the rich source of Vitamin A, B (Thiamine), riboflavin and minerals like calcium, phosphorus and iron. *Prunus salicina* grows up to 10 metres (33 ft) tall, and has reddish-brown shoots. The leaves are 6-12 cm long and 2.5-5 cm broad, with a serrated margin. The flowers are produced in early spring, 2 cm diameter with five white petals. The fruit is a drupe 4-7 cm in diameter with yellow-pink flesh; it can be harvested in the summer. When fully ripe it can be eaten raw (Anonymous, 2015–16).

19.4.1 Center of origin and distribution

The center of origin for plums is extended from Asia, Europe to America for different biotypes/landraces. In India, it is widely cultivated in the state of Himachal Pradesh, Jammu and Kashmir, Uttarakhand, Punjab and Uttar Pradesh.

Centers of origin for major plum species has been given below (Crane and Lawrence, 1956; Watkins, 1976)

Table 1: Recommended varieties of plums in temperate regions of different states of India

Sl.No.	States	Early	Mid season	Late
1.	**Himachal Pradesh**			
	High hills	Sweet early, Methley, Kelsey	Starking delicious, Satsuma, Burbank, Elephant heart	Mariposa
	Mid hills	Santa rose, Beauty, Early red beauty, Transparent cage	Frontier, Kanto 5	Maripose, Tarrol, Red Ace
	Low hills	Alucha purple, Titron	-	-
2.	**Uttar Pradesh**			
		Jamuni, Titron, Settler, Cloth of gold, Ramgarh maynard	Howe, Alubulkhara sharbati, Zardula, Zardula chittidar, Burbunk, Elephant heart, Satsuma, Black chamba, Victoria	Late yellow, Kanto 5, Sweet early, Santa rose, Mariposa
3.	**Jammu and Kashmir**			
		Saharanpuri white, Plum first Ramgarh, Maynard sharps, Early subza	Formosa, Maynard, Bryanstones ganze, Burbank, Santa rose	Satsuma, Grand duke, Silver wickson, Beauty, Cloth of gold, Reine claude bavy golden cage
4	**Punjab**	Satluj purple	Titron, Kala Amritsari, Kataru chak	Alubukhara

(Chadha, 2001)

Table 2:

S.No	Species	Origin
1.	Japanese plum (*Prunus salicina* Lindl.),	China
2.	European plums (*Prunus domestica* L.)	Southern Europe or western Asia
3.	American plum (*Prunus americana* Marsh.),	North America
4.	Damson plum (*P. insititia* L.),	Western Asia or Europe
5.	cherry plum (*P. ceracifera* Ehrh.).	Western and central Asia
6.	Italian plum (*P. cocomilia* Ten.),	Italy, Greece and Yugoslavia
7.	Blackthorn sloe (*P. spinosa* L.),	Europe and Asia
8.	Apricot plum (*P. simonii* Carr.),	North China
9.	Manchurian plum (*P. ussuriensis* Kov. and Kost.)	Ussuri river
10.	Canada plum (*P. nigra* Ait.)	Canada and US

19.4.2 Varieties and Cultivars

There are 280 varieties of plums available in the country from that few of them are grown commercially. Some Plum germplasms are collected in the gene bank at NBPGR, Regional Station, Shimla (India) viz., Au Rosa, Au Cherry, Grand Prize, Black Amber, Queen Ann, Satsuma, Fortune, Monarch, Plum Beauty, Tarrol, Red Plum, Santa Rosa, Yellow Plum, Black Plum, Late Plum, Methley, Prune, Frontier, etc. Cultivars namely: Pershore, French Damson, Stanley, Golden Transparent, Giant Prune, Victoria etc. are self-fruitful. Cultivars namely: Satsuma, Black Champa, Raine Claude, Transparent, Golden Drop, Frogmore, Italian Prune, Red Beaut, President, Mariposa, Burmosa, Kelsay, Frontier etc are self-unfruitful. Whereras, cultivars Santa Rosa, Beauty, Early Orleans etc are partially self-fruitful. Recommended cultivars in India (Anonymous, 2003) are Sweet Early, Methley, Red Beaut, Santa Rosa, Beauty, Frontier, Satsuma, Burbank and Alucha Purple.

19.4.3 References

Anonymous. 2003. Package of practices for fruit crops. Directorate of Extension Education, Dr. YSP University of Hort.& Forestry, Solan. H.P., India.

Crane, M.B and Lawrence, W.J.C. 1956. The genetics of garden plants. 4th ed., MacMillan Company, London.

Watkins, R. 1976. Cherry, plum, peach, apricot and almond. Prunus spp. In Simmonds NW (ed.) Evolution of Crop Plants, Longman, London, pp. 242–247.

Milosevic, T, Milosevic, N and Mratinic, E. 2010. Morphogenic variability of some autochthonous plum cultivars in western Serbia. *Brazilian Archives of Biology and Technology,* **53**: 1293–1297.

Anonymous. 2015–16. Area and Production of Horticultural Crops-All India Dinesh Kumar, K. K. Srivastava and S. R. Singh. 2018. Morphological and horticultural diversity of plum varieties evaluated under Kashmir conditions. *Tropical plant research.* **5**(1): 77-82.

Chadha, K.L. 2001. Hand Book of Horticulture. Indian Council of Agricultural Research, New Delhi. pp. 324-325.

19.5. Apricot

Botanical name: *Prunus armeniaca L.*

Family: Rosaceae

Scientific classification

Kingdom	:	Plantae- Plants
Division	:	Magnoliophyta–Flowering plants
Class	:	Magnoliopsida – Dicotyledons
Subclass	:	Rosidae
Order	:	Rosales
Family	:	Rosaceae – Rose family
Genus	:	*Prunus L. – plum*
Species	:	*armeniaca L.*

Prunus armeniaca (Apricot) is belonging to family Rosaceae is one of the most cultivated stone fruits in the world which is grown in Central Asia, Pakistan, Turkey, Iran, Iraq, Afghanistan, Pakistan, and Syria. Apricots are the deciduous plant which grows up to 9 m height. The leaves are oval and finely serrated with 5-petaled white flowers growing together in clusters. The fruit's color varies from yellow to orange to deep purple and ripens in late summer. The plant is best grown in deciduous climatic condition with low temperature (Shakir *et al.*, 2017)

19.5.1 Center of diversity and distribution

Prunus armeniaca L., the cultivated and wild apricot is native to Asia and Caucasus including north and north eastern China (Mratinic *et al.*, 2011). The main areas of distribution of wild apricots are generally reported to be north China, Manchuria, north Korea, Khingan mountains and north eastern Mongolia. Vavilov (1951) reported three centers of origin namely: 1) Chinese centre (north eastern, central and western China), 2) Central Asian centre (Hindu Kush to Kashmir mountains) and 3) Near eastern center (north eastern Iran to Caucasus and central Turkey). The richest diversity is available in the central Asian region comprising Sinkiang (China), Afganistan, Baluchistan, Pakistan and northern India. In India, cold arid region exhibits a great extent of variability in seedling population of apricot genotypes found in wild, semi wild and cultivated forms (Mir, 2000; Das *et al.*, 2004).

It is mostly cultivated in the Mediterranean countries, Central Asia, Russia, USA, Iran, Iraq, Afghanistan, Pakistan, Syria and Turkey. United States produces almost 90% of the world's apricot crop (Asma *et al.*, 2005 and Mohamed *et al.*, 2011). In India it is mainly cultivated in North West Hills Region, Jammu & Kashmir, Himachal Pradesh and Uttar Pradesh and also in North Eastern Hills Region comprising the state of Arunachal Pradesh, Nagaland, Meghalaya, Sikkim and Manipur (Varsha *et al.*, 2012).

19.5.2 Cultivars and Varieties

Generally, there are two types of apricot, namely, sweet kernel type and bitter kernel type. About 81 exotic accessions and 19 indigenous cultivars were collected and evaluated. Major cultivars in India are namely: Kaisha, New Castle, Early Shipley, Saffaida, Charmagaz, Shakarpara, Gilgati Sweet, St Ambroise, Moorpark, Nari Halman, Rakcha Karpu, Narmo are commonly grown in, India (Anonymous, 2003; Das *et al.*, 2004).The local types Halman and Rakhaikarpo have been popular, whereas exotic introductions namely, Nari, Kaisha, Shakarpa and New Castle are promising. These cultivars are recommended for dry cold areas. The USA variety Nugget is self-fruitful and bears sweet and attractive colored fruits.

Table 1: Promising cultivars in major production regions of India

Sl.No.	Regions	Cultivars
1.	Jammu and Kashmir	Charmagz, Halman, Rachkaikarpo, Nari, Shakarpara
2.	Himachal Pradesh	Kaisha, Nugget, Castle, Saffeida, Charmagz
3.	Uttar Pradesh	Sharmagz Kaisha, Moorpark, Turkey, St. Ambrose

19.5.3 References

Anonymous, 2003. Package of practices for fruit crops. Directorate of Extension Education, Dr. YSP University of Hort.& Forestry, Solan. H.P., India.

Asma, B.M., and K. Ozturk. 2005. Analysis of morphological, pomological and yield\characteristics of some apricot germplasm in Turkey. *Genetic Resources and Crop Evolution*, **52**:305–313.

Das, B, Sharma, A.K, Srivastava, K.K and Mir, M.S. 2004. Studies on phenology and fruit characters of some commonly grown apricot cultivars/landraces in Ladakh. *The Hort. J.*, **17**(3): 205-211.

Mir, M.S. 2000. Potential and problems of fruit crop production in Ladakh. In Sharma JP, Mir AA (eds.) Dynamics of Cold Arid Agriculture, Kalyani Publishers, Ludhiana, pp.188-213.

Mohamed, F.G, Mohamed, S.S, Khalil, K.H.S, Hussein, M.S.A and Kamil, M.M. 2011. Application of FT-IR spectroscopy for Rapid and simultaneous quality Determination of some fruit products, *Nature and Science*, **9**: 21-31.

Mratinic, E., Popovski, B., Milosevic, T and Popovska. 2011. Analysis of morphological pomological characteristics of apricot germplasm in FYR Macedonia. *Journal of Agricultural Science and Technology*, **13**: 1121–34.

Shakir Ullah, Aish Muhammad, Iqbal Hussian, Hafeez-UrRahman, Muhammad Zeeshan Hyder, Muhammad Din and Nizamud Din. 2017. Morphological variations in apricot (*Prunus armeniaca*) cultivars grown in Gilgit Baltistan Pakistan. *Pakistan J. Agric. Res.* **30**(1): 1-16.

Varsha Raj, Akash Jain and Jasmine Chaudhary. 2012. Prunus Armeniaca (Apricot): An Overview. *Journal of Pharmacy Research,* **5**(8): 3964-3966.

Vavilov, N.I. 1951. Phytogeographical basis of plant breeding. In: CHESTER, K. S. (trans): The origin, variation, immunity and breeding of cultivated plants *Chronical Botany.***13**: 13–54.

19.6. Cherry

Botanical name: *Prunus avium (L.) L.*

Family: Rosaceae

Scientific classification

Kingdom	:	Plantae- Plants
Division	:	Magnoliophyta–Flowering plants
Class	:	Magnoliopsida – Dicotyledons
Subclass	:	Rosidae
Order	:	Rosales
Family	:	Rosaceae – Rose family
Genus	:	*Prunus L. – plum*
Species	:	*avium (L.) L.*

Sweet cherry (*Prunus avium* L., Rosaceae, 2n = 16) is cultivated in temperate regions of the world for its edible fruits. The sweet cherry trees are large (30 feet to 40 feet tall) and usually pyramid shaped. Branches grow upright. The fruits are large, have a deep stem cavity, vary in colour from light yellow to dark red to purplish black, and the stems or pedicels are about 1.5 inches long. Flowers arise from clusters of 2 to 5 flowers on short spurs with multiple buds at tips; the distal bud develops into a leafy shoot. The flesh ranges in texture from tender to firm, and is sweet. Mostly sweet cherries are consumed as fresh fruit.

19.6.1 Center of diversity and distribution

Most of the cherry species are indigenous to Europe and Asia (Rehder, 1974). There are two types of cherries viz., Sweet (*Prunus avium* L.) and sour (*P. cerasus* L.) cherries which are native to Near East centre which includes Asia Minor, Iran, Iraq and Syria (Vavilov, 1951). More specifically, sour cherry is native to Carpathian Basin (Anonymous, 2002a). Sweet cherries (mazzard) have been grown in southern Russia, north of the Caucasian mountains to the north of France for a long time. Natural range of sweet cherry also includes the temperate regions of Europe, from the northern part of Spain to the south eastern part of Russia (Hedrick, 1915). Cherry was introduced in England by Romans from Persia in the first Century (Layne *et al*., 1996). Thereafter, its cultivation started in other European countries.

In India it is widely distributed in the state of Jammu and Kashmir, Himachal Pradesh and a very little acreage in Uttarakhand (Srivastava *et al.* 2014). There are 90 per cent of the total cherry production of India is confined to Kashmir valley of Jammu and Kashmir due to the temperate climatic condition of Kashmir valley (Ahmed *et al.* 2010).

19.6.2 Varieties and Cultivars

Many cultivars of sweet cherry have been introduced from Europe, USSR and British Columbia. Promising exotic cultivars like Bigarrean Napoleon, Black Heart, Guigne Noir for J&K and cultivars like Black Tartarian Bing, Napoleon (white) Sam, Sue (White), Shella for H.P. have been identified. For warmer climate, cultivars like Summit, Sunburst, Lapins, Compacat and Stella have been found to be promising.

Table 1: Promising cultivars in major production regions of India

Sl.No.	Regions	Cultivars
1.	Jammu and Kashmir	Bigarreau Napoleon, Black heart
2.	Himachal Pradesh	Black Tartarian, Napoleon, Sue Sam, Lambert Bing
3.	Uttar Pradesh	Bedford Prolific, Black Heart, Governor's Wood

19.6.3 References

Anonymous. 2002a. Consensus Document on the Biology of Prunus spp. (Stone fruits), Environment Directorate. Org. Eco. Co-op. and Dev. Paris., pp. 1-42.

Hedrick, U.P. 1915. The cherries of New York. *J. B. Lyon, Albany,* New York.

Layne, R.E.C, Bailey, C.H and Hough, L.F. 1996. Apricots. In Janick J, Moore JN (eds.) Fruit Breeding Vol. I: Tree and Tropical Fruits, John Wiley and Sons, Inc. New York, pp. 79-111.

Rehder, A. 1954. Manual of cultivated trees and shrubs. MacMillan Company, New York.

Vavilov, N.I. 1951. Phytogeographic basis of plant breeding. In Chester, K. S. (ed.) The Origin, Variation, Immunity and Breeding of Cultivated Plants, *Chronica Bot.*, **13**(1/3): 13-54.

19.7. Strawberry

Botanical name: *Fragaria* x *ananassa Duch.*

Family: Rosaceae

Scientific classification

Kingdom	:	Plantae- Plants
Division	:	Magnoliophyta–Flowering plants
Class	:	Magnoliopsida – Dicotyledons
Subclass	:	Rosidae
Order	:	Rosales
Family	:	Rosaceae – Rose family
Genus	:	*Fragaria L.*
Species	:	*ananassa Duch.*

Strawberry is one of the most popular soft fruit and an attractive, luscious, tasty and nutritious fruit with a distinct and pleasant aroma, and delicate flavor. It is cultivated in plains as well as in hills up to an elevation of 3000 meters in humid or dry regions, widely grown under protected and open condition in temperate and subtropical countries with maximum temperature of 22° -25° C in the day and 7° -13° C at night. It has rich in vitamin C, iron, proteins, minerals Ca, P and K. Strawberry is mainly consumed as fresh fruits and also processed for making wine, jam, jelly, ice cream and soft drinks etc. Strawberries are low-growing herbaceous plants with a fibrous root system and a crown from which arise basal leaves. The leaves are compound, typically with three leaflets, sawtooth-edged, and usually hairy. The flowers, generally white, rarely reddish, are borne in small clusters on slender stalks arising, like the surface-creeping stems, from the axils of the leaves.

19.7.1 Center of origin and distribution

The garden strawberry (*Fragaria* x *ananassa* and related cultivars) is the most common variety of strawberry cultivated worldwide. It originated in Europe in the early eighteenth century, and represents the cross of *Fragaria virginiana* from eastern North America (the main native strawberry in the United States), which was noted for its fine flavor, and *Fragaria chiloensis* from Chile, noted for its large size. Cultivars of *Fragaria* x *ananassa* have replaced in commercial production the woodland strawberry (*Fragaria vesca*), which was cultivated in the early seventeenth century. It is cultivated in tropical and sub-tropical areas round the year. The leading strawberry producing countries are United States, followed by China and Spain. In India, it is commercially cultivated in Himachal Pradesh, Uttar Pradesh, Maharashtra, West Bengal, Nilgiris hills, Delhi, Haryana, Punjab and Rajasthan (Chadha, 2001). Nainital and Dehradun districts of Uttarakhand, Mahabaleshwar (Maharashtra), Kashmir valley, Bangalore and Kalipong (West Bengal) are the main centers of strawberry cultivation in India (Tyagi *et al.*, 2015).

Table 1: Worldwide distribution of native strawberries (Fragaria spp.) (Hummer and Hancock 2009)

Sl.No	Species	Geographic Distribution
1.	*F. bucarica* Losink.	Asia (W. Himalayas)
2.	*F. daltonia* J. Gay	Asia (Himalayas)
3.	*F. nubicola* Lindl.	Asia (Himalayas)
4.	*F. gracilis* Losink.	Asia (N. China)
5.	*F. mandshurica* Staudt	Asia (N. China)
6.	*F. pentaphylla* Losink.	Asia (N. China)
7.	*F. corymbosa* Losink.	Asia (N. China)
8.	*F. moupinensis* (French.) Card	Asia (N. China)
9.	*F. gracilis* Losink.	Asia (N.W. China)
10.	*F. tibeticia* spec. nov. Staudt	Asia (China)
11.	*F. innumae* Makino	Asia (Japan)
12.	*F. yezoensis* Hara.	Asia (Japan)
13.	*F. nipponica* Lindl.	Asia (Japan)
14.	*F. iturupensis* Staudt.	Asia (Iturup Island)
15.	*F. nilgerrensis* Schlect.	Asia (S.E. Asia)
16.	*F. orientalis* Losink. syn. = *F. corymbose* Losink.	Asia (China, Far Eastern Russia)
17.	*F. Americana* (Porter) Britton syn. =*F. vesca* (Porter) Staudt.	Europe, Asia, N. America
18.	*F. viridis* Duch.	Europe, Asia
19.	*F. moschata* Duch.	Europe, Asia
20.	*F. chilonensis* (L.) Miller	America, S. America (Western N. America, Hawaii, Chile)
21.	*F. virginiana* Miller	N. America

Table 2: Types of Strawberry

Sl.No	Chromosome numbers	Species
1	Diploid species	*Fragaria daltoniana* *Fragaria iinumae* *Fragaria nilgerrensis* *Fragaria nipponica* *Fragaria nubicola* *Fragaria vesca* (Woodland Strawberry) *Fragaria viridis* *Fragaria yezoensis*

Contd.

2	Tetraploid species	*Fragaria moupinensis* *Fragaria orientalis*
3	Hexaploid species	*Fragaria moschata* (Musk Strawberry)
4	Octoploid species and hybrids	*Fragaria x ananassa* (Garden Strawberry) *Fragaria chiloensis* (Beach Strawberry) *Fragaria iturupensis* (Iturup Strawberry) *Fragaria virginiana* (Virginia Strawberry)
5	Decaploid species and hybrids	*Fragaria* × *Potentilla* hybrids *Fragaria* × *vescana*

(Darrow, 1966)

19.7.2 Important varieties and cultivars

Chandler - Fruit is having high dessert quality with outstanding colour and flavor. It is very resistant to physical damages caused by rain. Plants are tolerant to viruses. It is suitable for fresh market and processing. The fruits having good TSS (12%), acidity (0.85%), vitamin C (55.5mg/100g) and sugar content (6.1%).

Tioga - An early maturing cultivar and it is tolerant to viruses. Fruits are very large, fleshy and skin firm. Average berry weighs about 9g and it contains TSS 12.2%, acidity 0.98% and sugar 6.2%.

Torrey- It is tolerant to viruses; it produces numerous runners. Fruits are large in size, fleshy and skin medium firm, dessert quality is excellent with good TSS (12%), acidity (0.97%) and sugars (6.1%). Average fruit weight is 6.9g.

Selva - It is a day neutral cultivar and it produce off season fruits. Fruits are large, flesh, skin firm, conic to blocky in shape, dessert quality is good. The average size of berry weighs 15-18 g with 11.1% TSS, 1% acidity and 5.5% sugar.

Belrubi - Fruits are large, conical, bright red skin, some what firm, less hallow at core, high quality, sweet, slightly sub acid and attractive red flesh. The average fruit weight is 15g with TSS 11.8%, acidity 0.98% and sugars 6%.

Fern - It is a day neutral, early ripening and over bearing cultivar. Fruits are large, conical, slid internally, slightly hollow, skin red, flesh red, firm, flavor excellent. It tastes sweet to slightly sub acidic. Average berry weighs 20-25g with TSS 11.2%, acidity 0.88% and sugars 6.1%.

Pajaro - It is very successful under summer season. Plant tolerant to virus. Fruit is quite susceptable to physical damage caused by rain. Friuts are large, flesh very firm and red in colour. The average berry weighs 7.6g with TSS 12.2%, acidity 0.97% and sugars 5.5%.

Sweet Charli (A.L.80-456 x pajaro) - It is early maturing variety and average fruit weight is 17g. It is resistance to anthracnose and susceptible to blast, powdery mildew and mites.

Premier, Red coat, Local jeolikot, Dilpasand, Bangalore, Florida 90, Katrain sweet, Pusa early dwarf and blackmore are some other cultivars of strawberry which is also grown.

19.7.3 References

Chadha, K.L. 2001. Hand Book of Horticulture. Indian Council of Agricultural Research, New Delhi. pp. 324-325.

Darrow, G.M. 1966. The Strawberry: History, Breeding and Physiology. New York. Holt, Rinehart and Winston.

Hummer, K., and J.F. Hancock. 2009. Strawberry genomics: botanical history, cultivation, traditional breeding, and new technologies. In: K. Folta and S Gardiner (eds.), Genetics and Genomics of Rosaceae. Springer Science + Business Media, New York: 413-436.

Tyagi, S, Mukhtar Ahmad, Sanjay Sahay, A.H. Nanher and Brajesh Nandan. 2015. Strawberry: A potential cash crop in India. *Rashtriyakrishi*, **10**(2): 57-59.

Appendix

Botanical name, family and origin of important fruit crops

S.No.	Common name	Synonyms	Botanical name	Family	Origin
1	Chinese gooseberry	Kiwi fruit	*Actinidia chinensis* Planch	Actinidiaceae	China
2	Bael	Bengal Quince, Vilvam, Bil Patram	*Aegle marmelos* Correa	Rutaceae	North India
3	Indian Horse Chestnut	Pangar, Bankhor	*Aesculus indica* Colebr (Syn. *A. glabra*)	Hippocastanaceae	Kashmir
4	Cashew nut	Kaju	*Anacardium occidentale* L.	Anacardiaceae	Tropical America
5	Pineapple	Ananas	*Ananas comosus* Merr.	Bromeliaceae	Brazil
6	Atemoya	Lakshmanphal	*Annona atemoya* Hort.	Annonaceae	Man made hybrid, also found in India
7	Cherimoya	Hanumanphal	*Annona cherimola* Mill	Annonaceae	In Israel, Florida, Egypt
8	Ilama	White annona	*Annona diversifolia* Safford	Annonaceae	Tropical America
9	Mamphal	Sour sop	*Annona muricata* L.	Annonaceae	America, West Indies
10	Ramphal	Bullock's heart	*Annona reticulata* L.	Annonaceae	Tropical America
11	Custard apple	Sweet sop, Sugar apple, Sharifa, Sitaphal	*Annona squamosa* L.	Annonaceae	Tropical America
12	Bread fruit	Vilayati phanas	*Artocarpus altilis*	Moraceae	Malaya, Pacific Island
13	Jackfruit	Katahal	*Artocarpus heterophyllus*	Moraceae	India
14	Monkey jack	Barhal	*Artocarpus lakoocha* Roxb.	Moraceae	India to Malacea
15	Carambola	Kamrakh	*Averrhoa carambola* L.	Oxalidaceae	India-China
16	Cucumber tree	Bilimbi, Tree sorrel	*Averrhoa bilimbi* L.	Oxalidaceae	India-Malaya
17	Mountain Papaya	-	*Carica candamarcensis* Hook F.	Caricaceae	Ecuador
18	Papaya	Papita	*Carica papaya L.*	Caricaceae	Tropical America
19	Karonda	-	*Carissa carandas* L.	Apocynaceae	India-Java
20	Natal plum	-	*Carissa grandiflora* A.D.C.	Apocynaceae	South Africa
21	Pecannut	Pecan	*Carya illinoensis* Koch	Juglandaceae	Southern United States
22	White Sapota	Kochii Sapota	*Casimirao edulis* Llav & Rex	Rutaceae	Mexico
23	Chestnut	-	*Castanea sativa* Mill	Fagaceae	North Africa Asia Minor

Contd.

24	Butter lime	West Indian star apple	*Chrysophyllum cainito* L.	Sapotaceae	West Indies and Central America
25	Lime	Nimbu	*Citrus aurantifolia* Swingle	Rutaceae	India
26	Sour orange	-	*Citrus aurantium* L.	Rutaceae	India
27	Pummelo	Shaddock	*Citrus grandis* (*Citrus maxima*)	Rutaceae	South East Asia
28	Indian wild orange	-	*Citrus indica* Tanaka	Rutaceae	India
29	Rough lemon	Jattikhatti	*Citrus jambhiri* Lush	Rutaceae	India
30	Karna khatta	Soh Sarkar	*Citrus karna* Raf	Rutaceae	India
31	Khasi papeda	Soh shyrkhoit	*Citrus latipes* Tanaka	Rutaceae	Eastern part of India
32	Indian sweet lime	-	*Citrus limettoides* Tanaka	Rutaceae	India
33	Lemon	-	*Citrus limon* Burma	Rutaceae	East Asia
34	Rangpur lime	Canton lemon	*Citrus limonia* Osbeck	Rutaceae	India
35	Kichili	-	*Citrus mederaspatara* Tanaka	Rutaceae	South India
36	Citron	Turanj	*Citrus medica* L.	Rutaceae	India
37	Sour pummelo	-	*Citrus megaloxycarpa* Lush	Rutaceae	India
38	Grape fruit	-	*Citrus paradisi* Macf.	Rutaceae	West Indies
39	Hill lemon	Soh- long	*Citrus pseudalimon* Tanaka (*Citrus limon* var. decumana)	Rutaceae	India
40	Mandarin	-	*Citrus reticulata* Blanco	Rutaceae	Southern China and Cochin-China
41	Indian grape fruit	-	*Citrus rugulosa* Tanaka	Rutaceae	India
42	Sweet orange	-	*Citrus sinensis* Osbeck	Rutaceae	China
43	Quince	-	*Cydonia oblonga* Mill	Rosaceae	South Asia
44	Tree tomato	-	*Cyphomandra betacea*	Solanaceae	Peru
45	Mabola persimmon	Butter fruit	*Diospyros discolor* Willd.	Ebenaeae	Philippines
46	Persimmon	Kaku	*Diospyros kaki* L.	Ebenaceae	China
47	Date plum	Amlok	*Diospyros lotus* L.	Ebenaceae	West Asia Himalayas China
48	Durian	Civet fruit	*Durio zibethinum* L.	Bombacaceae	East Indies Malaya
49	Indian gooseberry	Aonla	*Emblica officinalis* Gaertn.	Euphorbiaceae	Tropical Asia
50	Loquat	Japanplum	*Eriobotrya japonica* Lindl.	Rosaceae	China
51	Surinam cherry	Brazil cherry	*Eugenia uniflora* L.	Myrtaceae	Tropical South America

Contd.

52	Longan	-	*Euphoria longan* Steud	Sapindaceae	India- Burma
53	Pineapple guava	New Zealand banana	*Feijoa sellowiana* Berg.	Myrtaceae	Brazil
54	Wood apple	Elephant apple, Kaitha	*Feronia limonia* (L.) Swing	Rutaceae	India
55	Fig	-	*Ficus carica* L.	Moraceae	West Asia
56	Gular	-	*Ficus glomerata* Roxb	Moraceae	East Indies to Australia
57	Governor's plum	Batoko plum	*Flacourtia indica* Merr.	Flacourtiaceae	India and Tropical Africa
58	Paniala	Puneala plum	*Flacourtia jangomas*	Flacourtiaceae	India
59	Kumquat	-	*Fortunella japonica* *Fortunella margarita Swingle*	Rutaceae	South China
60	Strawberry	-	*Fragaria chilaensis* x *Fragaria virginiana* = *Fragaria* x *ananas*	Rosaceae	-
61	Cowphal	Kau	*Garcinia cowa* Roxb.	Guttiferae	India
62	Kokam butter	Kokum	*Garcinia indica* Choicy	Guttiferae	India
63	Dampel	Tamal	*Garcinia xanthochymus* Hook. F.	Guttiferae	India-Burma
64	Phalsa	-	*Grewia subinaequalis* D. C.	Tiliaceae	India
65	Walnut	-	*Juglans regia*	Juglandaceae	South East Europe
66	Litchi	-	*Litchi chinensis* Sonn.	Sapindaceae	China
67	Macadamia nut	-	*Macadamia ternifolia*	Proteaceae	-
68	Mahua	-	*Madhuka indica* J.F.G. Mel. (Syn. *Madhuka latifolia*)	Sapotaceae	India
69	Barbados Cherry	-	*Malpighia glabra* L.	Malpighiaceae	South America
70	West Indian Cherry	-	*Malpighia punicifolia* L.	Malpighiaceae	South America
71	Crab apple	-	*Malus baccata* var. Himalacca	Rosaceae	Himalayas
72	Apple	-	*Malus* x *domestica*	Rosaceae	Asia Minor to Western Himalayas
73	Mango	-	*Mangifera indica* L.	Anacardiaceae	Indo Burma region
74	Khirnee	-	*Manilkara hexandra* Roxb	Sapotaceae	Probably India
75	Ceriman	-	*Monstera deliciosa* Liebm.	Araceae	Mexico Guatemala
76	White Mulberry	Safed shahtut	*Morus alba* L.	Moraceae	China
77	Black Mulberry	Kali shahtut	*Morus nigra* L.	Moraceae	West Asia (Iran)

Contd.

78	Banana	Apple of paradise	*Musa paradisiaca* L.	Musaceae	Malaysia-India
79	Plantain banana	-	*Musa sapientum*	Musaceae	-
80	Rambutan	-	*Nephelium lappaceum* L.	Sapindaceae	Malaya
81	Olive	Zaitun	*Olea europaea* L.	Oleaceae	Mediterranean region
82	Avocado	Alligator pear	*Persea americana Mill*	Lauraceae	Mexico and West Indies
83	Passion fruit	-	*Passiflora edulis* Sims	Passifloraceae	Brazil
84	Date palm	-	*Phoenix dactylifera* L.	Palmae	West Asia and Arabia
85	Wild date	East Indian wine palm	*Phoenix sylvestris* Roxb	Palmae	India-Madagascar
86	Stargoose berry	-	*Phyllanthus acidus* Skeels	Euphorbiaceae	India
87	Jam berry	Tomatillo	*Physalis ixocarpa* Brat.	Solanaceae	Mexico Southern USA
88	Cape gooseberry	Husk tomato	*Physalis peruviana* L.	Solanaceae	Tropics
89	Pistachionut	-	*Pistacia vera* L.	Anacardiaceae	West Asia
90	Trifoliate orange	-	*Poncirus trifoliata* Raf.	Rutaceae	China
91	Almond	-	*Prunus amygdalus* Batsch	Rosaceae	Persia Afghanistan
92	Apricot	-	*Prunus armeniaca* L.	Rosaceae	China
93	Sweet cherry	-	*Prunus avium* L.	Rosaceae	Europe-West Asia
94	Sour cherry	-	*Prunus cerasus* L.	Rosaceae	Europe-West Asia
95	Plum	-	*Prunus domestica* L.	Rosaceae	Europe-West Asia Europe
96	Bird cherry		*Prunus padus* L.	Rosaceae	Europe
97.	Peach	-	*Prunus persica* Batsch.	Rosaceae	China
98	Japanese plum	-	*Prunus salicina* Lindl.	Rosaceae	China
99	Strawberry guava	-	*Psidium cattleianum*	Myrtaceae	Brazil
100	Guava	Amrood	*Psidium guajava* L.	Myrtaceae	West Indies to Peru
101	Pomegranate	Anar	*Punica granatum*	Punicaceae	South East Europe to Himalayas
102	Pear	Nashpati	*Pyrus communis* L.	Rosaceae	Europe
103	Chinese sand pear	-	*Pyrus pyrifolia* var. culta nakai	Rosaceae	China
104	Orange raspberry	-	*Rubus ellipticus* Smith	Rosaceae	Probably Yunan
105	Blackberry	-	*Rubus fruticosus* L.	Rosaceae	West Asia
106	Hog plum	-	*Spondias cythera* Sonn.	Anacardiaceae	Polynesia
107	Indian hog plum	-	*Spondias pinnata* Kurz	Anacardiaceae	Tropical Asia

Contd.

			(*Spondias mangifera*)		
108	Java plum	Jamum	*Syzygium cuminii* Skeels	Myrtaceae	East India-Malaya
109	Rose apple	-	*Syzygium jambos* Alston	Myrtaceae	East Indies
110	Tamarind	-	*Tamarindus indica* L.	Leguminosae	Tropical Africa-India
111	Indian almond	-	*Terminalia catappa* L.	Combretacae	India (Tropical Asia)
112	Grape	-	*Vitis vinifera* L.	Vitaceae	Caucasus region
113.	Indian ber	-	*Ziziphus mauritiana* Lam.	Rhamnaceae	India
114.	China jujuba	-	*Ziziphus jujuba* Lam.	Rhamnaceae	China
115	Mangosteen		*Garcinia mangostana* L.	Guttiferae	Malaya Peninsula